D1233815

NUTRACEUTICALS:
DESIGNER FOODS III
GARLIC, SOY AND LICORICE

F
  N
    P
PUBLICATIONS IN
FOOD SCIENCE AND NUTRITION

Books

NUTRACEUTICALS: DESIGNER FOODS III, P.A. Lachance
DESCRIPTIVE SENSORY ANALYSIS IN PRACTICE, M.C. Gacula, Jr.
APPETITE FOR LIFE: AN AUTOBIOGRAPHY, S.A. Goldblith
HACCP: MICROBIOLOGICAL SAFETY OF MEAT, J.J. Sheridan *et al.*
OF MICROBES AND MOLECULES: FOOD TECHNOLOGY AT M.I.T., S.A. Goldblith
MEAT PRESERVATION: PREVENTING LOSSES AND ASSURING SAFETY,
     R.G. Cassens
S.C. PRESCOTT, PIONEER FOOD TECHNOLOGIST, S.A. Goldblith
FOOD CONCEPTS AND PRODUCTS: JUST-IN-TIME DEVELOPMENT, H.R. Moskowitz
MICROWAVE FOODS: NEW PRODUCT DEVELOPMENT, R.V. Decareau
DESIGN AND ANALYSIS OF SENSORY OPTIMIZATION, M.C. Gacula, Jr.
NUTRIENT ADDITIONS TO FOOD, J.C. Bauernfeind and P.A. Lachance
NITRITE-CURED MEAT, R.G. Cassens
POTENTIAL FOR NUTRITIONAL MODULATION OF AGING, D.K. Ingram *et al.*
CONTROLLED/MODIFIED ATMOSPHERE/VACUUM PACKAGING, A.L. Brody
NUTRITIONAL STATUS ASSESSMENT OF THE INDIVIDUAL, G.E. Livingston
QUALITY ASSURANCE OF FOODS, J.E. Stauffer
THE SCIENCE OF MEAT AND MEAT PRODUCTS, 3RD ED., J.F. Price and
     B.S. Schweigert
HANDBOOK OF FOOD COLORANT PATENTS, F.J. Francis
ROLE OF CHEMISTRY IN PROCESSED FOODS, O.R. Fennema *et al.*
NEW DIRECTIONS FOR PRODUCT TESTING OF FOODS, H.R. Moskowitz
ENVIRONMENTAL ASPECTS OF CANCER: ROLE OF FOODS, E.L. Wynder *et al.*
FOOD PRODUCT DEVELOPMENT AND DIETARY GUIDELINES, G.E. Livingston,
     R.J. Moshy and C.M. Chang
SHELF-LIFE DATING OF FOODS, T.P. Labuza
ANTINUTRIENTS AND NATURAL TOXICANTS IN FOOD, R.L. Ory
UTILIZATION OF PROTEIN RESOURCES, D.W. Stanley *et al.*
POSTHARVEST BIOLOGY AND BIOTECHNOLOGY, H.O. Hultin and M. Milner

Journals

JOURNAL OF FOOD LIPIDS, F. Shahidi
JOURNAL OF RAPID METHODS AND AUTOMATION IN MICROBIOLOGY,
     D.Y.C. Fung and M.C. Goldschmidt
JOURNAL OF MUSCLE FOODS, N.G. Marriott, G.J. Flick, Jr. and J.R. Claus
JOURNAL OF SENSORY STUDIES, M.C. Gacula, Jr.
JOURNAL OF FOODSERVICE SYSTEMS, C.A. Sawyer
JOURNAL OF FOOD BIOCHEMISTRY, N.F. Haard, H. Swaisgood and B. Wasserman
JOURNAL OF FOOD PROCESS ENGINEERING, D.R. Heldman and R.P. Singh
JOURNAL OF FOOD PROCESSING AND PRESERVATION, D.B. Lund
JOURNAL OF FOOD QUALITY, J.J. Powers
JOURNAL OF FOOD SAFETY, T.J. Montville and D.G. Hoover
JOURNAL OF TEXTURE STUDIES, M.C. Bourne and M.A. Rao

Newsletters

MICROWAVES AND FOOD, R.V. Decareau
FOOD INDUSTRY REPORT, G.C. Melson
FOOD, NUTRITION AND HEALTH, P.A. Lachance and M.C. Fisher

NUTRACEUTICALS:
DESIGNER FOODS III
GARLIC, SOY AND LICORICE

Edited by

Paul A. Lachance, Ph.D., FACN

Professor and Chair
Department of Food Science
Rutgers - The State University
New Brunswick, New Jersey

FOOD & NUTRITION PRESS, INC.
TRUMBULL, CONNECTICUT 06611 USA

Library of Congress Catalog Card Number: 97-060572
ISBN: 0-917678-40-0

Printed in the United States of America

DEDICATION

This book is dedicated to the memory of Dr. Herbert F. Pierson Jr., born 7 August 1952 in Trenton, New Jersey, who earned degrees in biochemistry and pharmacognosy.

As a scientist in the Diet and Cancer Prevention Branch, National Cancer Institute, he was Project Leader of the novel initiative to fingerprint the cancer-preventing compounds found in food. With boundless enthusiasm and energy, in 1989, he coined the descriptor "Designer Foods" to not only recognize the role of phytochemicals in thwarting disease pathogenesis, but to stimulate the food and pharmaceutical industries to research and to market existing and new foods "designed" to benefit the health of humans.

He was very closely allied to the three "Designer Foods" short courses held at Rutgers. By the time of the third Designer Foods short course, he had resigned his position at NCI to establish his own consulting business "Preventive Nutrition Consultants" in Woodinville, WA. In addition, he became involved, along with his wife, with the continuing education of pharmacist in this emerging field of herbals and designer or functional foods, which in the USA are grouped as Nutraceuticals.

He fought valiantly with lymphoma but his Creator called him home 14 March 1996.

CONTRIBUTORS

AWAZU, SHOJI, Department of Biopharmaceutics, Tokyo University of Pharmacy and Science, 1432-1 Horinouchi, Hachioji, Tokyo 192-03, Japan

BEECHER, C., Program for Collaborative Research in the Pharmaceutical Sciences, Department of Medicinal Chemistry and Pharmacognosy, College of Pharmacy (m/c 781), University of Illinois at Chicago, Chicago, IL.

BENNINK, MAURICE R., Departments of Medicine, and Food Science and Human Nutrition, Michigan State University, East Lansing, MI.

BLAKELY SHIRLEY, A.R., Food and Drug Administration, Office of Policy, Planning and Strategic Initiatives (HFS-19), 200 C Street, SW, Washington, DC.

BOEHM-WILCOX, CHRISTA, Department of Veterinary Pathology, University of Sydney, NSW 2006, Australia.

BOLAÑOS-JIMENEZ, FRANCISCO, Unite de Pharmacologie Neuro-Immuno-Endocrinienne, Institut Pasteur, 28 rue du Dr. Roux, F75015 Paris, France.

BOSNIC, MEIRA, Department of Veterinary Pathology, University of Sydney, NSW 2006, Australia.

CLARK, JAMES P., Henkel Corporation, 5325 South 9th Avenue, LaGrange, IL.

CONSTANTINOU, ANDREAS, Chemoprevention Program, Department of Surgical Oncology, University of Illinois, Chicago.

DAVIS, DEVRA L., U.S. Department of Health and Human Services, Public Health Service, Assistant Secretary for Health, Office of Disease Prevention and Health Promotion, DHHS/PHS, 330 C Street S.W., Room 2132, Washington, DC.

DIMITROV, NIKOLAY V., Departments of Medicine, and Food Science and Human Nutrition, Michigan State University, East Lansing, MI.

DUKE, JAMES, U.S. Department of Agriculture, Agricultural Research Service, Building 003, Room 227, 10300 Baltimore Avenue, Beltsville, MD.

ERDMAN, JOHN W. JR., Division of Nutritional Sciences, University of Illinois, 451 Bevier Hall, 905 S. Goodwin, Urbana, IL.

FILLION, GILLES M., Unite de Pharmacologie Neuro-Immuno-Endocrinienne, Institut Pasteur, 28 rue du Dr. Roux, F75015 Paris, France.

FILLION, MARIE-PAULE, Unite de Pharmacologie Neuro-Immuno-Endocrinienne, Institut Pasteur, 28 rue du Dr. Roux, F75015 Paris, France.

GEARY, RICHARD S., Department of Applied Chemistry and Chemical Engineering, Southwest Research Institute, San Antonio, TX.

GRIMALDI, BRIGITTE, Unite de Pharmacologie, Neuro-Immuno-Endocrinienne, Institut Pasteur, 28 rue du Dr. Roux, F75015 Paris, France.

GUHR, GERDA, Department of Food Science, Rutgers — The State University, New Brunswick, NJ.

HARTMAN, THOMAS, G., Center For Advanced Food Technology, Cook College, Rutgers University, New Brunswick, NJ.

HASLER, CLARE M. Director, Functional Foods for Health Program, Department of Food Science and Human Nutrition, 103 Agricultural Bioprocess Lab, 1302 West Pennsylvania Avenue, Urbana, IL.

HATONO, SHUNSO, Wakunaga Pharmaceutical Co., Ltd., 1624 Shimokotachi, Koda-cho, Takata-gun, Hiroshima Pref. 739-11, Japan.

HENLEY, E.C., Protein Technologies International Inc., Checkerboard Square, St. Louis, MO.

HORIE, TOSHIHARU, Department of Biopharmaceutics, Faculty of Pharmaceutical Sciences, Chiba University, 1-33 Yayoi-cho, Inage-ku, Chiba-shi, Chiba 263, Japan.

JENKS, BELINDA H., Protein Technologies International Inc., Checkerboard Square, St. Louis, MO.

KAMADA, TAKAATSU, The Second Department of Internal Medicine, Hirosaki University School of Medicine, 5 Zaifu-cho, Hirosaki, Aomori, Japan 036.

KANAZAWA, TAKEMICHI, The Second Department of Internal Medicine, Hirosaki University School of Medicine, 5 Zaifu-cho, Hirosaki, Aomori, Japan 036.

KATSUKI, HIROSHI, Department of Chemical Pharmacology, Faculty of Pharmaceutical Sciences, The University of Tokyo, 7-3-1 Hongo, Bunkyo-Ku, Tokyo 1 13, Japan.

KODERA, YUKIHIRO, Wakunaga Pharmaceutical Co., Ltd., 1624 Shimokotachi, Koda-cho, Takata-gun, Hiroshima Pref. 739-11, Japan.

KOJINRA, RYUSUKE, Philadelphia Biomedical Research Institute, 100 Ross and Royal Roads, King of Prussia, PA.

KRISHNARAJ, RAJABATHER, Geriatric Medicine (787), University of Illinois at Chicago, Chicago, IL.

LACHANCE, PAUL A., Department of Food Science, Rutgers — The State University, New Brunswick, NJ.

LEWANDOWSKI, C., Program for Collaborative Research in the Pharmaceutical Sciences, Department of Medicinal Chemistry and Pharmacognosy, College of Pharmacy (m/c 781), University of Illinois at Chicago, Chicago, IL.

MATSUURA, HIROMICHI, Institute for OTC Research, Wakunaga Pharmaceutical Co., Ltd., Hiroshima, Japan.

MEHTA, RAJENDRA G., Chemoprevention Program, Department of Surgical Oncology, University of Illinois, Chicago.

METOKI, HIROBUMI, Reimeikyo Rehabilitation Hospital, 30 Ikarigaseki-mura, Aomori, Japan 038-01.

MILLER, MICHAEL A., Department of Applied Chemistry and Chemical Engineering, Southwest Research Institute, San Antonio, TX.

MOON, RICHARD C., Chemoprevention Program, Department of Surgical Oncology, University of Illinois, Chicago.

MORIGUCHI, TORU, Department of Chemical Pharmacology, Faculty of Pharmaceutical Sciences, The University of Tokyo, 7-3-1 Hongo, Bunkyo-Ku, Tokyo 113, Japan.

NISHINO, HOYOKU, Cancer Prevention Division, National Cancer Center Research Institute, 1-1, Tsukiji 5-chome, Chuo-ku, Tokyo 104-, Japan.

NISHIYAMA, NOBUYOSHI, Department of Chemical Pharmacology, Faculty of Pharmaceutical Sciences, The University of Tokyo, 7-3-1 Hongo, Bunkyo-Ku, Tokyo 113, Japan.

NIXON, DANIEL W., Associate Director, Cancer Prevention and Control, Hollings Cancer Center and Folk Professor of Experimental Oncology, Medical University of South Carolina, Charleston, SC.

OHNISHI, S. TSUYOSHI, Philadelphia Biomedical Research Institute, 100 Ross and Royal Roads, King of Prussia, PA.

OIKE, YASABURO, Reimeikyo Rehabilitation Hospital, 30 Ikarigaseki-mura, Aomori, Japan 038-01.

ONODERA, KOGO, The Second Department of Internal Medicine, Hirosaki University School of Medicine, 5 Zaifu-cho, Hirosaki, Aomori, Japan 036.

OSANAI, TOMOHIRO, The Second Department of Internal Medicine, Hirosaki University School of Medicine, 5 Zaifu-cho, Hirosaki, Aomori, Japan 036.

PIERSON, HERBERT F., (Deceased), Vice President for Research and Development, Preventive Nutrition Consultants, Inc., 19508 189TH Place NE, Woodinville, WA.

PINTO, JOHN T., Clinical Nutrition Research Unit & Nutrition Research Laboratory, Department of Medicine, Memorial Sloan-Kettering Cancer Center & New York Hospital — Cornell Medical Center, New York, NY.

REEVE, VIVIENNE E., Department of Veterinary Pathology, University of Sydney, NSW 2006, Australia.

RIVLIN, RICHARD S., Clinical Nutrition Research Unit & Nutrition Research Laboratory, Department of Medicine, Memorial Sloan-Kettering Cancer Center & New York Hospital — Cornell Medical Center, New York, NY.

ROSEN, ROBERT T., Center For Advanced Food Technology, Cook College, Rutgers University, New Brunswick, NJ.

ROZINOVA, EMILIA, Department of Veterinary Pathology, University of Sydney, NSW 2006, Australia.

SAITO, HIROSHI, Department of Chemical Pharmacology, Faculty of Pharmaceutical Sciences, The University of Tokyo, 7-3-1 Hongo, Bunkyo-Ku, Tokyo 1 13, Japan.

SARHAN, HALA, Unite de Pharmacologie Neuro-Immuno-Endocrinienne, Institut Pasteur, 28 rue du Dr. Roux, F75015 Paris, France.

SCHMEISSER, DALE D., Department of Nutrition and Medical Dietetics, University of Illinois, Chicago, Chicago, IL

SIEBERT, BRIAN D., CSIRO Division of Human Nutrition, Adelaide, SA, Australia.

STEELE, VERNON, Chemoprevention Branch, The National Cancer Institute, Bethesda, MD.

TESTA, LUCIA C.A., MacAndrews and Forbes Co., 3rd Street and Jefferson Avenue, Camden, NJ.

UEMURA, TSUGUMICHI, The Second Department of Internal Medicine, Hirosaki University School of Medicine, 5 Zaifu-cho, Hirosaki, Aomori, Japan 036.

VORA, PETER S., MacAndrews and Forbes Co., 3rd Street and Jefferson Avenue, Camden, NJ.

WAGGLE, DOYLE H., Protein Technologies International Inc., Checkerboard Square, St. Louis, MO.

WARGOVICH, MICHAEL J., Section of Gastrointestinal Oncology and Digestive Disease, University of Texas, M.D. Anderson Cancer Center, Houston, TX.

WEINBERG, DAVID S., Southern Research Institute, 2000 Ninth Avenue, South, Birmingham, AL.

ZHANG, YONGXIANG, Department of Chemical Pharmacology, Faculty of Pharmaceutical Sciences, The University of Tokyo, 7-3-1 Hongo, Bunkyo-Ku, Tokyo 113, Japan.

ZHOU, JIN-R., Nutrition & Metabolism Laboratory, Beth Israel Deaconess Medical Center, 194 Pilgrim Road, Boston, MA.

PREFACE

The term "Designer Foods" was the programmatic and funding term of the National Cancer Institute and can be briefly defined as the study of one or more phytochemical components(s) acting individually, additively or synergistically, usually as component(s) of whole food, that have the characteristic of providing protective, preventative and possibly curative roles in the pathogenesis of cancer and other chronic disease progressions. The observation of benefits may be based on the results of one or more epidemiological, *in vitro* and/or *in vivo* cellular, organ system or whole organism assays, and may include human clinical experiences and trials. Other terms have emerged to describe "designer foods." The University of Illinois has evolved a coordinated phytochemical research effort entitled "Functional Foods for Health." Dr. Stephen DeFelice of the Foundation for Innovative Medicine coined the term "Nutraceutical" to describe beneficial phytochemicals. This term has gained considerable acceptance because it is unique, concise and somewhat descriptive of the intended field of endeavor.

Rutgers held the first short course on Designer Foods in January 1990, and the second in March 1993, with the third being held 23-25 May 1994 at Georgetown University of which these are the proceedings.

In contrast to the field of Pharmacognosy and the discovery of drugs from natural products, nutraceutical research discovers substances that are considered a food or a part of a food and which provide health or even medical benefits. Such substances are invariably phytochemicals and may range from being isolated nutrients to other non-nutrient phytochemicals that act directly, or may exert biochemical activity directly, or by modulating genomic materials. Food, be it commodity, or herb, or spice, is any substance that is eaten or otherwise taken into the body (e.g., ostomy) to sustain psychological and physiological life, provide energy and promote the biochemical and physiological interactions that are nutrition. The substance of food is a vast array of chemicals of which a very small number are essential to life and are recognized and classified as nutrients. In contrast a vast array of food chemicals provide flavor, color and texture attributes, while others are key to the metabolism of the food (e.g., enzymes). Of the functions of food, nine are psychological and only three are physiological. Nutraceuticals are naturally-occurring and if determined to be beneficial could be biotechnologically-enhanced.

A paradigm I have evolved to give dimension to the breadth and potential of the implications of nutraceuticals is shown in Fig. P.1. There are four major components to the paradigm. First is the recognition that a major consideration in the risk of any disease developing is our genetic make-up revealed to some degree by our family history. Secondly, there is no question

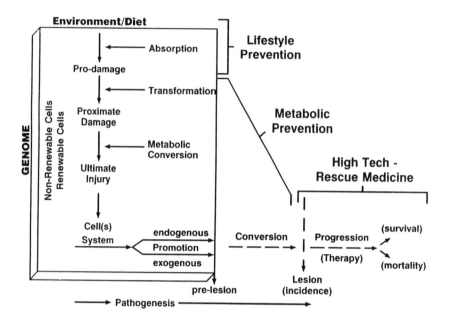

FIG. P.1. PARADIGM OF METABOLIC PREVENTION ROLE FOR NUTRACEUTICALS

The paradigm utilizes concepts of phases in cancer induction recognizing that a genetic predisposition exists. In addition renewable cells are more prone to pathology. The predominate philosophy is treatment *after* recognition of the pathology at the high costs associated with high-tech-rescue medicine. Lifestyle practices, such as exposure to smoking and antioxidant dietary composition, can induce, accelerate or thwart partially or completely the initiation of the pathogenesis. Nutraceuticals function at the metabolic prevention level to block, inhibit or suppress promotion, thus thwarting pathogenesis and the cost of pathology and its consequences.

that our exposure to the environment, including smoking, and to various favorable and unfavorable diet components can be partially counterbalanced by our choice of lifestyle practices, such as consuming no less than five servings of fruit and vegetables per day and regular exercise. Thirdly, today's medical care system is devoted to the remediation of disease that is already pathological, and thus we experience a very costly practice of high-tech-rescue medicine. The fourth dimension encompasses the overlooked numerous metabolic steps in the transformation of normal cellular events into damaged events (e.g., oxidized or precancerous lesions), which nutraceuticals at the proper dose can block or suppress, or metabolically transform, thus setting the stage to prevent, or even revert pathology. The mechanisms of action of the various nutraceuticals may be one or several. For example, free radical scavenger and antioxidant nutraceuticals can nullify damage by any number of biochemical mechanisms, but some also exert benefit by enhancing immune function. Table P.1 lists the broad classes of compounds that have nutraceutical activity, and there are very likely more classes of compounds that have nutraceutical activity. It is very likely there are many other nutraceutical components, as well as interactions, yet to be discovered.

TABLE P.1
PHYTOCHEMICALS: THE NEXT FRONTIER

Phytochemical Family	Food Sources
Allyl Sulfides	Onions, garlic, leeks, chives
Indoles	Cruciferous vegetables (broccoli, cabbage, kale, cauliflower)
Isoflavones	Soybeans (tofu, soy milk)
Isothiocyanates	Cruciferous vegetables
Phenolic Acids (ellagic acid, ferulic acid)	Tomatoes, citrus fruits, carrots, whole grains, nuts
Polyphenols	Green tea, grapes, wine
Saponins	Beans and legumes
Terpenes (perillyl alcohol, limonene)	Cherries, citrus fruit peel

A very conservative 1993 economic analysis of solely hospital care costs and the role that three nutrient antioxidants could exert on cardiovascular disease, breast cancer and cataracts is presented in Table P.2. It is common knowledge that billions of dollars are expended each year for hospital cost relating to breast, lung, stomach and other cancers, cardiovascular disease and cataracts. The study[1] considers the potential impact of only three antioxidants, namely vitamin C, vitamin E and beta carotene; the annual savings in hospital care costs alone would exceed 8 billion dollars. Expert public health physicians[2] believe that as much as 70% of disease is in fact preventable!

TABLE P.2
U.S. HOSPITALIZATION EXPENDITURES

	Total Expenditures ($ Mil)	Percentage Preventable from Increased Antioxidants (Vits. C,E, Beta-Carotene)	Estimated Medicare Cost Savings ($ Mil)	Estimated Total Savings ($ Mil)
Cancer				
Breast	1,088.3	16	27.4	174.1
Lung	3,263.7	21	125.4	685.4
Stomach	538.0	30	43.6	161.4
Cardiovascular Disease	30,823.7	25	1,500.0	7,705.9
Cataracts	98.5	50	7.1	49.3
TOTAL	35,812.3		1,676.6	8,776.1 Annual

Source: Pracon Inc. Economic Analysis (1993)

The papers in this symposium were organized to reveal the existing and emerging knowledge of the nutraceuticals found in garlic, soy and licorice. The lead paper discusses the epidemiological evidence, and the papers following present and discuss chemical or biochemical evidence at the cellular level; and some clinical data are presented.

I'm indebted to the late Dr. Herbert Peirson, who first headed up the Designer Foods Program for the National Cancer Institute and later had the courage and foresight to start a consulting business in this area. His suggestions as to relevant investigators were invaluable, as was his assistance in gaining funding to support the symposium/short course. He served as co-chair of the meeting and these proceedings which I have chosen to dedicate to his memory. We in turn are indebted to the authors who provided succinct texts of their presentations to serve as a status summary of the current state of the available knowledge.

A major conclusion of the overall effort is that the science is very incomplete, but that the findings to date have great promise. The problem is that marketing persons are trying to capitalize on the "promise" of the existing knowledge, but often to a degree that the existing data cannot support. In the interim, the consumer needs to increase their intakes of conventional fruits and vegetables and some cereal grain products, but otherwise should "beware" of the false and misleading promises.

REFERENCES

[1]   Anon. 1993. The potential of antioxidant vitamins in reducing health care costs. Pracon Inc., Reston, VA. Commissioned by the Council for Responsible Nutrition, Washington, D.C.
[2]   James F. Fries, C. Everett Koop — The health project. The Sciences (NYAS) 34(1):19,1994.

PAUL A. LACHANCE
NEW BRUNSWICK, NEW JERSEY

CONTENTS

Section IV. Phytopharmacology of Soy Food Forms

Section V. Phytopharmacology of Licorice Food Forms

Section VI. Bridging the Gaps in Knowledge for Designer Food Applications

CHAPTER 1

PLANT FOODS IN DISEASE PREVENTION — STUDY OF PROCEDURE LESSONS FROM EASTERN VERSUS WESTERN DIET

DEVRA L. DAVIS

U.S. Department of Health and Human Services
Public Health Service
Assistant Secretary for Health
Office of Disease Prevention and Health Promotion
DHHS/PHS
330 C Street. S.W., Room 2132
Washington, D.C. 20201

ABSTRACT

A synopsis is presented of two publications concerning disease prevention associated with plant food constituents. Human breast cancer is increasing worldwide. Total lifetime exposure to estrogen is a risk factor; 16 alpha hydroxyestrone increases tumor growth. Indole-2-carbinol and genestein promote an alternate favorable pathway to 2-hydroxyestrone. Natural products are evidently sources of potent anticarcinogens but require research.

Some Clues about Plant Foods in Disease Prevention

These remarks introduce reprints of two recent papers pertinent to the subject of opportunities for disease prevention through plant foods. Before presenting these papers, I want to recapitulate the theory that some portion of breast cancer today may be due to avoidable exposures to some xenoestrogenic materials, and that other xenoestrogenic materials, which are natural products and plant foods, may help to prevent breast cancer. This work has been developed with H. Leon Bradlow, of Cornell Medical Center, Mary Wolff of Mt. Sinai Medical Center, and other colleagues who are part of the Breast Cancer Prevention Collaborative Research Group (BCPCRG). The BCPCRG is an informal network of clinical, toxicologic, and epidemiologic researchers who work chiefly on the Internet and by Fax to exchange late-breaking developments pertinent to the identification of avoidable causes of breast cancer.

Breast Cancer

Though breast cancer mortality has recently fallen in younger white women, mortality and incidence have risen substantially for African American women of all ages and for all women over 65 in the U.S. and U.K. Worldwide patterns also reveal growing breast cancer incidence and mortality. Geographic and time trend analysis provide an especially intriguing part of the breast cancer puzzle, and bolster the notion that a substantial portion of breast cancer should be avoidable. Women and men in Japan and Asia have rates of new cases and deaths from breast cancer that are consistently about five times lower than those in other industrial countries. Breast cancer mortality for women ages 45 to 84 in 15 major industrial countries from 1968 to 1986, grouped into six major global regions, rose between 0.2 and 2% annually. While rates are highest in U.K. and U.S., the fastest rate of growth is evident in East Asia and Eastern Europe, two regions where mammographic screening cannot generally be conducted. Thus, breast cancer deaths in all countries appear to be on the rise, growing most in regions where improved diagnostic technology does not play a role.

For breast cancer, and possibly other hormonally mediated cancers, such as testicular, prostate, and endometrium, the common factor which links most known risk factors is total lifetime exposure to estrogen. Based on epidemiologic studies and on previous experimental research that found that many of the same materials which increase breast tumor development in animals are also estrogenic, we have recently hypothesized that some breast cancer cases are due to avoidable environmental exposures that increase estrogen levels and alter overall metabolism. We have termed these materials "xenoestrogens." They include some organochlorine pesticides, pharmaceuticals, fuels, plastics, and natural products — all of which have been found to affect the amounts and types of estrogen produced.

Experimental studies on the ability of a wide variety of materials to affect estrogen production have been conducted for more than two decades in several medical laboratories. For breast and other hormonally mediated cancers, agents which affect the body's natural levels of estrogen may thereby affect the risk of breast cancer. With respect to PCBs, a number of isomers are anti--estrogenic, while others are potent estrogens. The picture is complex and incomplete, but clearly indicates that a number of widely used compounds affect estrogen production and metabolism and other hormonally mediated biologic activities.

We have hypothesized that two competing enzyme systems can alter metabolism of the 18-carbon estradiol by contending for two mutually exclusive pathways. Pathway I inserts a hydroxyl (OH•) radical at the 2-carbon position and yields the catechol estrogen 2-hydroxyestrone (2-OHE$_1$), a weakly anti-estrogenic metabolite. Pathway II adds an OH• at the 16 position and yields 16α-hydroxyestrone (16α-OHE$_1$); this creates a fully potent metabolite which increases tumor growth, DNA damage and cell proliferation. While many of the synthetic materials identified as xenoestrogens to date promote the cell proliferative pathway, several natural products, such as indole-2-carbinol and genistein, appear to promote the beneficial pathway. Studies underway in laboratories across the country will clarify the relative role of these materials in helping reduce the risk of recurrence in women with breast cancer, as well as in reducing the risk of the disease overall. David Zava and others have recently speculated that genistein and related compounds may work by keying into estrogen receptors, and blocking its activation by more potent estradiol. In fact, most xenoestrogens identified to date are thousands of times less potent than the body's own estradiol. But, materials such as genistein can occur with circulating levels that are thousands of times higher than those of the body's own estradiol, especially in Asian women whose diets are rich in soy products.

Papers Published

The first paper[1] reviews in vitro and in vivo evidence that a number of common natural products function as potent anticarcinogens. The second paper[2] assesses patterns of cancer in farmers as clues to preventable causes of the disease.

Both papers[1,2] present evidence that current patterns of cancer may be linked, in part, to avoidable exposures. We know that it is feasible to prevent cancer because the rates of the disease vary substantially, even within the industrial world. The rates of brain cancer, multiple myeloma, and breast cancer are all three to five times higher in the United States for males and females than in Japan. This variation in patterns between countries suggests that there are preventable factors.

Consider what is generally known about cancer. Genetics, which gets a lot of front-page coverage, explains, in some sense, all about cancer. But, only a small proportion of cancer is in fact inherited. Most cancer is acquired in a lifetime through interaction of normal genes with the environment, which obviously includes nutrition as a major component. We have to ask what in the environment interacts with the inherited genes that most people have to cause the development of cancer. Food, air, and water sustain life and are also critical components of the environment to which we are all exposed every day of our lives. This essential microenvironment contains both natural and synthetic carcinogens and anticarcinogens.

As to other major causes of cancer, smoking is the single most important preventable cause in industrial countries. Food is a very important factor, both in terms of preventing and causing cancer. Contamination with mycotoxins and fungal molds in the developing world is an important factor, which explains some of the high rates of esophageal, liver and stomach cancer that are endemic in parts of Asia and Africa.

Chemicals, as dusts and fumes, have been clearly characterized as important causes of cancer in highly exposed workers. Sexual practices are relevant to AIDS, but are also relevant to cervical cancer. Viruses, except for the AIDS virus, have not yet been identified as clear viral causes of human cancer, but have been implicated for stomach cancer, as well as cervix. And of course, radiation — both ionizing and non-ionizing — is important, as it can ionize cells and lead to direct mutations.

With my colleagues, David Hoel and Gregg Dinse, I published in 1994[3] a relevant article in the *Journal of the American Medical Association*. We found that baby-boomer males had three times higher rates of cancer not related to smoking than we estimate occurred for their grandfathers. Many of the tumors that have occurred at greater rates in these men, also have occurred more often in farmers; the latter are reviewed in reference number 2. And baby-boomer females have a 50% higher rate of new cases of cancer compared to their grandmothers. We took away the smoking-related cancers to do this analysis, as we wanted to see whether causes other than smoking appeared to be operating.

Taken together with the theory of xenoestrogens, these three publications provide evidence that some segment of cancer is preventable. Variations in cancer rates may very well be linked to parts of the environment, including foods as protective factors and as sources of contaminants. We have to ask what we control as a society that will reduce the risk of cancer? We all recognize that food policy is a vital part of preventative public health programs. In the past, ketchup was proposed as a vegetable in the school lunch program. New guidelines have been developed that provide for more consumption of fruits and vegetables, less consumption of salt and animal fat. Other ways to effect positive food policy changes will include programs or subsidies that we can change. Thus, beef is now graded as to fat content. And subsidies could be developed to reward those who produce healthier produce. Rather than subsidize and stockpile butter, we could develop economic incentives to encourage the production of more healthful crops, such as soy, and to discourage tobacco, which could prove quite valuable in advancing public health.

The rationale for pursuing opportunities to prevent cancer and other diseases grows stronger day by day. New cases of cancer in young persons continue to increase for reasons that are unclear. And rates of deaths from cancer have increased in persons over 65. Along with other efforts to promote health and prevent disease, the development of foods rich in anticarcinogens could play a critical role in ensuring that fewer cases of cancer occur in the first place.

REFERENCES

[1] Davis, D.L. and Babich, H. Natural Anticarcinogens and Mechanisms of Cancer. *In* Handbook of Hazardous Materials, (Corn, M., ed.) Baltimore. Academic Press, San Diego, 1993:463–474.

[2] Davis, D.L., Blair, A. and Hoel, D.G. Agricultural Exposures and Cancer Trends in Developed Countries. Environmental Health Perspectives 1992; 100:39–44.

[3] Davis, D.L., Dinse, G.E. and Hoel, D.G. Decreasing Cardiovascular Disease and Increasing Cancer Among Whites in the United States From 1973 through 1987. J Am. Med. Assoc. 1994; 271:431–437.

CHAPTER 2

SAPONINS IN *ALLIUM* VEGETABLES

C. LEWANDOWSKI and C. BEECHER

Program for Collaborative Research in the Pharmaceutical Sciences
Department of Medicinal Chemistry and Pharmacognosy
College of Pharmacy (m/c 781)
University of Illinois at Chicago
Chicago, IL 60612

ABSTRACT

The saponins of the genus Allium *have been little explored either chemically or biologically, possibly because of the intense interest in the more obvious classes of* Allium *compounds, the thiols and flavonoids. This brief survey demonstrates the chemical complexity of* Allium *saponins in an attempt to direct researchers' attention to this little recognized class of dietary compounds.*

INTRODUCTION

Allium species are among the most widely used of all of the flowering plants. Consider first that one or more members of this widespread genus are an integral component in almost every cuisine. As Table 2.1 shows, the most prominent members of the genus are recognized as foods by almost all of the races of man. A less commonly acknowledged, but almost as universal, fact is the use of these plants for a variety of medical conditions. In the United States this unacknowledged fact has developed into a large and lucrative but somewhat underground market. Though openly sold for "medical" uses, to the tune of millions of dollars per year, the Food and Drug Administration (FDA) has yet to sanction, or conduct a detailed investigation into the use of, any *Allium* product for any specific medical condition. This is unfortunate in view of the extensive peer-reviewed, yet highly variable, experimental/medical literature that surrounds these plants and the chemical compounds that they contain. In this brief review, we will examine in some detail the chemical, taxonomic and biological range of one of the minor classes of compounds that is found in these plants, namely, saponins. This class of compounds may hold the key to some of the medical claims and may explain the variability of the published results, since different preparation protocols will either concentrate or remove them from the final product.

Allium Taxonomy

The genus *Allium*, in the family Amaryllidaceae, is a genus of perennial herbs usually with bulbous rootstocks. Their leaves are simple and more-or-less linear. There are some 300 defined species, many of which are human dietary components (see Table 2.1). Some are horticulturally ornamental. Only a small portion of the known species have been chemically examined.

TABLE 2.1
COMMONLY CONSUMED *ALLIUM* SPECIES

Species	Common name
Allium cepa	Onion
Allium sativum	Garlic
Allium ascalonium	Shallot
Allium ampelorasum	Leek
Allium fistulosum	Scallion
Allium schoenprasum	Chive

Allium Chemistry

The chemistry (and medicine) of alliums has been dominated by the various thiols that contribute most tellingly to the pungency of individual species. These recently reviewed compounds (Fig. 2.1a) provide the pronounced and distinctive odors to onions, garlic and their like.[1] Possibly because of high (olfactory) profile, they have been the subject of much investigation and have been shown to contribute to some of the recognized biology of the crude extracts of the plant material. Thus, compounds such as ajoene and allicin have been widely investigated and shown to be antitumor-promoter[2], antifungal[3], and anti-inflammatory, possibly due to an inhibition of phospholipase.[4] A second class of compounds that are present in many of these species, the flavonoids (Fig. 2.1b) are well known as antioxidants and anti-inflammatory compounds.[5] Their presence in some *Allium* species has recently been noted as onions may be the single biggest source of flavonoids in the human diet.[6]

Because of the prominence of the above two groups, it is at first surprising to recognize that saponins are commonly present throughout the genus, since saponins are generally assumed to be associated with toxicity or

1a. Organosulfur compounds

Onion

Garlic

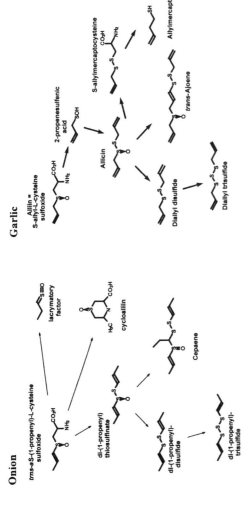

1b. Flavonoids

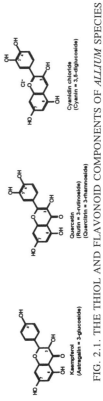

FIG. 2.1. THE THIOL AND FLAVONOID COMPONENTS OF *ALLIUM* SPECIES

hemolysis. It is even more surprising when it is recognized that they are present in sufficiently high levels, in some cases reportedly as high as 0.46% (see below), and that some *Allium* species were originally examined as possible commercial sources for the steroid precursor, diosgenin, until the yam was found to be more economical. The biological activity of these compounds is totally unknown[7], but the reported activity of some of their structural cousins is tantalizing and prompts this analysis. These closely related compounds have been shown to be anti-inflammatory[8], decrease blood cholesterol[9], strengthen veins, and increase their permeability.[10] The clear parallel to some of the reported activities of the extract is obvious. The previously reported concentrations of onion and garlic saponins are shown in the table below.[11]

TABLE 2.2
THE PUBLISHED CONCENTRATIONS OF KNOWN SAPONINS IN
ONIONS AND GARLIC

Alliospiroside A	0.46%	onion
Alliospiroside B	0.05%	onion
Alliospiroside C	0.049%	onion
Alliospiroside D	0.007%	onion
Alliofuroside A	0.022%	onion
Gitonin F	0.03%	garlic
Sativoside R-1	0.05%	garlic
Sativoside R-2	0.03%	garlic
Sativoside B-1	0.003%	garlic
Degalactotigonin	0.04%	garlic
Protoeruboside B	0.01%	garlic
Eruboside B	0.0013%	garlic

While these numbers must be accepted with a significant dose of skepticism, it can be extrapolated from this table that if the average person minimally consumes approximately one-half pound of onion per week (200 gm), then they could potentially be consuming almost a whole gram of these compounds. (It is probable that some portions of the population greatly exceed this.) Since these saponins could augment or strongly modify the biological activity of the other classes of compounds in these vegetables, it would seem important to determine their actual range of concentration and biological activity.

The alliaceous saponins are all of the steroidal class and are typically based on two different skeletal types, the spirostane and furostane nuclei (see Fig. 2.2a). The spirostane-type saponins, the more commonly found, are based on the compound spirostane. This molecule can undergo a number of modifications including hydroxylations and dehydrations to give rise to a variety of agylcones that form the basis of all of the *Allium* saponins. The furostan-type

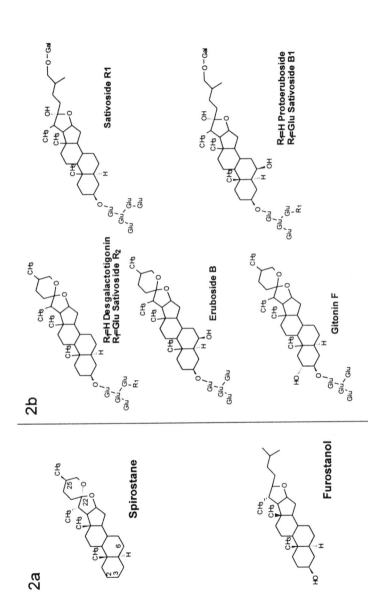

FIG. 2.2. THE AGLYCONE PORTIONS OF SAPONINS OF *ALLIUM* ARE ALL BASED ON ONE OF THESE TWO STEROIDAL NUCLEI (2a) Saponins from other plants may be based on very different fundamental structures. The complete structures of the major saponins found in *A. cepa* (2b). The placement and type of sugars is highly variable within the family.

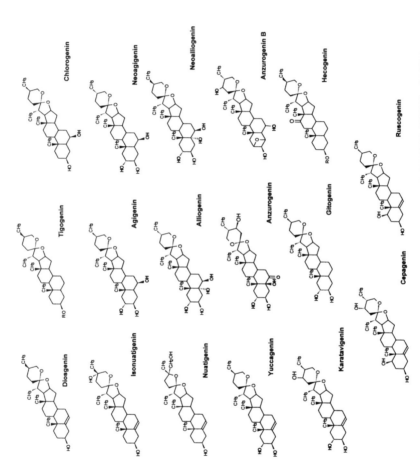

FIG. 2.3. THE FUNDAMENTAL GENINS FOUND IN THE GENUS *ALLIUM*

saponins are based on one such modification, specifically the reduction of the spiro ring; thereby, opening the F ring. By definition, all of these saponins are glycosolated, generally on the 3 position. The spirostane-based saponins are all monodesmosidic, while some of the furostane-based have two sets of sugar linkages and are thus bidesmosidic. Finally, the classes of saponins found differ from one another in the additional modification, i.e., hydroxylation, double bond, or carbonyl, added to the parent ring system to yield the various genins that are actually found. The most important genins of *Allium cepa* are shown in Fig. 2.2b. This figure serves to show the complexity of an individual plant saponin collection. The number of saponins found in all of the *Allium* species examined to date grows proportionately. Furthermore, it is not clear that we have any better than a random perspective on the saponins isolated.

The genin structures of the known *Allium* saponins are found in Fig. 2.3. The majority of these are found as multiple glycosides, although the aglycone is sometimes isolated from some species. The most commonly found saponins in this genus appear to be alliogenin, yuccagenin, and diosgenin, respectively. The rest of these genins are currently reported isolated only a few times in few species, but this may represent a lack of consistent looking rather than have great chemotaxonomic significance. In the final analysis, this is a complex collection of compounds, very probably with significant biological activity, that is little understood, and yet forms a substantial portion of our diet. They need to be carefully considered, along with the thiols and flavonoids, when the possible biology of *Allium* vegetables is investigated.

REFERENCES

[1] Block, E. The organosulfur chemistry of the genus *Allium* — Implications for the organic chemistry of sulfur. Ange. Chem. Int. Ed. Eng. 1992; 31:1135–1178.

[2] Belman, S., Solomon, J., Segal, A., Block, E. and Barany, G. Inhibition of soybean lipoxygenase and mouse skin tumor promotion by onion and garlic components. J. Biochem Toxicol 1989; 43:151–160.

[3] Singh, U.P., Pandey, V.N., Wagner, K.G. and Singh, K.P. Antifungal activity of ajoene, a constituent of garlic. Can J Bot 1990; 686:1354–1356.

[4] Gargour, Y., Moreau, H., Jain, M.K., De Hass, G.H. and Verger, R. Ajoene prevents fat digestion by human lipase in vitro. Biochem. Biophys Acta 1989; 10061:137–139.

[5] Middleton, E. and Kandaswami, C. Effects of flavonoids on immune and inflammatory cell functions. Biochem Pharmacol. 1992; 43:1167–1179.

[6] Hertog, M.G.L., Feskens, E.J.M., Hollman, P.C.H., Katan, M.B. and Kromhout, D. Dietary antioxidant flavonoids and risk of coronary heart disease: the Zutphen Elderly Study. Lancet 1993; 342:1007–1011.

[7] A recent search of NAPRALERT and MEDLINE failed to show any papers that explored their biology.

[8] Syrov, V.N., Kravets, S.D., Kushbaktova, Z.A., Nabiev, A.N., Vollerner, Y.S. and Gorovits, M.B. Steroid genins and glycosides as anti-inflammatory compounds. Khim-Farm Zh 1992; 26:71–76.

[9] Potter, S.M., Jimenez-Flores, R., Pollack, J., Lone, T.A. and Berber-Jimenez, M.D. Protein-saponin interaction and its influence on blood lipids. J. Agric and Food Chem. 1992; 41:1287–1291.

[10] Hostettmann, K. and Marston, A. Saponins: Chemistry and Pharmacology of Natural Products. Cambridge University Press, 1995.

[11] Beecher, C.W.W. and Farnsworth, N.R. eds. Napralert database, July 1995 edition.

THE CHEMICAL AND BIOLOGICAL PRINCIPLES OF DESIGNER FOODS

CHAPTER 3

OVERVIEW OF ANALYTICAL TECHNIQUES FOR THE FINGERPRINTING OF PHYTOCHEMICALS IN DESIGNER FOODS

ROBERT T. ROSEN and THOMAS G. HARTMAN

Center For Advanced Food Technology
Cook College, Rutgers University
New Brunswick, NJ 08903

ABSTRACT

Particle Beam (EI) HPLC-MS and Short Path Thermal Desorption GC-MS are useful for the identification and quantitation of phytochemicals in Designer Foods and natural products. Both techniques utilize electron ionization (EI)-MS, which is more useful for structural elucidation of unknowns than mass spectrometry-mass spectrometry (MS-MS). This chapter discusses experimental conditions and uses for the two methods with an emphasis on obtaining complete EI mass spectra with 70 eV fragmentation patterns.

INTRODUCTION

Chromatography and mass spectrometry are routinely utilized at the Center for Advanced Food Technology for the determination of beneficial phytochemicals in food and in natural products. Thermal desorption gas chromatography-mass spectrometry (GC-MS) and liquid chromatography-mass spectrometry (LC-MS) have been especially useful in identifying and quantifying important organics. The Mass Spectrometry and Chromatography Facility at the Center For Advanced Food Technology (CAFT) at Rutgers University has a LC-MS capability, which is unique in being one of a few facilities to utilize an electron ionization (EI) source with a particle beam interface for structural work in food, natural products and the environment. Patents are held by Dr. Thomas Hartman of CAFT for the development of a short-path thermal desorption device for GC and GC-MS, now marketed by Scientific Instrument Services, Inc. of Ringoes, New Jersey.

Overview of Purge and Trap Thermal Desorption

Volatile and semi-volatile organics are purged with moderate heat by a stream of inert gas from the matrix of interest and the organics trapped by a bed of preconditioned Tenax-TA and/or activated carbon. After trapping the organics, the traps are first purged with nitrogen or helium at room temperature to dry them, then attached to the thermal desorption device and rapidly heated to thermally desorb the trapped volatile and semi-volatile compounds and deliver them into the GC column. The gas chromatograph separates the compounds and the mass spectrometer then identifies the separated individual compounds. A detailed description of this methodology is presented in the next section.

Selection of Trapping Agents for Organics of Interest

The two most popular trapping agents for this type of investigation are Tenax-TA and Carbotrap. These are registered trademarks for a porous polymer resin based on 2,6-diphenyl-p-phenyene oxide and an activated form of graphitized carbon, respectively. Both trapping agents have a high affinity for nonpolar and polar organic compounds and a very low affinity for water and low molecular weight polar compounds, such as alcohols (less than C-3). Therefore, water will pass through the traps and organic compounds will be effectively retained, thereby greatly concentrating the semi-volatile organics on the trap. Tenax is ideal for aromatics, heterocyclics, aldehydes, ketones, alcohols, etc., as long as the alkyl chain lengths are C-3 or longer. Carbotrap is good for the same type of compounds, but also effectively traps aliphatic, olefinic and other types of paraffinic hydrocarbons, which lack any other type of functionality. For these reasons we often make combination traps containing two or more trapping agents combined, which then provide us with a broad spectrum general purpose trap.

Preparation and Conditioning of Adsorbent Traps

Traps are prepared by packing 10 cm × 4 mm i.d. silanized borosilicate glass tubes with 25 mg of Tenax-TA 60/80 mesh and/or 25 mg of Carbotrap 20/40 mesh. The ends of the tubes are plugged with silanized glass wool approximately one centimeter on each end. Traps are then temperature programmed from ambient temperature to 320C at a rate of 2C per minute, while purging with nitrogen at a flow rate not less than 20 ml per minute. The traps are held at the upper temperature limit for not less than four hours. After conditioning, the tubes are promptly sealed on both ends using column end caps fitted with graphite ferrules. Traps are then fitted with aluminum column tags which are used to inscribe identification. Traps prepared in this manner exhibit excellent adsorptive capacity and contain no organic background (bleed or artifact peaks) when analyzed by GC-MS.

Drying of the Tenax or Carbon Tubes

After the semi-volatile marker compounds have been trapped in the adsorbent, the trapping materials are placed in a tube and purged for approximately ten minutes (at room temperature) with dry nitrogen gas to remove any traces of water vapor. This is very important because in the next step of analysis the sample is thermally desorbed into a capillary column gas chromatograph and a low temperature is used to focus the semi-volatile compounds at the head of the column. If appreciable water vapor is left on the traps, then the vapor will freeze on the capillary column and block the flow of carrier gas, thus affecting the overall performance of the system. The entire apparatus is easily disassembled for cleaning between samples. Of course in our experimental design, we always include a blank or negative control, which is treated the same as the sample, but contains distilled water instead. This allows us to determine if any artifacts or impurities are from the isolation apparatus or from the sample itself.

Thermal Desorption GC-MS Analysis of Adsorbent Traps

Charged adsorbent traps which now contain the volatile and semi-volatile compounds are thermally desorbed into the GC-MS system. To accomplish this our lab utilizes a short path thermal desorption apparatus of our own unique design (Fig. 3.1). The apparatus provides for ballistic heating of the adsorbent trap and efficiently transfers the desorbed semi-volatiles into the GC injection port via a short (4 cm) fused silica-lined adaptor-needle assembly. The features of this system are numerous. The entire heating zone and transfer lines exposed to the sample flow path are either borosilicate glass or fused silica-lined, thereby providing an extremely inert environment, which minimizes degradation of labile compounds, which are often decomposed upon contact with hot catalytic metallic surfaces. A second major advantage is that each sample utilizes its oven new adsorbent trap and the thermal desorber inlet and outlet adapters are changed for each separate sample. This establishes a completely independent pneumatic circuit for each adsorbent trap. The advantage of this is that it eliminates the possibility of cross contamination from sample to sample. Also since a separate adsorbent trap is used for each sample it prevents a "memory effect" from occurring, which is usually due to overloading. An additional feature is the ability to utilize the split ratio valve of the GC injection port to adjust the sample loading onto the column. Thermal desorption is carried out at 250C for 10-15 minutes during which time the GC column oven temperature is held at an

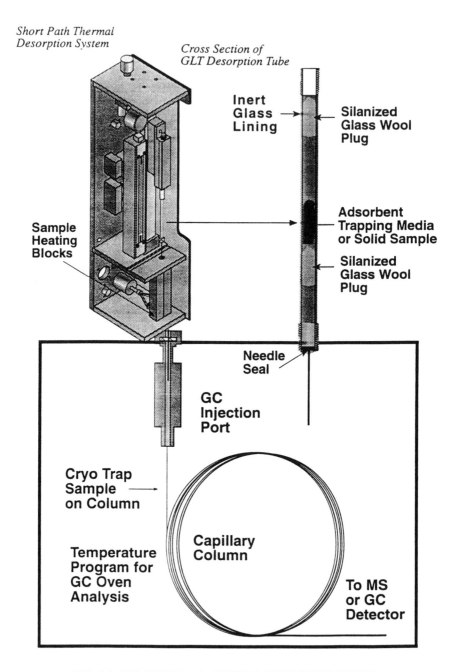

Short Path Thermal Desorption System

Cross Section of GLT Desorption Tube

Inert Glass Lining →

← Silanized Glass Wool Plug

Sample Heating Blocks

Adsorbent Trapping Media or Solid Sample

← Silanized Glass Wool Plug

Needle Seal

GC Injection Port

Cryo Trap Sample → on Column

Capillary Column

Temperature Program for GC Oven Analysis

To MS or GC Detector

FIG. 3.1. THE SHORT PATH THERMAL DESORPTION SYSTEM

appropriate temperature. This provides for focusing of the desorbed semi-volatiles during the thermal desorption interval resulting in excellent resolution and peak shape in the capillary chromatography.

We have utilized this methodology hundreds of times with excellent results. The performance of our system greatly exceeds that of any other commercial equipment currently available. Other commercial units which combine the purge and trap, thermal desorption, transfer lines and cryotraps into one unit suffer the following disadvantages. In these systems the adsorbent trap is a permanent integral part of the apparatus which is not easily removed or replaced. They operate on a purge and trap cycle, then thermal desorption, and then a backflush to clear the adsorbent trap for the next sample. Although these systems work well for EPA priority pollutant analyses, they do not work well for higher boiling compounds or for other general purpose semi-volatiles screening. Furthermore, they are at high risk for sample-to-sample cross contamination, since they utilize the same adsorbent trap over and over for hundreds of cycles. In addition, the transfer lines are long and often constructed of metalS, thereby causing a loss of resolution and introducing catalytic active sites which may cause degradation of labile samples.

Liquid Chromatography-Mass Spectrometry (LC-MS)

LC-MS permits separation of compounds too polar or thermally unstable for GC, while the mass spectrometer confers much greater selectivity and sensitivity than uv detectors. Initially, the thermospray technique referred to the production of ions from interaction with a solution of ammonium acetate. No filament is used in pure "thermospray" to generate ions. Ammonium acetate is an ideal buffer for LC/MS applications, because it is volatile and does not contaminate the mass spectrometer ion source and it is used as a source of ions. Reaction with the ammonium ion to give an MH^+ or $M + NH4^+$, where M is the molecular weight of the analyte, gave thermospray its initial success. Many HPLC separations, however, do not include ammonium acetate as a buffer. More recent models have filament and discharge ionization modes not requiring ammonium acetate, as well as atmospheric chemical ionization sources, which also utilize a discharge electrode to produce electrons. The definition of thermospray has now been changed, however, to reflect the technique whether ions result from ammonium acetate, a filament, or from discharge ionization. Also new is a technique called electrospray ionization mass spectrometers, where ions in solution may be directly analyzed by the mass spectrometer. This has been extremely useful for the determination of peptides and proteins. Another technique, Atmospheric Pressure Chemical Ionization (APCI) has now replaced thermospray for LC/MS applications.

The Center for Advanced Food Technology mass spectrometry facility is unique in utilizing a particle beam interface with an electron ionization (EI) source. This new technique is important as it gives fragmentation information, so as to ascertain structure, and allows comparison of spectra with NIH/EPA and NBS mass spectral libraries. An example of the chromatography obtained with an extract of sage is shown in Fig. 3.2, and the particle beam (EI) mass spectrum of rosmanol, which is one of the peaks in the chromatogram is shown in Fig. 3.3.

In our laboratory, particle beam electron ionization (EI) mass spectra are obtained on a Vestec model 201 mass spectrometer equipped with a Vestec particle beam interface, which is called a Universal Separator. The mass spectrometer is a single stage quadrupole equipped with a standard (open) EI source. Mass spectra are obtained with a filament operated at 70eV and 200 microamps. The Universal Separator operates by using a heated probe (operating temperature 120-130 degrees C) which nebulizes and partially vaporizes the LC eluent. Helium is added immediately after nebulization to carry the particles through the system. Nebulization produces particles, vapor and droplets. The droplets sink to the bottom of a U-tube and the condensate is removed by a peristaltic pump. The vast majority of solvent vapor is removed by a warmed (30 degrees C) countercurrent flow gas diffusion membrane separator. The remaining particles containing the analyte are carried to the mass spectrometer through a momentum separator where helium is removed. Analysis is then performed by the mass spectrometer.

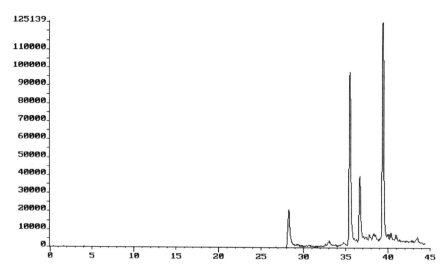

FIG. 3.2. HPLC-MS CHROMATOGRAM OF EXTRACT FROM SAGE

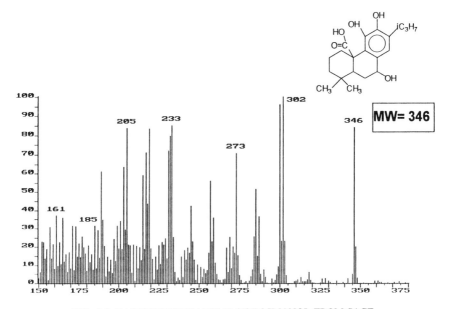

FIG. 3.3. PARTICLE BEAM EI-MS OF ROSMANOL FROM SAGE

REFERENCES

[1] Ho, C.T., Ferraro, T., Chen, Q., Rosen, R.T. and Huang, M.T. Phytochemicals in Teas and Rosemary and Their Cancer Preventive Properties. *In* Food Phytochemicals For Cancer Prevention, (Huang, M.T., Osawa, T., Ho. C.T. and Rosen, R.T., eds.) Washington DC:ACS Symposium Series, 1994; 547:2–19.

[2] Rosen, R.T., Roshdy, T.H., Hartman, T.G. and Ho, C.T. Determination of Free and Glycosidically Bound Organics in an Umbelliferous Vegetable Drink. *In* Food Phytochemicals For Cancer Prevention, (Huang, M.T., Osawa, T., Ho, C.T. and Rosen, R.T., eds.) Washington DC:ACS Symposium Series, 1994; 546:249–257.

[3] Ruiz, R., Hartman, T.G., Karmas, K., Lech, J. and Rosen, R.T. Breath Analysis Of Garlic Borne Phytochemicals in Human Subjects by Combined Adsorbent Trapping, Short Path Thermal Desorption GC-MS. *In* Food Phytochemicals For Cancer Prevention, (Huang, M.T., Osawa, T., Ho, C.T. and Rosen, R.T., eds.) Washington DC:ACS Symposium Series 1994; 546:102–119.

[4] Hartman, T.G., Lech, J., Karmas, J., Salinas, J., Rosen, R.T. and Ho, C.T. Flavor Characterization Using Adsorbent Trapping - Thermal Desorption or Direct Thermal Desorption-Gas Chromatography and Gas

Chromatography-Mass Spectrometry. *In* Flavor Measurement, (Ho, C.T. and Manley, C.H., eds.) New York:Marcel Dekker, Publisher, 1993:37–60.

[5]   Roshdy, T.H., Rosen, R.T., Hartman, T.G., Lech, J., Clark, L.B., Fukuda, E. and Ho, C.T. Glycosidically Bound Phenolics and Other Compounds in an Umbelliferous Vegetable Beverage. *In* Phenolic Compounds in Food and Their Effects on Health, (Ho, C.T., Lee, C.Y. and Huang, M.T., eds.) Washington DC:ACS Symposium Series 506, American Chemical Society, 1992:85–92.

[6]   Ho, C.T., Sheen, L.Y., Wu, P., Kuo, M.C., Hartman, T.G. and Rosen, R.T. Glycosidically Bound Aroma Compounds in Pineapple and Peach. *In* Proceedings of the 6th Weurman Flavour Research Symposium, (Thomas, A.F. and Bessier, Y., eds.) Chichester, UK: John Wiley & Sons, 1990:77–80.

[7]   Hartman, T.G., Ho, C., Rosen, J.D. and Rosen, R.T. Modern Techniques for the Analysis of Non-volatile and Thermally Labile Flavor Compounds. *In* Thermal Generation of Aromas, (Parliment, T.H., McGorrin, R.J., Ho, C.T., eds.) ACS Symposium Series, American Chemical Society, 1989; 8: 73–92.

CHAPTER 4

CHEMICAL EFFECTS OF PROCESSING AND FOOD PREPARATION ON CAROTENOIDS AND SOY AND GARLIC PHYTOCHEMICALS

JIN-R. ZHOU[1] and JOHN W. ERDMAN, JR.

Division of Nutritional Sciences
University of Illinois
451 Bevier Hall, 905 S. Goodwin
Urbana, IL 61801

ABSTRACT

Food processing procedures may play important roles in the biological functions of medicinal foods by altering the chemical composition and/or the bioavailability of the phytochemicals in those foods. Among extensively-studied phytochemicals, carotenoids in fruits and vegetables, isoflavones and saponins in soybean products, and sulfur compounds in garlic are reviewed in this chapter in regard to their reported biological functions and the effects of processing and food preparation on their content in prepared foods. The existing evidence reviewed demonstrates the importance of food processing on the potential medicinal functions of those phytochemical-containing foods. It is clear that much more research is needed to better understand the effects of processing on phytochemical composition in foods and the related biological functions of foods as consumed.

INTRODUCTION

A growing body of scientific data suggest that consumption of fruits, vegetables, and grains is inversely related to reduced risks of chronic diseases, such as cancer and heart disease, and that non-nutritive, naturally-occurring chemicals, known as phytochemicals, in those foods contribute to the disease-preventive properties. It has been noted that food processing procedures play a very important role in medicinal functions of phytochemicals in those foods. Many food processing and preparation procedures result in degradation, oxidation, geometric changes, and/or the formation of secondary phytochem-icals; all of these changes alter the bioactivity of specific components of foods.

---

[1] Current address: Nutrition & Metabolism Laboratory, Beth Israel Deaconess Medical Center, 194 Pilgrim Road, Boston, MA 02215

In this review, we will discuss effects of commonly-applied food processing and preparation procedures on the composition of phytochemicals in foods, particularly carotenoids in fruits and vegetables, isoflavones and saponins in soybeans and soy products, and sulfur compounds in garlic.

Carotenoids

Classification of Carotenoids

Carotenoids are a large family of compounds, with more than 500 naturally-occurring carotenoids in nature. They are synthesized by bacteria, higher plants, fungi, and algae. Food carotenoids are divided into two classes: the non-polar hydrocarbons called carotenes and their oxygenated derivatives called xanthophylls. Because of the presence of several sets of conjugated double bonds, carotenoids can specifically absorb light in the UV and/or visible region of the spectrum and thus produce color. In nature, carotenoids are predominantly found in the all-trans-isomer forms with only small amounts of mono-, or poly-cis isomers. Under the conditions of experimental treatment, food processing, or light exposure, carotenoids may isomerize and/or oxidize.[1,2]

Biological Functions of Carotenoids

Historically, carotenoids have been primarily recognized for their biological function as precursors of vitamin A. Among the naturally-occurring carotenoids, only about 10% of them have provitamin A activity. Besides their provitamin A activity, some carotenoids are now recognized to have other functions in biological systems. Many carotenoids are excellent antioxidants and participate in free radical-trapping and singlet oxygen-quenching reactions, and thereby have the potential to protect cells and organisms from oxidative damage.[3,4] The antioxidant property of carotenoids has been suggested to be responsible for their potential roles in reducing risks of certain types of cancer[5], protecting against heart disease[6], and enhancing immune response.[7] Carotenoids have also been shown to enhance the gap-junctional communication between cells by stimulating the synthesis of the gap-junction protein, connexin-43, and its mRNA[8,9], which does not appear to be due to their provitamin A or antioxidant properties.

Effects of Food Processing

Humans have to obtain their carotenoids from food systems because they are unable to synthesize these compounds. The presence of conjugated double bonds makes carotenoids very susceptible to thermal processing, even moderate heating, to light and to other processing procedures, which can result in isomerization of double bonds and/or oxidation of carotenoids. Table 4.1 lists the effects of major procedures of processing and food preparation on carotenoid content, and the extent of carotenoid isomerization and oxidation in a variety of fruits and vegetables.

Most processing and food preparation procedures involve heat treatment. Exposure of carotenoid-containing foods to moderate heating or light can result in isomerization of carotenoids. Severe heat treatment may also cause the oxidation of carotenoids. It has been reported that processing procedures, such as canning[10,12], blanching[11], baking[11], storage[13,14], light[14], and food preparation procedures, such as microwave and conventional heating[15] all resulted in increased cis-isomers and subsequent decrease of all-trans carotenoids with epoxy-containing carotenoids being more susceptible than other carotenoids.[15] For example, fresh sweet potato and carrots have 100% of β-carotene (βC) in all-trans configuration, while canned sweet potato and carrots contain 24.6 and 27.2% cis isomers mainly in the 13- and 9-cis isomer forms.[10] Thermal treatment primarily forms 13-cis βC[11,14], while light exposure favors 9-cis βC formation.[14,17]

Some mild treatments, however, such as blanching of sweet potatoes for 2 minutes[11], resulted in an increased carotenoid content most likely due to the release of carotenoids from the food matrix and/or increased digestibility of the food. Our laboratory has reported[18] a small enhancing effect of mild heat treatment on the serum and tissue accumulation (bioavailability) of α- and β-carotene from carrot slurries using the preruminant calf model, most likely due to release of carotenes from the carrot matrix.

Phytochemicals in Soy

Classification of Soy Phytochemicals

Soybean seeds contain various types of phytochemicals, including isoflavones, saponins, protease inhibitors, phytic acid, and tannins. Among these phytochemicals, isoflavones and saponins will be discussed in this review.

Isoflavones are the major phenolic compounds in soybeans. The main isoflavones which exist as glycosides are genistin, daidzin, and small amounts of glycitin. The enzymatic cleavage of isoflavone glycosides by β-glycosidase produces the isoflavone aglucones called genistein, daidzein, and glycitein, respectively. Isoflavones are also present in acylated forms, especially in the forms of malonyl isoflavones and acetyl isoflavones, with the malonylated-derivatives representing about 60% of the isoflavones.[21]

Saponins are steroid or triterpenoid glycosides which occur in a wide variety of plants. Among edible legume seeds, saponins are highest in soy. Soybean saponins are divided into three groups: soyasaponins A, B, and E, based on their aglucones of soyasapogenols A, B, and E, respectively. Soyasaponins I-V and A_1-A_6 have been isolated and their structure elucidated.

TABLE 4.1
EFFECTS OF PROCESSING AND FOOD PREPARATION ON THE CAROTENOID
CONTENT, CAROTENOID ISOMERIZATION, AND OXIDATION IN FRUITS
AND VEGETABLES

Processing Procedures	Effect on Carotenoid Content	References
Canning (vs. fresh)		
fruits & vegetables	decreased all-trans, increased cis-isomers	(10-12)
Pasteurization (vs. fresh)		
tomato juice	decreased carotenoid content 21%	(13)
Storage (vs. fresh)		
tomato juice (canned)	decreased carotenoid content 40%	(13)
carrot juice		
without light	increased 13-cis βC	(14)
with light	increased 9-cis βC	(14)
Cooking		
Microwave cooking (vs. fresh)		
garland chrysanthemum	decreased all-trans, increased cis-isomers, more loss of epoxy-containing carotenoids	(15)
sweet potatoes	decreased βC 23%, increased 13-cis βC 15%	(11)
Brussel sprouts	decreased xanthophylls	(19)
Conventional cooking (vs. microwave)	less all-trans,	
garland chrysanthemum	more cis-isomer (except cis-violaxanthin)	(15)
Blanching (vs. fresh)		
sweet potatoes	increased total carotene, increased 13-cis βC	(11)
Baking (vs. fresh)		
sweet potatoes	decreased total carotene 31% increased 13-cis βC 25%	(11)
Dehydration (vs. fresh)		
sweet potatoes	decreased total carotene, increased 13-cis βC	(11)
pepper	increased carotenogenesis, increased degradation	(16)
carrots, broccoli, & spinach	decreased carotene	(20)
Lighting (vs. fresh)		
Carrot juice	increased 9-cis βC, increased oxidation	(14,17)
pepper	increased carotenogenesis	(16)

Abbreviation: βC, β-carotene

Soyasaponins I-V possess soyasapogenol B as the common aglucone and are the monodesmosides of soyasapogenol B.[22] Soyasaponins A_1-A_6 possess soyasapogenol A as the common aglucone and are the bisdesmosides of soyasapogenol A.[23] Soyasaponins A_1-A_6 are present in partially acetylated forms.[24,25]

Biological Actions of Soy Phytochemicals

One biological effect of phytochemicals in soybean is their adverse sensory characteristics. The soyasaponin A group was reported to be responsible for an undesirable bitter and astringent taste. Further studies indicated that partially acetylated soyasaponins A_1-A_6 had bitter and astringent taste, while their parent soyasaponins A_1-A_6 did not.[24,25] This objectionable flavor was also reported for daidzein and genistein.

Isoflavones have been shown to function as antioxidants[26], antifungal agents[27,28], and anticarcinogens.[29,30] It has been suggested that isoflavones play a role in the prevention of estrogen-dependent breast cancer. Genistein was shown to inhibit the growth of human breast cancer[29] and prostate cancer[30] cell lines in culture. The possible mechanism may involve the inhibitory effect of isoflavones on the activity of tyrosine protein kinase.[31]

Soyasaponins possess a diversity of biological properties, such as hemolytic and hypocholesterolemic[32-34], antioxidative, antitumor-promoting, and anti-HIV-infecting properties.[35]

Effects of Food Processing

The common processing procedures used to produce soy products include germination, fermentation, cooking and a variety of commercial processes to produce soy protein concentrates and isolates. The effects of processing and food preparation on the content of isoflavones and saponins in soybeans and soy products are shown in Table 4.2.

Germination of soybean results in the overall increase and redistribution of soyasaponins in different parts of the germinating soybean. Upon germination in the dark, soyasaponins S-I and S-II levels in seed cotyledon and sprouts, and S-V and acetyl-soyasaponin A_4 (AS-A_4) levels in sprouts are elevated, compared to the levels in non-germinated soybeans.[36] Germination under light results in an even higher increase of S-I in sprouts.[36]

Compared to soybean seeds, fermented soybean products have decreased isoflavone[40] and saponin[41,42] contents, and increased conversion of isoflavone β-glycosides to their aglucones[40] due to enzymatic hydrolysis of isoflavone β-glycosides. The enzymatic activity of β-glycosidase is influenced by heating and pH. Boiling soybean products under alkaline condition substantially inactivated the activity of the enzyme.[43]

TABLE 4.2

EFFECTS OF PROCESSING AND FOOD PREPARATION PROCEDURES ON
PHYTOCHEMICALS IN SOYBEANS AND SOY PRODUCTS

Processing Procedures	Effect on Phytochemical Content	References
Germination (vs. soybean seed)		
without light		
seed cotyledons	increased S-I, S-II	(36)
	increased isoflavones (vs, roots & hypocytyls)	(37)
sprouts	increased S-I, S-II, S-V, and AS-A$_4$	(36)
	increased genistein	(38)
epicotyls	increased S-I, S-III	
hypocotyls	increased S-I, AS-A$_4$, concentrated isoflavones	(36,39)
roots	increased S-I, S-II, S-III, S-V, AS-A$_4$	(36)
with light		
cotyledons	increased S-I	(36)
sprouts	greatly increased S-I, increased S-II & S-V	(36)
Fermentation (vs. soybean seed)		
fermented	decreased isoflavone, increased isoflavone	
	aglucones	(40,45)
	decreased saponins	(41,42)
Soaking		(43)
50°C water, 6 hr	mainly daidzein and genistein	
50°C 0.25% NaHCO$_3$, 6 hr	decreased genistein and daidzein (10%)	
boiling, 0.25% NaHCO$_3$, 30 min	decreased genistein and daidzein (90%)	
Soy Products		
soy flour (vs. soybean)	maintained isoflavones	(40,45)
	maintained saponins	(47)
soy oil	trace isoflavones	(40,45)
soy protein concentrate (vs. soy flour)		
alcohol extraction	decreased isoflavones 10-20 fold	(40,45)
	traces of saponins retained	(47)
hot water (pH ~7)	maintained isoflavones	(40,45)
soy protein isolate (vs. soy flour)	decreased isoflavones 4-6 fold and	
	decreased daidzein/genistein ratio	(40)
	increased saponins	(47)
tofu (vs. soy flour)	increased genistein	(38)
	maintained saponins	(48)
soy sauce (vs. soybean)	decreased isoflavones	(40)
tempeh	glycitein converted to	
	6,7,4'-tri(OH)-isoflavone	
	(factor 2)	(40)
	daidzein converted to glycitein	
	and factor 2	(40)
	decreased saponins	(41)

Abbreviation: S-I to S-V, soyasaponin I to soyasaponin V; AS-A$_4$, acetyl-soyasaponin A$_4$.

Most of the isoflavone content of raw soybeans resides in the hypocotyl of the bean.[43] Hulls or soy oil are essentially devoid of isoflavones. There is considerable variation in the contents of isoflavones between varieties of beans as well as location and year of production.[43,44] Wang and Murphy[44] reported over a 3-fold range of total isoflavone contents (from 1,176 to 4,216 mg/g for 12 isoflavones) in 11 varieties, planted in different years and at different locations. Variety and crop year had a greater influence in the variation than did location. The same authors[45] compared concentration and distribution of the same 12 isoflavones in 29 commercial soybean foods. They found that heat processing, enzymatic hydrolysis, and fermentation all significantly altered the isomer distribution of isoflavones. In regard to concentration of total isoflavones, high protein soy ingredients (concentrates and isolates) contained similar levels compared to unprocessed soybeans, except for alcohol-leached soy concentrate where the majority of isoflavones are removed by alcohol leaching.[40,45] The so-called "second generation" soy foods, such as soy yogurt, soy bacon and soy or tempeh burgers, all have reduced concentrations of isoflavones due to dilution of the soy with non-soy ingredients.

As with isoflavones, the highest concentration of saponins is in the hypocotyl of soybeans. Hypocotyls are about 8-fold higher in saponin concentration than are cotyledons. Hulls are essentially devoid of saponins.[46] Since saponins are strongly surface active and bind to proteins, it is understandable that saponins in soy protein products, such as soy flour[40], soy protein isolate[47], and tofu[48], were maintained or even increased. However, soy protein concentrate produced from alcohol extraction showed only a trace amount retained of saponins.[47] Saponins are heat-stable under normal food processing procedures.

Yu et al.[49] recently investigated the bioavailability of isoflavones from soy milk in adult women. They determined that about 85% of the isoflavones were degraded in the intestine, but an average recovery of 21% and 9%, of daidzein and genistein, respectively, was noted in the urine. Thus, daidzein appears to be more bioavailable than genistein.

Phytochemicals in Garlic

Classification of Garlic Phytochemicals

The intact and undisturbed bulb of garlic contains only a few biologically active compounds, mainly alliin (S-allyl-L-cysteine S-oxide), a colorless and odorless compound. When garlic cloves are chopped or crushed, several diallyl thiosulfinates are rapidly formed by the action of the enzyme alliin alliinase, where alliin is enzymatically converted into thiosulfinates with allicin predominating. Allicin is the principal strong smelling compound in chopped garlic. Most of the thiosulfinates, especially allicin, are very unstable and convert to polysulfides such as diallyl sulfide (DAS), diallyl disulfide (DADS), diallyl trisulfide (DAT), allyl methyl disulfide (AMD), allyl methyl trisulfide (AMT), and lesser amounts of other compounds, such as ajoene and vinyldithiins.[50,51]

Biological Activities of Garlic Phytochemicals

Garlic has been used as a folk medicine since ancient time. Consisting of various physiologically active compounds, garlic has been reported to be used for prevention of stroke, coronary thrombosis and atherosclerosis, as well as for treatment of various diseases including infections and vascular disorders.

Epidemiological investigations suggest that the risk of stomach cancer decreases with increasing dietary intake of garlic.[52] Animal and in vitro studies indicate that garlic sulfur compounds have anticarcinogenic effects. For example, DAS has been shown to reduce chemically-induced colon and esophageal cancers[53,54], forestomach tumors[55], lung tumors[56] and skin tumors[57], while DADS reduced colon and renal cancers[58]; AMT, AMD, allyl trisulfide, and allyl sulfide reduced forestomach neoplasms and lung cancer.[59]

Besides their anticarcinogenic effects, garlic sulfur compounds have also been shown to have other biological actions including: ajoene[60], allicin, DADS, AMT, and DAT[61] in inhibiting platelet aggregation; aged garlic extract or alcohol extract of garlic[62,63] in antibacterial activity; ajoene[64,65] in antifungal activity; and garlic extract in antivirus activity[66], in enhancing immune responses[67,68], and prevention of lipid peroxidation.[69]

Effects of Food Processing

Table 4.3 lists the primary products formed in garlic during different processing procedures. A variety of compounds are produced depending on the methods of treatment. The intact bulb of garlic contains only a few bioactive compounds. When chopped, steamed or processed as an ingredient in food, garlic's chemistry is very much altered. In these processes at least 100 sulfur-containing compounds are produced, many of which may relate to garlic's medicinal use.

When garlic is allowed to ferment, S-allyl cysteine and mercapto cysteine are the two major water-soluble compounds formed, together with minor amounts of other sulfur compounds.[70] Maceration of garlic in vegetable oil resulted in the formation of vinyldithiins as major compounds, as well as smaller amounts of ajoene, DAT and AMT.[73,75]

Processing by way of cooking results in a variety of compounds which are reported to be biologically active. Oven- or microwave-baking of garlic produces DADS and DAT as major compounds, while frying, oil-cooking, or microwave-cooking garlic mainly forms DADS, AMD, and vinyldithiins.[76] Garlic juice mainly contains 2-vinyl 4H-1,3-dithiin and 3-vinyl-4H-1,2-dithiin. Upon heating the juice (40°C) the amount of 2-vinyl-4H-1,3-dithiin was increased.[77]

TABLE 4.3
EFFECTS OF PROCESSING AND FOOD PREPARATION ON SULFUR
PHYTOCHEMICALS IN GARLIC AND GARLIC PRODUCTS

Processing Procedures	Primary Sulfur Products Produced During Processing	References
Fermentation	S-allyl cysteine, S-allyl mercapto cysteine, other sulfur-containing amino acids	(70)
Steam distillation		
distillate oil	diallyl, methyl allyl, dimethyl, and allyl 1-propenyl oligosulfides	(71-73)
	DADS, AMT, and DAT	(74)
Maceration	vinyl dithiins (major), ajoene, DAT, AMT	(73,75)
Cooking		
oven- or microwave-baking	DADS and DAT	(76)
fried, oil-cooked,		
microwave-fried	DADS, AMD, vinyldithiins	(76)
Heating (40°C)		
garlic juice (vs. unheated juice)	increased 2-vinyl-4H-1,3-dithiin, 3-vinyl-4-H-1,2-dithiin	(77)
pH effects (raw, blended in different pH, heated)		(78)
pH 5.5	2-vinyl-4H-1,3-dithiin, and 3-vinyl-4H-1,2-dithiin	
neutral or weak acid	DAT, AMT, cis-1-propenyl allyl disulfide, isobutyl isothiocyanate, 2,4--dimethylfuran, 1,3-dithiane, aniline, AMS, DMDS	
pH 9.0	DADS, DAS, AMD, propenylthiol, propyl allyl disulfide, 1,2-epithiopropane	
Extraction		(50)
100°C steam	DADS	
ethyl alcohol and water (25°C)	allicin	
ethyl alcohol (<0°C)	alliin	
Commercial products		
spice garlic	allicin	(79)
"health food"	allicin	(79)
aged garlic extract	diallyl polysulfides (tri-, tetra-, penta-, hexa-, and hepta-)	(69)

Abbreviation: DADS, diallyl disulfide; DAT, diallyl trisulfide; DAS, diallyl sulfide; AMD, allyl methyl disulfide; AMS, allyl methyl sulfide; AMT, allyl methyl trisulfide; DMDS, dimethyl disulfide.

The formation of sulfur-containing compounds from heating garlic is also pH-dependent. Blanching raw garlic in acidic condition followed by heating results in the formation of 3-vinyl-4H-1,2-dithiin and 2-vinyl-4H-1,3-dithiin as major compounds, while similar treatments in neutral or weak acidic or basic conditions resulted in the formation of a variety of very different compounds.[78]

Several methods have been used for extraction of sulfur-containing compounds from garlic. Steam distillation, the harshest technique, yields a variety of oligosulfides including DADS, AMT, and DAT.[50, 71-74] The gentler method of extraction using ethyl alcohol and water at room temperature mainly yields allicin, while the application of ethyl alcohol at a subzero temperature, the gentlest technique, primarily extracts the parent compound alliin.[50]

To summarize, many individual food processing and preparation procedures influence not only the total phytochemical content, but the profile of derivative phytochemicals formed. Because each phytochemical may have a specific bioactivity, food processing procedures may play an important role in influencing medicinal functions of processed foods. For example, one garlic preparation may have a totally different array of sulfur compounds than another (see Table 4.3). Therefore, researchers must be particularly aware of prior processing history and the phytochemical profile of the products they are testing.

REFERENCES

[1] Quackerbush, F.W. Reversed-phase HPLC separation of cis- and trans-carotenoids and its application to beta-carotene in food materials. J. Liquid Chrom. 1987; 10:643–53.

[2] Erdman, J.W. Jr., Poor, C.L., and Dietz, J.M. Factors affecting the bioavailability of vitamin A, carotenes and vitamin E. Food Technol. 1988; 42:214–221.

[3] Krinsky, N.I. Beta-carotene: Functions. In New Protective Roles for Selected Nutrients, (Spiller, G.A. and Scala, J., eds.) Alan R. Liss, New York, 1989:1–15.

[4] Lieber, D.C. Antioxidant reactions of carotenoids. In Carotenoids in Human Health (Canfield, L.M., Krinsky, N.I. and Olson, J.A., eds.) Annals NY Acad. Sciences, New York, 1993; 691:20–31.

[5] Ziegler, R.D. A review of epidemiologic evidence that carotenoids reduce the risk of cancer. J. Nutr. 1989; 119:116–122.

[6] Hennekens, C.H. and Eberlein, K. A randomized trial of aspirin and β-carotene among U.S. physicians. Prev. Med. 1985; 14:165–168.

[7] Bendich, A. Carotenoids and the immune response. J. Nutr. 1989; 119:112-115.

[8] Rogers, M., Berestecky, J.M., Hossain, M.E., Guo, H., Kadle, R., Nicholson, B.J. and Bertram, J.S. Retinoid-enhanced gap junctional communi-

cation is achieved by increased levels of connexin 43 mRNA protein. Mol. Carcinogenesis 1990; 3:335–343.

[9] Wolf, G. Retinoids and carotenoids as inhibitors of carcinogenesis and inducers of cell-cell communication. Nutr. Rev. 1992; 50:270–274.

[10] Chandler, L.A. and Schwartz, S.J. HPLC separation of cis-trans carotene isomers in fresh and processed fruits and vegetables. J. Food Sci. 1987; 52:669–672.

[11] Chandler, L.A. and Schwartz, S.J. Isomerization and losses of trans-β-carotene in sweet potatoes as affected by processing treatments. J. Agric. Food Chem. 1988; 36:129–133.

[12] Edwards, C.G. and Lee, C.Y. Measurement of provitamin A carotenoids in fresh and canned carrots and green peas. J. Food Sci. 1986; 51:534–535.

[13] Dietz, J.M. and Gould, W.A. Effects of process stage on retention of beta carotene in tomato juice. J. Food Sci. 1986; 51:847–848.

[14] Pesek, C.A. and Warthesen, J.J. Kinetic model for photoisomerization and concomitant photodegradation of β-carotene. J. Agric. Food Chem. 1990; 38:1313–1315.

[15] Chen, B.H. Studies on the stability of carotenoids in garland chrysanthemum (Ipomoea spp.) as affected by microwave and conventional heating. J. Food Prot. 1992; 55:296–300.

[16] Minguez-Mosquera, M.I., Jaren-Galan, M. and Garrido-Fernandez, J. Competition between the processes of biosynthesis and degradation of carotenoids during the drying of peppers. J. Agric. Food Chem. 1994; 42:645–648.

[17] Clydesdale, F.M., Ho, C.-T., Lee, C.Y., Mondy, N.I. and Shewfelt, R.L. The effects of postharvest treatment and chemical interactions on the bioavailability of ascorbic acid, thiamin, vitamin A, carotenoids, and minerals. Critical Rev. Food Sci. Nutr. 1991; 30:599–638.

[18] Poor, C.L., Bierer, T.L., Merchen, N.R., Fahey, G.C., Jr. and Erdman, J.W., Jr. The accumulation of α- and β-carotene in serum and tissues of preruminant calves fed raw and steamed carrot slurries. J. Nutr. 1993; 123:1296–1304.

[19] Khachik, F., Beecher, G.R. and Whittaker, N.F. Separation, identification, and quantification of the major carotenoid and chlorophyll constituents in extracts of several green vegetables by liquid chromatography. J. Agric. Food Chem. 1986; 34:603–616.

[20] Park, Y.W. Effect of freezing, thawing, drying, and cooking on carotene retention in carrots, broccoli and spinach. J. Food Sci. 1987; 52:1022–1025.

[21] Fleury, Y., Welti, D.H., Philippossian, G. and Magnolato, D. Soybean (malonyl) isoflavones: characterization and antioxidant properties. In Phenolic Compounds in Food and Their Effects on Health, (Huang, M.-T., Ho, C.-T. and Lee, C.Y., eds.) Am. Chem. Soc., Washington, D.C., 1992:98–113.

[22] Burrows, J.C., Price, K.R. and Fenwick, G.R. Saponin, IV, an additional monodesmosidic saponin isolated from soybean. Phytochem. 1987; 26:1214-1215.

[23] Curl, C.L., Price, K.R. and Fenwick, G.R. Soyasaponin A_3, a new monodesmosidic saponin isolated from the seeds of Glycine max. J. Natural Prod. 1988; 51:122-124.

[24] Kitagawa, I., Taniyama, T., Nagahama, Y., Okubo, K., Yamauchi, F. and Yoshikawa, M. Saponin and sapogenol. XLII. Structures of acetyl-soyasaponins A_1, A_2, and A_3, astringent partially acetylated bisdemosides of soyasapogenol A, from American soybean, the seeds of Glycine max Merrill. Chem. Pharm. Bull. 1988; 36:2819-2828.

[25] Taniyama, T., Nagahama, Y., Yoshikawa, M. and Kitagawa I. Saponin and sapogenol. XLIII. Acetyl-soyasaponins A_4, A_5, and A_6, new astringent bisdesmosides of soyasapogenol A, from Japanese soybean, the seeds of Glycine max Merrill. Chem. Pharm. Bull. 1988; 36:2829-2839.

[26] Wei, H.C., Wei, L.H., Frenkel, K., Bowen, R. and Barnes, S. Inhibition of tumor promoter — induced hydrogen peroxide formation in vitro and in vivo by genistein. Nutr. Cancer 1993; 20:1-12.

[27] Naim, M., Gestetner, B., Zilkah, S., Birk, Y. and Bondi, A. Soybean isoflavones, characterization, determination and antifungal activity. J. Agric. Food Chem. 1974; 22:806.

[28] Kramer, R.P., Hindorf, H. and Jha, H.C. Antifungal activity of soybean and chickpea isoflavones and their reduced derivatives. Phytochem. 1984; 23:2203-2205.

[29] Peterson, T.G. and Barnes, S. Genistein inhibition of the growth of human breast cancer cells: independent from estrogen receptors and the multi-drug resistance gene. Biochem. Biophys. Res. Commun. 1991; 179:661-667.

[30] Peterson, T.G. and Barnes, S. Genistein and biochanin A inhibit the growth of human prostate cancer cells but not epidermal growth factor receptor tyrosine autophosphorylation. Prostate 1993; 22:335-345.

[31] Akiyama, T., Ishida, J., Nakagawa, S., et al. Genistein, a specific inhibitor of tyrosine-specific protein kinase. J. Biol. Chem. 1987; 262:5592-5595.

[32] Oakenfull, D.G. Dietary fiber, saponins and plasma cholesterol. Food Tech. Aust. 1981; 33:4-32.

[33] Potter, J.D., Illman, R.D., Calvert, G.D., Oakenfull, D.G. and Topping, D.L. Soya saponins, plasma lipids, lipoproteins and fecal bile acids: a double blind cross-over study. Nutr. Rep. Int. 1980; 22:521-528.

[34] Price, K.R., Johnson, I.T. and Fenwick, G.R. The chemistry and biological significance of saponins in foods and feedingstuffs. Crit. Rev. Food Sci. Nutr. 1987; 26:27.

[35] Nakashima, H., Okubo, F., Honda, Y., Tamura, T., Matsuda, S. and Yamamoto, N. Inhibitory effect of glycosides-like saponin from soybean on

the infectivity of HIV in vitro. AIDS 1989; 3:655–658.

[36] Shimoyamada, M. and Okubo, K. Variation in saponin contents in germinating soybean seeds and effect of light irradiation. Agric. Biol. Chem. 1991; 55:577–579.

[37] Suganuma, N. and Takaki, M. Changes in amounts of isoflavones in seeds during germination of soybean and role in the formation of root nodules. Soil Sci. Plant Nutr. 1993; 39:661–667.

[38] Murphy, P.A. Phytoestrogen content of processed soybean products. Food Technol. 1982; 16:60.

[39] Eldridge, A.C. and Kwolek, W.F. Soybean isoflavones: effect of environment and variety on composition. J. Agric. Food Chem. 1983; 31:394–396.

[40] Coward, L., Barnes, N.C. and Setchell, D.R. Genistein, daidzein, and their β-glycoside conjugates: antitumor isoflavones in soybean foods from American and Asian diets. J. Agric. Food Chem. 1993; 41:1961–1967.

[41] Fenwick, D.E. and Oakenfull, D. Saponin content of food plants and some prepared foods. J. Sci. Food Agric. 1983; 34:186–191.

[42] Kitagawa, J., Yoshikawa, M., Hayashi, T. and Tanayama, T. Characterization of saponin constituents in soybeans of various origins and quantitative analysis of soyasaponin by gas-liquid chromatography. Yakugaku Zasshi 1984; 104:162.

[43] Ha, E.Y.M., Morr, C.V. and Seo, A. Isoflavone aglucones and volatile organic compounds in soybeans: effects of soaking treatments. J. Food Sci. 1992; 57:414–417.

[44] Wang, H.-J. and Murphy, P.A. Isoflavone composition of American and Japanese soybeans in Iowa: Effect of variety, crop year, and location. J. Agric. Food Chem. 1994; 42:1674–1677.

[45] Wang, H.-J. and Murphy, P.A. Isoflavone content in commercial soybean foods. J. Agric. Food Chem. 1994; 42:1666–1673.

[46] Taniyama, T., Yoshikawa, M. and Kitagawa, I. Saponin and sapogenol. XLIV. Soyasaponin composition in soybeans of various origins and soyasaponin content in various organs of soybean. Structure of soyasaponin V from soybean hypocotyl. Yakugaku Zasshi 1988; 108:562–571.

[47] Ireland, P.A., Dziedzic, S.Z. and Kearsley, M.W. Saponin content of soya and some commercial soya products by means of high-performance liquid chromatography of the saponins. J. Sci. Food Agric. 1986; 37:694–698.

[48] Fenwick, D.E. and Oakenfull, D. Saponin content of soya beans and some commercial soya bean products. J. Sci. Food Agric. 1981; 32:273–278.

[49] Xu, X., Wang, H.-J., Murphy, P.A., Cook, L. and Hendrick, S. Daidzein is a more bioavailable soy milk isoflavone than is genistein in adult women. J. Nutr. 1994; 124:825–832.

[50] Block, E. The chemistry of garlic and onions. Sci. Am. 1985; 252:114–119.

[51] Raghavan, B., Abraham, K.O. and Shankaranarayana, M.L. Chemistry of garlic and garlic products. J. Sci. Ind. Res. 1983; 42:401–409.

[52] You, W.C., Blot, W.J., Chang, Y.S., et al. Diet and high risk of stomach cancer in Shandong, China. Cancer Res. 1988; 48:3518.

[53] Wargovich, M.J. Diallyl sulfide, a flavor component of garlic (Allium sativum), inhibits dimethylhydrazine-induced colon cancer. Carcinogenesis 1987; 8:487–489.

[54] Wargovich, M.J. New dietary anticarcinogens and prevention of gastrointestinal cancer. Dis. Colon Rectum 1988; 31:72–75.

[55] Hadjiolov, D., Fernando, B.C., Schmeiser, H.H., Wiebler, M., Hadjiolov, N. and Pirajnov, G. Effect of diallyl sulfide on aristolochic acid-induced forestomach carcinogenesis in rats. Carcinogenesis 1993; 14:407–410.

[56] Hong, J.-Y., Lin, M.C., Wang, Z.Y., Wang, E.-J. and Yang, C.S. Inhibition of chemical toxicity and carcinogenesis by diallyl sulfide and diallyl sulfone. In Food phytochemicals for cancer prevention I. Fruits and vegetables. (Huang, M.T., Osawa, T., Ho, C.T. and Rosen, R.T., eds.) Maple Press, New York. 1994:97–101.

[57] Dwivedi, C., Rohlfs, S., Jarvis, D. and Engineer, F.N. Chemoprevention of chemically-induced skin tumor development by diallyl sulfide and diallyl disulfide. Pharm. Res. 1992; 9:1168–1170.

[58] Takahashi, S., Hakoi, K., Yada, H., Hirose, M., Ito, N. and Fukushima, S. Enhancing effects of diallyl sulfide on hepatocarcinogenesis and inhibitory actions of the related diallyl disulfide on colon and renal carcinogenesis in rats. Carcinogenesis 1992; 13:1513–1518.

[59] Sparnins, V.L., Barany, G. and Wattenberg, L.W. Effects of organosulfur compounds from garlic and onions on benzo[a]pyrene-induced neoplasia and glutathione S-transferase activity in the mouse. Carcinogenesis 1988; 9:131.

[60] Srivastava, K.C. and Tyagi, O.D. Effects of a garlic-derived principle (ajoene) on aggregation and arachidonic acid metabolism in human blood platelets. Prostaglandins, Leukotrienes and Essential Fatty Acids 1993; 49:587–595.

[61] Ariga, T., Oshiba, S. and Tamada, T. Platelet aggregation inhibitor in garlic. Lancet 1981; i:150–151.

[62] Deshpande, R.D., Khan, M.B., Bhat, D.A. and Navalkar, R.G. Inhibition of mycobacterium-avium complex isolates from AIDS patients by garlic (Allium sativum). J. Antimicrobial Chemotherapy 1993; 32:623–626.

[63] Farbman, K.S., Barnett, E.D, Bolduc, G.R and Klein, J.O. Antibacterial activity of garlic and onions — A historical perspective. Ped. Inf. Dis. J. 1993; 12:613–614.

[64] Singh, U.P., Pandey, V.N. and Wagner, K.G. Antifungal activity of ajoene, a constituent of garlic (Allium sativum). Canad. J. Botany 1990; 68:1354–1356.

[65] Singh, U.P., Chauhan, V.B. and Wagner, K.G. Effect of ajoene, a compound derived from garlic (Allium sativum), on phtophthora drechsleri f. sp. cajani. Mycologia 1992; 84:105-108.

[66] Esanu, V. and Prehoveanu, E. The effect of garlic, applied as such or in association with NaF, on experimental influenza in mice. Rev. Roum. Med. Virol. 1983; 34:11-17.

[67] Kandil, O.M., Abdullah, T.H. and Elkadi, A. Garlic and the immune system in humans: its effect on natural killer cells. Fed. Proc. 1987; 46:441.

[68] Lau, B.H.S., Yamasaki, T. and Gridley, D.S. Garlic compounds modulate macrophage and T-lymphocyte functions. Mol. Biother. 1991; 3:103-107.

[69] Horie, T., Awazu, S., Itakura, Y. and Fuwa T. Identified diallyl polysulfides from an aged garlic extract which protects the membranes from lipid peroxidation. Planta Med. 1992; 58:468-469.

[70] Gruber, P. Dags for en ny syn pa vitloksprodukter. Svensk Farmacevtisk Tidskrift 1992; 96:26.

[71] Vernin, G., Metzger, S., Fraisse, D. and Scharff, C. GC-MS (EI, PCI, NCI) computer analysis of volatile sulfur compounds in garlic essential oils: application of mass fragmentometry SIM technique. Planta Med. 1986; 52:96.

[72] Yu, T.H., Wu, C.M. and Liou, Y.C. Volatile compounds from garlic. J. Agric. Food Chem. 1989; 37:725.

[73] Lawson, L.D., Wang, Z.Y.J. and Hughes, B.G. Identification and HPLC quantitation of the sulfides and dialk(en)yl thiosulfinates in commercial garlic products. Planta Med. 1991; 57:363-370.

[74] Yan, X.J., Wang, Z.B. and Barlow, P. Quantitative estimation of garlic oil content in garlic oil based health products. Food Chem. 1992; 45:135-139.

[75] Brodnitz, M.H, Pascale, J.V. and van Derlice, L. Flavor components of garlic extract. J. Agric. Food Chem. 1971; 19:273.

[76] Yu, T.H., Wu, C.M. and Ho, C.T. Volatile compounds of deep-oil fried, microwave-heated, and oven-baked garlic slices. J. Agric. Food Chem. 1993; 41:800-805.

[77] Yu, T.H. and Wu, C.M. Stability of allicin in garlic juice. J. Food Sci. 1989; 54:977-981.

[78] Yu. T.H., Wu, C.M. and Liou, Y.C. Effects of pH adjustment and subsequent heat treatment on the formation of volatile compounds of garlic. J. Food Sci. 1989; 54:632-635.

[79] Saito, K., Horie, M. and Hoshino, Y. Determination of allicin in garlic and commercial garlic products by gas chromatography with flame photometric detection. J. Assoc. Off. Anal. Chem. 1989; 72:917-920.

CHAPTER 5

PHYTOCHEMICALS: BIOCHEMICAL MARKERS OF INGESTION, ABSORPTION AND METABOLISM USING FLAXSEED AS A MODEL

CLARE M. HASLER

Director, Functional Foods for Health Program
Department of Food Science and Human Nutrition
103 Agricultural Bioprocess Lab
1302 West Pennsylvania Avenue
Urbana, IL 61801

ABSTRACT

Phytochemicals will play an increasingly important role in optimal nutrition in the future. Flaxseed is the richest source of a unique class of phytochemicals — the mammalian lignans; enterodiol and enterolactone. Lignans are significantly elevated in human urine following flaxseed consumption, and are thought to have potential as a chemopreventive agent because of their ability to modulate estrogen metabolism. Flaxseed may serve as a model to examine the ingestion, absorption and metabolism of phytochemicals. Reliable markers of phytochemical metabolism are necessary to ascertain the safety and efficacy of additional designer/functional foods as this new field in the food and nutrition sciences continues to develop.

INTRODUCTION

As we approach the year 2000, we are faced with an entirely different array of health problems than those which challenged nutrition researchers in the early part of this century. There is currently an epidemic of diet-related chronic diseases — particularly cancer — which is predicted to become the number one killer within five years.[1] Phytochemicals will play an increasingly important role in optimal nutrition as this new era in the food and nutrition sciences develops. Thus, there is a need for reliable biochemical markers of phytochemical ingestion, absorption and metabolism.

Phytochemicals in Cancer Prevention

In 1982, the National Research Council advised that cancer risk might be reduced by increasing the consumption of citrus fruits, and carotene-rich and cabbage family vegetables.[2] That recommendation continues to be substantiated by strong epidemiological evidence including a recent review of 200 studies, which demonstrated that persons with a low fruit and vegetable intake experienced twice the risk of cancer at most bodily sites.[3] The chemopreventive effects of fruits and vegetables is thought to be due to the presence of non-nutritive, physiologically-active secondary metabolites called phytochemicals[4], of which more than a dozen classes have been identified (Table 5.1). Phytochemicals have recently been a topic of intense research efforts[5] and popular press attention.[6]

TABLE 5.1
MAJOR CLASSES OF PHYTOCHEMICALS POTENTIALLY INVOLVED
IN CANCER PREVENTION

Phytochemical Class	Food Source
Allium compounds	Garlic, onions, chives, leeks
Carotenoids	Yellow and orange vegetables and fruits, dark green leafy vegetables
Coumarins	Vegetables and citrus fruits
Dithiolthiones	Cruciferous vegetables
Flavonoids	Most fruits and vegetables
Glucosinolates, indoles	Cruciferous vegetables, particularly Brussels sprouts, rutabaga, mustard greens, dried horseradish
Inositol hexaphosphate	Plant foods, particularly soybeans and cereals
Isoflavones	Soybeans
Isothiocyanates	Cruciferous vegetables
Lignans	Flaxseed
Limonene	Citrus fruits
Phenols	Nearly all fruits and vegetables
Protease inhibitors	Plant foods, particularly seeds and legumes, including soybeans
Saponins	Plant foods, particularly soybeans
Sterols	Most vegetables, soybeans

Lignans as Markers for Flaxseed Ingestion and Absorption

Flaxseed is the most abundant plant source of precursors for the mammalian lignans, enterolactone (ENL), and its primary reduction product, enterodiol (END), producing up to 800 times more than other foods.[7] Following flaxseed ingestion, ENL and END (Fig. 5.1), are produced due to the action of intestinal bacteria on the lignan precursor, secoisolariciresinol diglycoside.[8] Lignans are an excellent marker for the ingestion and absorption of flaxseed, as they are readily excreted at significant levels in the urine.[9,10] In this regard, flaxseed is similar to soy, which is extremely rich in another unique group of polyphenolic phytochemicals — the isoflavones.[11] Total urinary isoflavones (diadzein, genistein and equol) have been shown to increase 1000-fold above baseline levels following soy consumption.[12] Unfortunately, the absorption of phytochemicals from other foods are not quantified as readily as the soy isoflavones and flaxseed lignans, and the identification of accurate and reliable biochemical markers for the ingestion and absorption of other food phytochemicals is necessary. Ideally, quantitative data on major classes of phytochemicals of particular relevance to disease prevention should be added to existing nutrient composition databases to aid in future epidemiological studies with designer/functional foods.

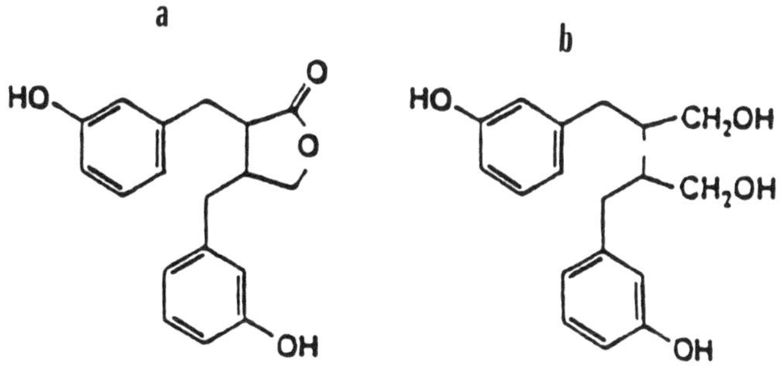

FIG. 5.1. CHEMICAL STRUCTURES OF THE TWO PRINCIPAL
MAMMALIAN LIGNANS
(a) enterolactone (ENT); *trans*-2,3-bis(3-hydroxybenzyl) butyrolactone and
(b) enterodiol (END); 2,3-bis(3-hydroxybenzyl) butane-1,4-diol.

The Effect of Lignans on Estrogen Metabolism

Estrogen metabolism is the primary biological endpoint modulated by flaxseed. Lignans possess both weak estrogenic and antiestrogenic activity[13], and are structurally similar to tamoxifen, a breast cancer chemopreventive agent.[14] Further, nonhuman primates excreting significant amounts of dietary-derived lignans rarely exhibit spontaneous[15] or chemically-induced[16] mammary cancer, thus a number of recent studies have attempted to clarify the role of lignans in the prevention of human breast cancer.

Lignans may be one of the dietary factors protecting those consuming a vegetarian or semi-vegetarian diet against the development of hormone-dependent cancers. Adlercreutz found urinary ENT to be the highest in women consuming a macrobiotic diet, while the lowest ENL excretion was seen in omnivores. In a separate study[17], he found the lowest ENT excretion in older women with breast cancer and highest in young vegetarians with no history of disease. The exact mechanism by which lignans modulate hormone metabolism is unclear, but it has been suggested that they reduce hormone bioavailability via the induction of hepatic sex hormone binding globulin.[13] Data from a recent clinical did not support that observation, however, instead suggesting that flaxseed may reduce ovarian dysfunction.[18] Further work is necessary to identify the mechanism(s) by which flaxseed lignans modulate estrogen metabolism.

SUMMARY

There is a great deal of excitement about the potential for phytochemicals in disease prevention and health promotion. As the field of designer/functional foods moves forward, we may see the development of many new foods with enhanced levels of physiologically-active phytochemicals, undoubtedly resulting in their elevated intake. Additional reliable, quantitative markers for the phytochemical ingestion, absorption and metabolism are urgently needed in order to assess the safety and efficacy of designer/functional foods.

REFERENCES

1   Panel says cancer fight needs to be overhauled. Wall Street Journal, September 30, 1994.
2   Diet, Nutrition, and Cancer Committee on Diet, Nutrition, and Cancer, Assembly of Life Sciences, National Research Council. Washington, DC: National Academy Press, 1982.
3   Block, G., Patterson, B. and Subar, A. Fruit, Vegetables, and Cancer Prevention: A review of the Epidemiological Evidence. Nutr. Cancer 1992; 18:1–28.
4   Steinmetz, K.A. and Potter, J.D. Vegetables, Fruit, and Cancer. II. Mechanisms. Cancer Causes and Control 1991; 2:427–442.
5   Dietary Phytochemicals in Cancer Prevention and Treatment. Adv. Exp. Med. Biol. Volume 401. American Institute for Cancer Research (ed.). Plenmum Press, New York, 1996.
6   Begley, S. Beyond Vitamins. Newsweek 1994:45–49.
7   Thompson, L.U., Robb, P., Serraino, M. and Cheung, F. Mammalian Lignan Production from Various Foods. Nutr. Cancer 1991; 16:43–52.

[8] Setchell, K.D.R., Borriello, S.P., Gordon, H., Lawson, A.M., Harkness, R. and Morgan, D.M.L. Lignan Formation in Man — Microbial Involvement and Possible Roles in Relation to Cancer. Lancet 1981; ii:5–7.

[9] Shultz, T.D., Bonorden, W.R. and Seaman, W.R. Effect of Short-term Flaxseed Consumption on Lignan and Sex Hormone Metabolism in Men. Nutr. Res. 1991; 11:1089–1100.

[10] Phipps, W.R., Martini, M.C., Lampe, J.W., Slavin, J.L. and Kurzer MS. Effect of Flaxseed Ingestion on the Menstrual Cycle. J. Clin. Endocrinol. 1993; 77:1215–1219.

[11] Messina, M., Messina, V. and Setchell, K. The Simple Soybean and Your Health. (B. Conner and E.W. Sparber, eds.) pp. 71–76. Avery Publishing Group, New York, 1994:Chap. 7.

[12] Cassidy, A., Bingham, S. and Setchell, K.D.R. Biological Effects of a Diet Soy Protein Rich in Isoflavones on the Menstrual Cycle of Premenopausal Women. Am. J. Clin. Nutr. 1994; 60:333–340.

[13] Adlercreutz, A., Hockerstedt, K., Bannwart, B., Bloigu, S., Hamalainen, E., Fotsis, T. and Ollus, A. Effect of Dietary Components, Including Lignans and Phytoestrogens, on Enterohepatic Circulation and Liver Metabolism of Estrogens and on Sex Hormone Binding Globulin (SHBG). J. Steroid Biochem. 1987; 27:1135–1144.

[14] Nayfield, S.G., Karp, J.E., Ford, L.G., Dorr, F.A. and Dramer, B.S. Potential Role of Tamoxifen in Prevention of Breast Cancer. J. Natl. Cancer Inst. 1991; 83:1450–1459.

[15] Adlercreutz, H., Musey, P.I., Fotsis, T., Bannwart, C., Wahala, K., Makela, T., Brunow, G. and Hase, T. Identification of Lignans and Phytoestrogens in Urine of Chimpanzees. Clin. Chim. Acta 1986; 158:147–154.

[16] Pfeiffer, C.A. and Allen, B. Attempts to Produce Cancer in Rhesus Monkeys with Carcinogenic Hydrocarbons and Estrogens. Cancer Res. 1948; 8:97–127.

[17] Adlercreutz, H., Fotsis, T., Bannwart, C., Wahala, K., Makela, T., Brunow, G. and Hase, T. Determination of Urinary Lignans and Phytoestrogens Metabolites, Potential Antiestrogens and Anticarcinogens, in Urine of Women on Various Habitual Diets. J. Steroid Biochem. 1986; 25:791–797.

[18] Lampe, J.W., Martini, M.C., Kurzer, M.S., Adlercreutz, H. and Slavin, J.L. Urinary Lignan and Isoflavonoid Excretion in Premenopausal Women Consuming Flaxseed Powder. Am. J. Clin. Nutr. 1994; 60:122–128.

[19] Adlercreutz, H., Honjo, H., Higashi, A., Fotsis, T., Hamalainen, E., Hasegawa, T. and Okada, H. Urinary Excretion of Lignans and Isoflavonoid Phytoestrogens in Japanese Men and Women Consuming a Traditional Japanese Diet. Am. J. Clin. Nutr. 1991; 54:1093–1100.

RESEARCH APPROACHES TO SPECIAL PRECLINICAL SAFETY AND TOXICOLOGICAL EVALUATIONS

SHIRLEY A.R. BLAKELY

Food and Drug Administration
Office of Policy, Planning and Strategic Initiatives (HFS-19)
200 C Street, SW, Washington, DC 20204

ABSTRACT

Foods which contain phytochemical components and that have been shown to exhibit activity in biological systems are termed designer foods. The toxicological and safety aspects of many designer foods are largely unknown. Therefore in order to be considered for human consumption in concentrated forms or in higher than usual amounts, systematic safety evaluations on designer foods are essential. Approaches to conducting systematic preclinical safety studies on designer foods begin with information gathering through literature searches and reviews of databases. Systematic evaluations of the in vivo effects of test materials on reproduction, bioavailability and tissue storage of essential nutrients, bone mineral homeostasis, and lipid metabolism will be needed. Many factors affect the safety evaluation of phytochemicals in extracts and crude mixtures, including components in the basal diet of the animals, dosage of test material, method and route of administration of the test material, adequacy of the material used as the control, the experimental design, measures taken to control bias, animal models, effects across age and gender ranges, endpoints, and interpretation of the results. The goals of these studies will be to determine toxicity, adverse effects on nutritional status, and to develop surrogate biomarkers of compliance and efficacy. The technology in the area of designer foods is advancing so that needs, such as producing active components in large quantities, developing stable test material and adequate standards for testing, developing analytical methodology, and appropriate animal models, will soon be addressed.

INTRODUCTION

Designer foods is a term that has been used to describe a category of foods grouped according to their phytochemical components.[1] The phytochemicals in these foods are generally non-nutrient components. They may, however, possess some biological functional properties whereby they act, for instance, as antioxidants or antiproliferative agents. Therefore, they have recently been categorized or referred to as candidate nutrients.[2] The list of suggested candidate nutrients includes isoflavones, tocotrienols, and carotenoids, particularly, certain non-provitamin A carotenoids. Many of these candidate nutrients are ubiquitous in foods that contain known nutrients and have been shown to exhibit synergistic interactions with essential nutrients. For the purpose of the discussion on research approaches, crude extracts, whole food items, and food components, in general, will be used to describe the physical form of the test materials.

Research methodologies for the complex mixtures just mentioned may be different from the traditional research approaches for purified chemicals and could present major challenges in conducting preclinical safety and toxicological evaluations. Because the test materials are crude extracts and mixtures of compounds, a combination of strategies and an interdisciplinary approach will be required. It is first necessary to understand as much as possible about the chemical properties of the test article. The biggest challenge, however, is to determine how to fit evaluation studies of complex mixtures into traditional research paradigms commonly used in studies of pure and fully characterized compounds. Finally, from this information a list of future needs for research approaches to special preclinical safety and toxicological evaluations can be developed.

Challenges

Chemical characterization and compositional studies of a particular food phytochemical are essential before preclinical evaluations are undertaken. Questions regarding the chemical structure of the active phytochemical, its functional groups, and the physical state of the test article must be answered. In addition, the stereochemical configuration and comparative structural activity of the phytochemical, including solubility, ionization constant, and particle size, must be known. This information can be used to predict how the test article might function in biological systems and *in vivo*. Databases depicting chemical composition and biological activity of the components can be useful in describing general components of the food or food groups that are abundant sources of a compound of interest. The Natural Products Alert (NAPRALERT)[3] database of foods, phytochemicals, and biochemical functions is possibly the most extensive natural products database in existence. Another, the carotenoids database[4], represents a source of consistent data on the carotenoid content of foods commonly consumed in the United States. Foods rich in carotenoids also contain other phytochemicals, such as isoflavones, flavonoids, and other polyphenols.[5]

Analytical methodology must be developed or refined to describe chemical properties of the components of interest and to produce sufficient quantities of compounds with a sufficient potency to support preclinical evaluations. In most cases the structure of the active constituent is not known; a series of extraction steps coupled with biological assays is necessary to determine the active fractions. Numerous examples exist in the literature in which the active food component may have been overlooked because of inadequate attention given to the initial search for the most active fractions. Recent evidence suggests, for example, that β-carotene may not be the most potent biological antioxidant among the family of carotenoids.[6] The greater the attention initially devoted to the separation of the most active fractions, the greater the likelihood of reducing erroneous conclusions about the properties of a given food. Newer analytical technologies such as supercritical fluid extraction may be useful. Computerized data analysis techniques such as pattern recognition and neural networks to compare chromatographic profiles can also be used.

After the richest source of the active constituent or target component has been identified, the next challenge is to determine stability under various conditions of use and storage. Tests will be needed to determine the shelf life of the product when it is stored in a freezer, protected from air and light, or when it is added to test diets. The fact that these are naturally-occurring plant products raises concerns about the levels of contaminants from various sources. In addition to the usual contaminants, such as pesticide residues and toxic elements, there may also be unintentional contaminants, such as *bis*phenol-A[7,8], which arise from autoclaveware and have been shown to exhibit estrogenic activity. If not avoided, such contaminants can interfere with the outcome of preclinical evaluations of various compounds and may lead to false positive or negative findings.

Major challenges exist in both *in vivo* and *in vitro* preclinical evaluations when the test article is a crude mixture. However, in this discussion, our evaluation of food components and crude extracts through traditional research paradigms will focus on *in vivo* studies that use whole animals. Information regarding effects of test components on reproduction, bioavailability, bone mineral homeostasis, lipid metabolism, and interaction with essential nutrients, such as lipid-soluble vitamins, can be obtained by using traditional methodology and whole animal systems. When the test article is a crude extract or a complex mixture of chemical compounds, some of the challenges of *in vivo* preclinical evaluations include determination of (1) which basal diet to use, (2) the amount and range of dosing, (3) the method and route of administration, (4) the control article and, if needed, the positive control article, (5) the experimental design, (6) the control of bias measures, (7) the suitable experimental animal models, (8) the gender and age effects, (9) which endpoints to measure, and (10) how to interpret the results.

Because findings have indicated that some ingredients in these diets may exhibit some of the variables under evaluation, the choice of diets is critical.

Making use of various standardized diets, such as the AIN-76A diet[9,10], or the more recent AIN-93 growth and maintenance, purified diets for laboratory rodents can be a starting point.[11] Although these diets provide a basal response and allow for altered responses from test substances, some problems have been identified. It may also be desirable to use a stock-type diet for carcinogenesis and other types of studies. Reports indicate that the outcome of the study may be influenced by the type of baseline diet used.[12,13] The stock diet contains factors which increase the activity of HMGCo-A reductase[14,15] and may be the preferred diet in cholesterol studies.

Determining the amount to administer depends upon the potency and purity of the mixture being administered, as well as the bioavailability. Administering the test component in levels that exceed normal amounts by 5 or 10 times might be a good starting point. The bioavailability of phytochemicals may differ in test animals and in humans, and thus presents a major challenge in finding the appropriate amount to administer. For example, in some of the earlier β-carotene studies in rats, average consumption was estimated at 4 mg per day, which on a body weight basis is equivalent to 1,400 mg/day for humans, or almost 400 times the average dose. At that level it has been shown to interfere with vitamin E status in rats.[16,17] The data on vitamin E interference in humans are mixed[18,19] because the test dosages for β-carotene are typically at levels of less than 50 mg per day. Other animal models for β-carotene studies are being evaluated. Estimating the amount to administer in the context of what is practicable for the human situation may be useful.

Exposing the test animal to the test article in the manner anticipated for human exposure and in the context of a normal diet would suggest the choice of an oral route of administration through feeding. Graded amounts administered in a standardized semipurified or a stock-type diet may be the desired approach. When added to the diet, test articles that are bulky and contain high amounts of inactive constituents tend to displace essential nutrients. Chemical definitions of bulk components, e.g., fiber, starch, and/or fat, can assist in determining which macro-ingredient in the basal diet to displace. However, in such cases, questions regarding maintenance of the isocaloric nature across all treatment groups become very important. It may be preferable to have the test article in a concentrated powder form because liquids increase the overall moisture content of the diet, which may in turn affect the stability of the other essential constituents in the diet during mixing, storage, and use.

The control article is used to establish a baseline response to the diet and to verify changes caused by the test compound. Ideally, the control article added to the basal diet is a placebo of the test article, and should resemble the test article in every respect except for the absence of the active constituent. Selecting a placebo can be a major challenge when the test article is a whole food or even a fractionated food containing complex mixtures of chemicals. One way to address this problem may be to start with the whole commodity and remove one or more of the components that might interfere with the outcome. A test article that is a whole ground seed, for example, might contain about 40% (by weight) fat and 10% fiber. As part of the experimental design, a defatted test group and/or a defatted + defibered group can be added along with the graded levels of the whole seed. The defatted and/or defibered foods can serve as additional control articles.

In addition to those added groups, the design could also include a positive control group. The positive control article should be one that is known to affect the endpoint expected for the test article. The use of a positive control helps to validate findings that show effects due to the test articles to be used in the comparisons.

Using animals to evaluate safety and toxicity should help to answer questions about the role of essential nutrients in bioavailability and metabolism and about the direct effects of the test article on chronic degenerative disease biomarkers. The choice of animal model should be based on whether the animal uses the test compound in a manner which mimics human metabolism. Models for various human diseases are available and should provide useful information about the effects of a component. Transgenic animal models for specific human conditions are being used with increasing success.

Preclinical evaluations should be conducted in both male and female animals. The effects of a test article in females cannot be predicted from data obtained from males as many of the functional indicators of safety may be gender specific. Similarly, one cannot predict effects in older animal models by examining effects only in young, immature animals. Neither can one predict the effects of a test article on chronic ingestion by carrying out short-term studies. All of these factors must be considered in comprehensive preclinical safety and toxicological evaluations.

The experimental design will be heavily influenced by the objectives of the study, which should be clearly defined. Accurate determination of the number of factors to include in the study is also critical. Limiting the number of factors to enhance the interpretation of the findings may be desirable in a factorial design. Moreover, including measures to control for bias in the design of the experiment is important. Identifying the proper animals, diets, and housing unit; randomly assigning animals to the treatment groups; randomizing analyses across all groups; and maintaining accurate records are good quality control measures that should be incorporated in every aspect of an evaluation. Adherence to the Guidelines for Good Laboratory Practices (GLP) will ensure appropriate quality control points.[20]

Protocols for standard toxicological evaluations are listed in the Redbook I.[21] Screening techniques that use surrogate biomarkers to determine if a specific fraction of a food tests positive for toxicity, such as hepatotoxicity,

DNA damage, and adverse reproductive outcomes, may be needed. However, whatever the evaluation, the challenges surrounding the use of test articles that are mixtures or crude extracts are the same.

In addition, the goal of safety studies will be to determine adverse effects on nutritional status and to develop surrogate biomarkers of compliance and efficacy. Determination of the test article's bioavailability by measuring blood levels of the test compound and its accumulation and biodistribution in tissues is one of the fundamental issues in preclinical studies of designer foods. The assumption that it will be possible to measure the ingested active constituent in the fluids and tissues of the animal may or may not be the case. The components in a test article may be derivatized to other compounds or may have limited bioavailability.

A systematic evaluation will be needed of the modulatory effects of the test article on blood lipids and lipid metabolism, including measurement of liver lipids, fecal bile acids, and cholesterol and triglycerides, and a fatty acid profile of blood and tissue. If the test article lowers blood lipid levels, for example, then it might also affect absorption, distribution, metabolism, and the function of lipid-soluble vitamins. Interactions of the constituents of the test article with essential vitamins, particularly lipid-soluble vitamins, will need to be examined, especially because some active components that are antioxidants may also be prooxidants. The effects of interactions of the test article with essential minerals, for instance, on bone and mineral homeostasis, should also be investigated. Other important endpoints that assess the impact of a test article on nutritional status include evaluation of its influence on antioxidant and detoxification enzyme systems and immune function. Post-mortem pathology assessments will be needed to answer key questions about the safety of the test article.

To interpret the results, the statistical analysis might include, in addition to analysis of variance of all factors, a test for homogeneity of variances. The response may be linear or non-linear. Crude mixtures consist of many compounds that increase concomitantly with the increase in dosage of the test article. The use of graded dosages of a test compound creates a paradox because the substances of no interest may cause non-linear results. Because these other components may attenuate the response observed at lower levels, graded levels that are equally spaced or placed logarithmically should be used.

Future Needs

A partial list of future needs can be developed from the foregoing discussion. Future needs are based on the number of new designer foods identified and the ability to isolate and purify the compounds in sufficient quantities for the conduct of research. A series of challenges for the investigator begins with determining the active constituent and then isolating and quantifying that component. Based on awareness of these challenges, the following is a partial list of needs for future work with designer foods:

(1)   Scaled-up capabilities to produce larger quantities of test materials are essential. One of the major obstacles to conducting comprehensive preclinical studies is the unavailability of sufficient amounts of purified materials, caused, in part, by a lack of knowledge about the identity of the active constituent in the food items under study. The interdisciplinary approach in which an analytical chemist is part of the research team will be invaluable in investigating designer foods.

(2)   A stable test material is needed for reasonable shelf-life and constant composition during preclinical evaluations.

(3)   Improved animal models are needed for evaluation of these compounds, i.e., models in which absorption, utilization, and metabolism of the compounds occur as they do in humans.

(4)   Standards are needed for chemical analyses. As stated above, these are essential for accurate preclinical studies. Several sources may be of assistance in this area including the Standard Reference Materials (SRM) program[22,23] at the National Institute of Standards and Technology (NIST), U.S. Pharmacopeia (USP), and AOAC International.

(5)   Standardized analytical methodology is needed to minimize interlaboratory variation. For many essential nutrients, NIST has established interlaboratory studies of analytical performance called "round robin" studies, which provide the investigator with standardized procedures for measuring key nutrients.[23] This program would also be of great value if it were extended to cover designer foods. AOAC International, the American Oil Chemists Society, and the American Association of Cereal Chemists are other organizations which can also be of assistance.

Challenges, as well as opportunities, are involved in conducting systematic preclinical studies of designer foods. Safety and toxicological evaluations of specific designer foods will provide sufficient information to determine if and how to proceed to the next stage of studies. Innovative approaches will be needed to begin systematic studies of new designer foods and to obtain the information needed to continue them.

REFERENCES

[1] Haumann, B.F. Designer foods — Designing and Manipulating Foods to Promote Health. INFORM (Intl. News Fats, Oils and Related Mat.) 1993; 4:343-375.

[2] Hendrich, S., Lee, K.W., Xu, X., Wang, H.J. and Murphy, P. Defining Food Components as New Nutrients. J. Nutr. 1994; 124:1789S-1792S.

[3] Loub, W.B., Farnsworth, N.R., Soejarto, D.D. and Quinn, M.L. NAPR-ALERT: Computer Handling of Natural Product Research Data. J. Chem. Inf. Comput. Sci. 1985; 25:99-103.

[4] Mangels, A.R., Holden, J.M., Beecher, G.R., Forman, M.R. and Lanza, E. Carotenoid Content of Fruits and Vegetables: An Evaluation of Analytic Data. J. Am. Diet. Assoc. 1993; 93(3):284-286.

[5] Gross, J. Pigments in Vegetables. Van Nostrand Reinhold, New York, 1991.

[6] Bendich, A. Biological functions of dietary carotenoids. In Carotenoids in Human Health, (Canfield, L.M., Krinsky, N.I. and Olson, J.A., eds.) Ann. N.Y. Acad. Sci. 1993; 691:61-67.

[7] Miller, S.C., Bottema, C.D.K., Stathis, P.A., Tokes, R.L. and Feldman, D. Unexpected Presence of Estrogens in Culture Medium Supplements: Subsequent Metabolism by the Yeast Saccharomyces cerevisiae. Endocrinology 1986; 119:1362-1369.

[8] Bisphenol-A: An Estrogenic Substance is Released from Polycarbonate Flasks During Autoclaving. Endocrinology 1993; 132:2279-2286.

[9] AIN. Report of the American Institute of Nutrition ad hoc Committee on Standards for Nutrition Studies. J. Nutr. 1977; 107:1340-1348.

[10] AIN. Second Report of the American Institute of Nutrition ad hoc Committee on Standards for Nutrition Studies. J. Nutr. 1980; 110:1726.

[11] Reeves, P.H., Nielsen, F.H. and Fahey, G.C., Jr. AIN-93 Purified Diets for Laboratory Rodents: Final Report of the American Institute of Nutrition ad hoc Writing Committee on the Reformation of the AIN-76A Rodent Diet. J. Nutr. 1993; 123:1923-1931.

[12] Cohen, L.A., Epstein, M., Saa-Pabon, V., Meschter, C. and Zang, E. Interactions Between 4-HPR and Diet in NMU-Induced Mammary Tumorigenesis. Nutr. Cancer 1994; 21:271-283.

[13] Fisher, M.J., Sakata, T., Tibbels, R.S., Smith, R.A., Patil, K., Dhachab, M., Johansson, S.L. and Cohen, S.M. Effect of Sodium Saccharin and Calcium on Urinary Parameters in Rats Fed Prolab 3200 or AIN-76 diet. Food Chem. Toxicol 1989; 27:1-9.

[14] O'Brien, B.C. and Reiser, R. Comparative Effects of Purified and Human-type Diets on Cholesterol Metabolism in the Rat. J. Nutr. 1979; 109:98-104.

15 Reiser, R., Henderson, G.R., O'Brien, B.C. and Thomas, J. Hepatic 3-Hydroxy-3-Methylglutaryl Coenzyme A Reductase of Rats Fed Semi-purified and Stock Diets. J. Nutr. 1977; 107:453–457.

16 Blakely, S.R., Mitchell, G.V., Jenkins, M.Y. and Grundel, E. Effects of β-carotene and Related Carotenoids on Vitamin E Status. *In* Vitamin E: Biochemistry and Molecular Function, Marcel Dekker, New York, 1992: 63–68.

17 Blakely, S.R., Mitchell, G.V., Jenkins, M.Y., Grundel, E. and Whittaker, P. Effects of Vitamin A, Beta-carotene and Canthaxanthin on Alpha-tocopherol, Carotenoid and Iron Status in Retired Breeder Rats. J. Nutr. 1991; 121: 1649–1655.

18 Xu, M.J., Plezia, P.M. and Alberts, D.S. Reduction in Plasma or Skin Alpha-tocopherol Concentration with Long-term Oral Administration of Beta-carotene in Humans and Mice. Natl. Cancer Inst. 1992; 84:1559–1565.

19 Nierenberg, D.W., Stukel, T.A., Mott, L.A. and Greenberg, E.R. Steady-state Serum Concentration of Alpha-tocopherol Not Altered by Supplementation with Oral Beta-carotene. J. Natl. Cancer Inst. 1994; 86: 117–121.

20 Code of Federal Regulations, Title 21, Part 58, Good Laboratory Practice for Nonclinical Laboratory Studies. U.S. Government Printing Office, Washington, DC, 1992.

21 Toxicological Principles for the Safety Assessment of Direct Food Additives and Color Additives Used in Food — Redbook I. Bureau of Foods. U.S. Food and Drug Administration, Washington, DC, 1980.

22 Gills, T. Basic Measurement System Components — SI units, Definitive Methods, Standard Reference Materials. *In* Sixth Conference for Federally Supported Human Nutrition Research Units and Centers, National Institutes of Health (NIH), Bethesda, MD, 1994.

23 U.S. Department of Commerce, Standard Reference Materials Program and the "Round Robin" Program, National Institute of Standards and Technology, Gaithersburg, MD 20899.

PHYTOPHARMACOLOGY OF GARLIC FOOD FORMS

CHAPTER 7

PHYTOCHEMISTRY OF GARLIC HORTICULTURAL AND PROCESSING PROCEDURES

HIROMICHI MATSUURA

Institute for OTC Research
Wakunaga Pharmaceutical Co., Ltd.
Hiroshima, Japan

ABSTRACT

Recently, many studies have evaluated raw garlic and its different types of preparations for both their chemical and biological characteristics. These studies, have revealed that different processing procedures may influence the amount of organosulfur compounds, such as alliin and γ-glutamyl-S-allylcysteine, found in garlic. Some studies have also revealed that allixin and steroidal glycosides may posses anti-tumor-promoting activity.

Furthermore, analyses of the allicin content and allicin-producing potential of garlic preparations in water and simulated digestive conditions revealed that allicin-producing potential in water may not be a meaningful chemical evaluation for garlic products. It was also revealed that there was no allicin in any of the commercially available garlic preparations tested.

Finally, studies revealing biological characteristics, such as antioxidant activities, anti-tumor-promoting activities and intestinal flora enhancing effects have revealed that aged garlic extract is more effective and beneficial than aqueous extracts of either raw or boiled garlic.

INTRODUCTION

Since ancient times garlic (*Allium sativum* L.) has been used worldwide for its pungency and flavoring, for its medical properties, such as a tonic or bactericide, and as a popular remedy for various ailments. The source plant is extensively cultivated in the world, and the total world production in 1989 was approximately 3 million metric tons with China (21.7%), Korea (13.3%) and India (10.0%), being the major producers.

Regarding the chemical constituents in garlic, the structure of diallyl disulfide, the principal component of garlic oil, was discovered by Wertheim in 1844.[1] This, and other odorous compounds, are not present in intact garlic, rather, they are mainly formed once the cellular tissue in intact garlic is disrupted. In 1944, Cavallito and Bailey isolated an oily and unstable compound, allicin, which exhibited a potent anti-microbial activity.[2] In 1949, Stoll and Seebeck showed that allicin was generated from a precursor, S-allylcysteine sulfoxide (alliin) by an enzymatic reaction.[3] After these studies, a number of studies on volatile organosulfur compounds and their precursors in garlic and other *Allium* plants have been reported.[4,5]

It is well-known that organosulfur compounds in garlic are transformed into a variety of compounds depending upon the processing method applied to the raw garlic. Therefore, different processing procedures result in the generation of various kinds of chemical constituents in garlic preparations.[6] A recent study on the medicinal efficacy of garlic has revealed that some of the constituents and their derivatives have biological effects.

The present paper describes the chemical and biological characteristics of raw garlic and its processed preparations. Furthermore, recent studies on the chemical constituents of garlic are also summarized.

Organosulfur Compounds

Sulfur-containing Precursors

One of the organosulfur components in intact garlic, the odorless amino acid, alliin, is enzymatically converted into allicin when raw garlic cloves are crushed. Allicin has antibacterial properties, but is extremely unstable. Therefore, it is readily converted into various compounds, such as diallyl polysulfides, ajoene, and vinyl dithiins as shown in Fig. 7.1. Allicin which is released by enzymatic reaction from alliin, has been termed "allicin-producing potential," but it is a transient chemical compound, and has shown no meaningful biological role *in vivo*, even though some studies have shown antiplatelet aggregation[7] and chemoprevention activities *in vitro*.

Since allicin is such an unstable compound, the former studies also probably measured the efficacy of other sulfur-containing compounds, rather than allicin. Since they did not classify the exact compounds used in their experiments and there have been no verifications of allicin's effectiveness *in vivo*, vague references to "allicin-producing potential" should be avoided. In addition, it was found that γ-glutamyl-S-alk(en)ylcysteines proposed as biosynthetic precursors of S-alk(en)ylcysteine sulfoxides[8] are gradually converted to S-alk(en)ylcysteines through an enzymatic process when raw garlic is crushed

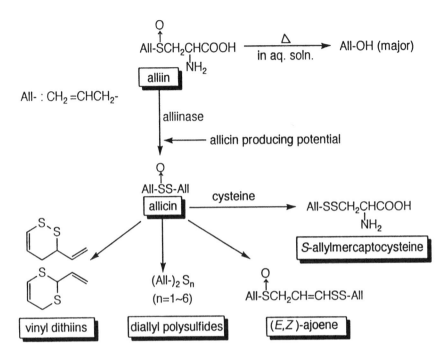

FIG. 7.1. TRANSFORMATION OF ALLIIN OF CRUSHED RAW GARLIC

or extracted in an aqueous medium (Fig. 7.2). S-Allylcysteine (SAC), the major compound transformed from γ-glutamyl-S-allylcysteine (γ-Glu-SAC) is a compound verified to be both biologically active[9,10] and bioavailable.[11] Therefore, it is useful to evaluate both the alliin and γ-Glu-SAC contents in raw garlic. The separation of alliin and its analogues using high performance liquid chromatography (HPLC) has been accomplished on an ion-exchange column followed by post-derivatization.[12] The alliin content in raw garlic measured by this method varied from 5.4 mg/g to 14.5 mg/g. Other ingredients, such as γ-glutamyl peptides and allicin, were determined by HPLC with a reversed phase column.[12] The amount of γ-Glu-SAC was determined to be in the range of 1.9 to 8.9 mg/g per fresh weight. The allicin content in raw garlic was determined after being finely-crushed and added to water at 25°C. It was determined to be in a range of 2.5 to 5.0 mg/g.

Variations in the above 3 or 4 compounds have been used to characterize the garlic collected from Japan, China, U.S.A. and several countries in South America.

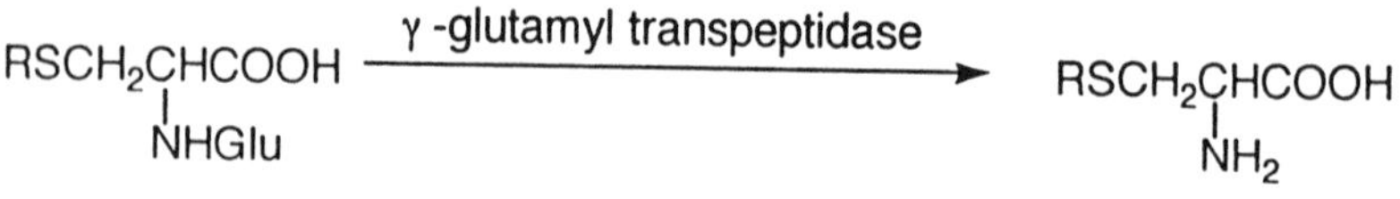

$$\text{RSCH}_2\text{CHCOOH} \xrightarrow{\gamma\text{-glutamyl transpeptidase}} \text{RSCH}_2\text{CHCOOH}$$

RSCH$_2$CHCOOH (NHGlu) → RSCH$_2$CHCOOH (NH$_2$)

R : CH$_2$=CHCH$_2$-   S-allylcysteine

CH$_3$CH=CH -   S-(trans-1-propenyl) cysteine

CH$_3$ -   S-methylcysteine

Glu : glutamic acid

FIG. 7.2. TRANSFORMATIONS OF γ-GLUTAMYL-S-ALLYLCYSTEINES OF CRUSHED RAW GARLIC

Cultural and Storage Effects

The effect of varying seasons on the amount of ingredients in intact garlic bulbs has not been well-studied. One of the very few studies as reported by Ueda et al.[13] has shown that the alliin content of garlic increased continuously until the bulb stopped growing.

Storage conditions also influence the amount of ingredients in freshly harvested garlic. It was found that garlic held at 4°C for 2 months showed a remarkable decrease in its concentration of γ-Glu-SAC accompanied by an increase in alliin and allicin. SAC content, on the other hand, was not influenced by such conditions. Lawson et al. also reported similar effects of storage conditions on intact garlic bulbs.[14] The increase in alliin and allicin content is considered to be a result of sprouting during storage, as a similar situation occurs in onion.[4]

Chemical Evaluation of Garlic Preparations and Processes

The procedures for processing garlic can be roughly classified into the following categories: (1) drying or dehydration without the inactivation of enzymes; (2) extraction with an aqueous or oily medium; (3) distillation, such as garlic oil; and (4) heating, including frying or boiling. Many results have been reported which elucidate the compounds generated from various garlic processing methods.[6,15-17] These studies have shown that variation in the amounts and types of the components found in the preparations is dependent upon both content of enzymes and natural decomposition during processing. These studies have also reported that no allicin was detected in any of the processed samples. Allicin was found in crushed raw garlic cloves. Ho and coworkers have reported changes in various volatile compounds in garlic due to assorted preparation and cooking methods, such as frying, oil-cooking, microwave-frying, oven-baking and microwave-baking.[18] In the present study, the garlic samples were prepared by Ho's methods, and alliin, γ-Glu-SAC, γ-glutamyl-S-(trans-1-propenyl)cysteine, SAC and allicin were analyzed. As shown in Table 7.1, the precursors, such as alliin, γ-Glu-SAC, and γ-glutamyl-S-(trans-1-propenyl)cysteine, and SAC were found in fried, oven-baked and microwave-baked garlic. These garlic preparations also contained similar amounts of volatile compounds as those reported by Ho et al. However, no allicin was detected in any of the cooked garlic used widely in foods. As mentioned before, allicin is a transient compound which is readily converted into other much more beneficial, important, and stable organosulfur compounds. Therefore, allicin itself does not appear to be a meaningful ingredient in garlic preparations. Furthermore, Freeman and Kodera have described the stability of allicin in solvents, blood and simulated digestive fluids.[19] They concluded that the allicin content in various commercial garlic preparations was less than 1 ppm and the possibility of producing allicin is severely suppressed in simulated digestive conditions.[19] Consequently, "allicin-producing potential" in water is meaningless as a marker for the chemical evaluation of garlic, even in preparations with an active alliin-alliinase system.

TABLE 7.1
CONTENT OF SOME AQUEOUS ORGANOSULFUR COMPOUNDS IN WHOLE RAW
GARLIC (A), FRIED GARLIC (B), OIL-COOKED GARLIC (C),
MICROWAVE-FRIED GARLIC (D), BAKED GARLIC (E), and
MICROWAVE-BAKED GARLIC (F)

Compound	Content (mg/g raw garlic)					
	A	B	C	D	E	F
Alliin	6.42	1.01	0.01	N.D.	0.91	0.80
γ-Glutamyl-S-allylcysteine	3.38	0.21	0.08	N.D.	1.60	1.58
γ-Glutamyl-S-(Trans-1-propenyl) cysteine	2.08	0.07	N.D.	N.D.	0.90	1.01
S-allylcysteine	N.D.	0.03	0.02	N.D.	0.05	0.05
Allicin	3.02†	N.D.	N.D.	N.D.	N.D.	N.D.
Total volatiles*	-	1.02	0.51	0.63	1.61	2.48

* These data were calculated from reference 18.
† In crushed raw garlic.
N.D.: Not detected. Detection limits were 0.002 mg/g (alliin), 0.05 mg/g [γ-glutamyl-S-allylcysteine, γ-glutamyl-S-(trans-1-propenyl) cysteine], and 0.01 mg/g (S-allyl cysteine, allicin), respectively.

Steroidal Glycosides

Chemical Studies

No investigation has been able to isolate and determine the structure of the steroidal glycosides from garlic, even though the presence of steroidal glycosides has been previously shown by thin-layer chromatography (TLC). In 1988, we first isolated a new furostanol glycoside named proto-eruboside-B from a crude glycoside fraction which we prepared from a metanolic extract of frozen garlic by reversed phase chromatography.[20] In this study, it was found that freezing was effective by depressing glucosidase activity during extraction in order to isolate genuine glycosides from raw materials such as garlic. Further studies on steroidal glycosides from bulbs of *A. sativum* led to the isolation of a new furostanol glycoside, named sativoside-B along with a small amount of proto-desgalactotigonin.[21] No spirostanol glycosides have been isolated from frozen garlic bulbs. In contrast, from the roots of this plant two new glycosides namely sativoside-R2 and sativoside-R1 have been isolated and their structures have been determined to be gluco-proto-desgalactotigonin and its corresponding spirostanol glycoside. In addition to these glycosides, three known glycosides, gluco-proto-desgalactotigonin, desgalactotigonin, and F-gitonin have been isolated and identified. No glycoside from β-chlorogenin has been isolated from roots. Moreover, on TLC analysis of the crude glycoside fraction and its hydrolysate from aerial parts of *A. sativum*, no corresponding glycosides and aglycones have been detected.

Other important Allium plants, such as the great headed garlic, the bulbs of *A. ampeloprasum* (common name: elephant garlic), which has large garlic-like bulbs, and the bulbs of *A. chinense* (common name: rakkyo) have been widely used for vegetables. From bulbs of *A. ampeloprasum*, collected in the U.S.A., several spirostanol glycosides, having agigenin as a aglycone, and their furostanol glycosides have been isolated.[22] The acid hydrolysates of the crude glycoside fraction show the presence of agigenin along with a small amount of β-chlorogenin, gitogenin, and tigogenin as aglycones by TLC analysis. Sometimes, it is difficult to distinguish between garlic and elephant garlic from the shape of their bulbs. Analyzing sapogenins by TLC, can confirm if a garlic-like bulb is real garlic or not and if a preparation is produced from real garlic or not. *A. chinense* is the original plant of the Chinese crude drug "Xiebei," which has been used for the treatment of thoracic pain and diarrhea in China. From bulbs collected in Japan, a new furostanol glycoside named chinenoside-I, which yielded laxogenin on acid hydrolysis, was isolated.[23] This was the first example of a furostanol glycoside having laxogenin as a spirostanol sapogenin.

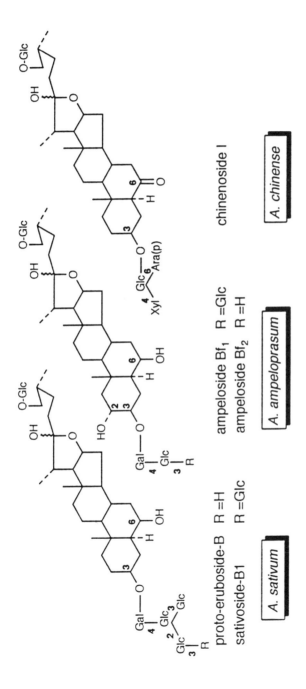

FIG. 7.3. STEROIDAL GLYCOSIDES FROM BULBS OF GARLIC AND RELATED PLANTS

Biological Activities

Antifungal activity of garlic has been attributed to allicin or ajoene.[24] Eruboside-B inhibits the growth of *Candida albicans*, and its activity (minimum inhibitory concentration: 25 μg/ml) is comparable to that of allicin and ajoene.[20] On the other hand, the genuine glycoside, proto-eruboside-B, does not show effective antifungal activity. It is interesting that the enzymatic transformation of eruboside-B from proto-eruboside-B when raw garlic bulbs are crushed is similar to that of allicin from alliin, followed by exhibiting antifungal activity.

The antitumor-promoting activity of steroidal glycosides and sapogenins in garlic has been examined by *in vitro* experiments using the tumor promoter, 12-O-tetradecanoylphorbol-13-acetate (TPA). As shown in Table 7.2, only eruboside-B showed an inhibition of TPA-enhanced ^{32}P-incorporation into the phospholipid portion of HeLa cells, and its activity was comparable to glycyrrhetinic acid, which has been shown to exhibit antitumor-promoting activity *in vivo*.[25]

TABLE 7.2
EFFECT OF STEROIDAL GLYCOSIDES FROM GARLIC ON TPA-ENHANCED
^{32}PI-INCORPORATION INTO PHOSPHOLIPID OF HELA CELLS

Compound		Inhibition (%)
Proto-eruboside-B	(50 μg/ml)	0
Eruboside B	(25 μg/ml)	39.4
	(10 μg/ml)	15.5
β-Chlorogenin	(50 μg/ml)	0
Glycyrrhetinic acid	(25 μg/ml)	30.9

TPA: 50nM

Flavonoids and Phenolics

It is well-known that onion contains a comparatively large amount of flavonoids, such as quercetin and its glycosides.[4] However, studies on the flavonoids and phenolics in garlic are much fewer. Very recently, we have isolated and identified a phenolic glucoside, coniferin from garlic (2 mg/100 g). Mizuno *et al.* have reported the quercitin content of garlic (8 μg/100 g) by HPLC.[26] Quercitin has been shown to have antitumor-promoting activity in a number of *in vitro* and *in vivo* studies using the tumor promoters, TPA and teleocidin.[27] In continuation of our garlic research for the prevention of cancer in humans, a new phenolic compound including a biosynthetically unusual n-pentyl group, named allixin, was isolated from garlic.[28] Allixin, 3-hydroxy-5-methoxy-6-methyl-2-pentyl-4H-pyran-4-one, has been shown to be a stress compound produced by garlic. Allixin inhibits the enhanced phospholipid metabolism of cultured cells induced by TPA. Furthermore, it was shown to suppress the promoting process of two-stage carcinogenesis *in vivo* and the promoting activity of TPA on skin tumor formation in DMBA initiated mice.[29] It was confirmed by an *in vitro* assay using cultured HeLa cells induced by TPA that the n-pentyl group from allixin is responsible for its antitumor-promoting activity due to the structure-activity relationship of derivatives of 3-hydroxy-4H-pyran-4-ones. In addition, Yamasaki *et al.* described the ability of allixin to inhibit aflatoxin B1-induced mutagenesis in *Salmonella typhimurium*.[30]

Biological Evaluation of Garlic Preparations

In general, crude drug extracts contain diverse compounds with certain biological activities, and therefore in many cases chemical analyses of several significant markers do not afford enough information to evaluate the efficacy of the extracts when taken as a whole. Specifically, organosulfur compounds in garlic have been shown to convert into a number of compounds and to yield variable components when they undergo different processing methods and preparations. This report describes the antitumor-promoting activity, antioxidant activity, and effects of various garlic preparations on the growth of *Lactobacillus acidophilus* and *Bifidobacterium bifidum*, which are useful enteric bacteria for humans.

The garlic preparations used in this present study were characterized by different organosulfur compounds and include: (1) a freshly prepared aqueous extract of raw garlic (RG) containing mainly allicin (9.9-11.4 mg/g, calculated as dry weight), (2) an aqueous extract of boiled garlic (BG) containing mainly alliin (14.8-16.7 mg/g), and (3) aged garlic extract (AGE) containing SAC (2.1 mg/g) and other organosulfur compounds quantified by Weinberg *et al.*[17]

Antitumor-promoting Activity

Tumor-inhibitory effects of garlic have been demonstrated in various experimental systems, and epidemiologic studies concerned with the preventative effects of garlic on human cancer have been reported.[31] Nishino *et al.* examined the effects of various garlic preparations by *in vitro* assays of TPA enhanced ^{3}H-choline incorporation into phospholipids in HeLa cells. They found that various kinds of chemicals which show inhibitory effects on the earliest phenomenon induced by a tumor promoter *in vitro*, also suppress carcinogenesis *in vivo* at the stage of promotion. As shown in Fig. 7.4, AGE inhibited TPA-stimulated ^{3}H-choline incorporation into the phospholipids of HeLa cells. In contrast, no effects were observed at 100 μg/ml of RG. RG showed an inhibitory effect on the growth of HeLa cells at the concentration of more than 100 μg/ml.

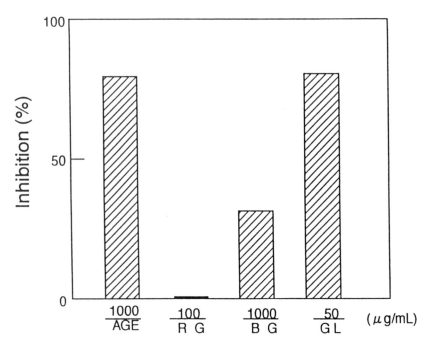

FIG. 7.4. INHIBITORY EFFECTS OF THE AQUEOUS EXTRACTS OF RAW GARLIC (RG) AND BOILED GARLIC (BG), THE AGED GARLIC EXTRACT (AGE) AND GLYCYRRHETINIC ACID (GL) ON TPA-ENHANCED ^{3}H-CHOLINE INCORPORATION INTO PHOSPHOLIPIDS INDUCED BY TPA IN HELA CELLS

Antioxidant Effects

Garlic has been shown to have antioxidant and free radical scavenging activities using *in vitro* systems such as iron-ascorbate acid in isolated rat liver membranes.[32] Lipid peroxidation has been reported to accompany the low level chemiluminescence that is closely correlated to other parameters of peroxidation, such as oxygen uptake and the formation of thiobarbituric acid reactive substances (TBA-RS).[33] In our previous study, we investigated the effects of garlic preparations on t-butyl hydroperoxide-induced chemiluminescence in a liver microsomal fraction.[12] At a concentration of 0.15 (w/v)% of the preparations, AGE demonstrated a decrease of about 30% in the light emission during lipid peroxidation, whereas aqueous extracts of raw or boiled garlic enhanced the light emission. Moreover, AGE suppressed the formation of TBA-RS at an early stage during lipid peroxidation, but no suppression was observed at the end point. The chemiluminescence in the early stage is generated from a self-reaction of t-BuOO•, whereas in the later stage it is generated from a self-reaction of t-BuOO• and its breakdown free radicals and lipid peroxides, which ultimately result in the formation of TBA-RS.[33] These results suggest that AGE mainly scavenges t-BuOO• as an antioxidant, whereas in contrast, RG and BG might act as oxidants.

Effects on the Growth of the Beneficial Enteric Flora, *L. acidophilus* and *B. bifidum*

Garlic has been known to contain soluble carbohydrates such as oligosaccharides and the group of fructose polymers called fructanes.[4] Recently, some kinds of oligofructanes have attracted considerable attention for their effects on the growth of *B. bifidum*.

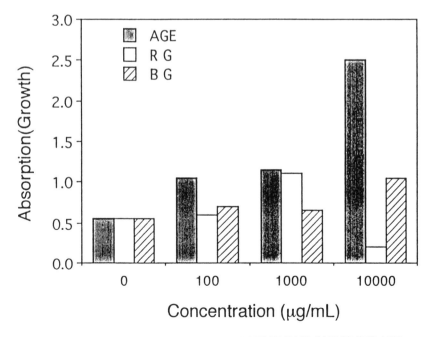

FIG. 7.5. EFFECT OF THE AQUEOUS EXTRACTS OF RAW GARLIC (RG) AND BOILED GARLIC (BG), AND THE AGED GARLIC EXTRACT (AGE) ON GROWTH OF *L. ACIDOPHILUS*

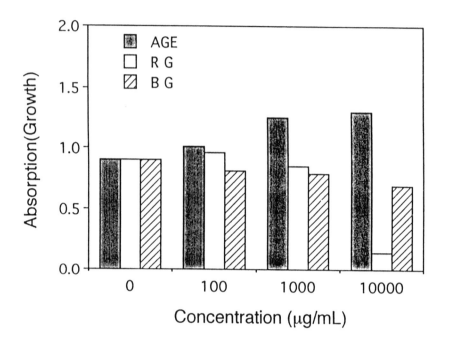

FIG. 7.6. EFFECT OF THE AQUEOUS EXTRACTS OF RAW GARLIC (RG) AND BOILED
GARLIC (BG), AND THE AGED GARLIC EXTRACT (AGE) ON GROWTH OF
B. BIFIDUM

In this present study, the effects of various garlic preparations were examined by *in vitro* assays of the growth. As shown in Fig. 7.5 and 7.6, AGE promoted the growth of both *L. acidophilus* and *B. bifidum* in a dose-dependent manner, whereas the other extracts showed an inhibition of the growth of *B. bifidum*. There was not a large difference in the quantity of fructanes present in the three garlic preparations, however, AGE contained a small amount of oligofructanes which the other preparations did not. One of our previous studies has shown that when AGE is subjected to an activity-guided fractionation, it affords F-4, its protein fraction.[34] The effects of AGE on the growth of *L. acidophilus* and *B. bifidum* can be attributed to F-4.

ACKNOWLEDGEMENTS

I am grateful to Dr. H. Nishino at the National Cancer Center Institute in Japan. Thanks are also due to Dr. Y. Itakura and the staff at the Institute for OTC Research at Wakunaga Pharmaceutical Co. Ltd., Japan.

REFERENCES

[1] Wertheim, T. Investigation of garlic oil. Annalen 1844; 51:289.

[2] Cavallito, C.J. and Bailey, J.H. Allicin, the antibacterial principle of *Allium sativum* L. Isolation, physical properties and antibacterial action. J. Am. Chem. Soc. 1944; 66:1950–1951.

[3] Stoll. A. and Seebeck, E. Alliin, the genuine mother compound of garlic oil. Helv. Chim. Acta. 1948; 31:189–210.

[4] Furia, T.E. Critical reviews in food science and nutrition: The genus *Allium*. Part 2. CRC Press, Boca Raton, FL, 1985.

[5] Block, E. The organosulfur chemistry of the genus *Allium* — Implication for the organic chemistry of sulfur. Angew. Chem. Int. Ed. Engl. 1992; 31:1135–1178.

[6] Iberl, B., Winkler, G. and Knobloch, K. Products of allicin transformation: ajoene and dithiins, characterization and their determination by HPLC. Planta Med. 1990; 56:202–211.

[7] Lawson, L.D., Ransom, D.K. and Hughes, B.G. Inhibition of whole blood platelet aggregation by compounds in garlic clove extracts and commercial garlic products. Thromb. Res. 1992; 65:141–156.

[8] Lancaster, J.E. and Shaw, M.L. γ-Glutamyl peptides in the biosynthesis of S-alk(en)yl-L-cysteine sulfoxides (flavor precursors) in *Allium*. Phytochem 1989; 28:455–460.

[9] Sumiyoshi, H. and Wargovich, M.J. Chemoprevention of 1,2-dimethylhydrazine induced colon cancer in mice by naturally occurring organosulfur compounds. Cancer Res. 1990; 50:5084–5087.

[10] Nakagawa, S., Kasuga, K. and Matsuura, H. Prevention of liver damage by aged garlic extract and its components in mice. Phytother. Res. 1989; 3:50–53.

[11] Nagae, S., Ushijima, M. and Hatono, S. *et al.* Pharmacokinetics of the garlic compound S-allylcysteine. Planta Med. 1994; 60:214–217.

[12] Imai, J., Ide, M., Nagae, S., Moriguchi, T., Matsuura, H. and Itakura, Y. Antioxidant and radical scavenging effects of aged garlic extract and its constituents. Planta Med. 1994; 60:417–420.

[13] Ueda, Y., Kawajiri, H., Miyamura, N. and Miyajima, R. Content of some sulfur containing components and free amino acids in various strains of garlic. J. Jpn. Soc. Food. Sci. Technol. 1991; 38:429–434.

[14] Lawson, L.D., Wang, Z.Y.J. and Hughes, B.G. γ-Glutamyl-*S*-alkylcysteines in garlic and other *Allium* spp; precursors of age-dependent *trans*-1- propenyl thiosulfinates. J. Nat. Prod. 1991; 54:436–444.

[15] Pentz, V.R., Guo, Z., Muller, B., Aye, R.D. and Siegers, C.P. Standardization of garlic preparations. Dtsch. Apoth. Ztg. 1992; 132:1779–1782.

[16] Lawson, L.D., Wang, Z.Y.J. and Hughes, B.G. Identification and HPLC quantitation of the sulfides and dialk(en)yl thiosulfinates in commercial garlic products. Planta Med. 1991; 57:363–370.

[17] Weinberg, D.S., Manier, M.L., Richardson, M.D. and Haibach, F.G. Identification and quantification of organosulfur compliance markers in a garlic extract. J. Agric. Food. Chem. 1993; 41:37–41.

[18] Yu, T.H., Wu, C.M. and Ho, C.T. Volatile compounds of deep-oil fried, microwave-heated, and oven-baked garlic slices. J. Agric. Food. Chem. 1993; 41:800–805.

[19] Freeman, F. and Kodera, Y. The stability of allicin in solvents, blood, and simulated physiological fluids. J. Agric. Food. Chem. 1995; 43:2332–2338.

[20] Matsuura, H., Ushiroguchi, T., Itakura, Y., Hayashi, M. and Fuwa, T. A furostanol glycoside from garlic, bulbs of *Allium sativum* L. Chem. Pharm. Bull. 1988; 36:3659–3663.

[21] Matsuura, H., Ushiroguchi, T., Itakura, Y. and Fuwa, T. Further studies on steroidal glycosides from bulbs, roots and leaves of *Allium sativum* L. Chem. Pharm. Bull. 1989; 37:2741–2743.

[22] Morita, T., Ushiroguchi, T., Hayashi, N., Matsuura, H., Itakura, Y. and Fuwa, T. Steroidal saponins from elephant garlic, bulbs of *Allium ampeloprasum* L. Chem. Pharm. Bull. 1988; 36:3480–3486.

[23] Matsuura, H., Ushiroguchi, T., Itakura, Y. and Fuwa, T. A furostanol glycoside from *Allium chinese* G. Don. Chem. Pharm. Bull. 1989; 37: 1390–1391.

[24] Yoshida, S., Kasuga, S., Hayashi, N., Ushiroguchi, T., Matsuura, H. and Nakagawa S. Antifungal activity of ajoene derived from garlic. Appl. Envir. Microbiol. 1987; 53:615–617.

[25] Nishino, H., Yoshioka, K. and Iwashima, A. *et al.* Glycyrrhetic acid inhibits tumor-promoting activity of teleocidin and 12-O-tetradecanoylphorbol-13-acetate in two stage mouse skin carcinogenesis. Jpn. J. Cancer Res. 1986; 77:33–38.

[26] Mizuno, W., Tsuchida, H., Kozukue, M. and Mizuno, S. Rapid quantitative analysis and distribution of free quercetin in vegetables and fruits. J. Jpn. Soc. Food Sci. Technol. 1992; 39:88–92.

[27] Nishino, H., Iwashima, A., Fujiki, H. and Sugimura, T. Inhibition by quercetin of the promoting effect on teleocidin on skin papilloma formation in mice initiated with 7,12-dimethylbenz[a]anthracene. Gann 1984; 75:113–116.

[28] Kodera, Y., Matsuura, H. and Yoshida, S. *et al.* Allixin, a stress compound from garlic. Chem. Pharm. Bull. 1989; 37:1656–1658.

[29] Nishino, H., Nishino, A. and Takayasu, J. *et al.* Antitumor-promoting activity of allixin, a stress compound produced by garlic. Cancer J. 1990; 3:20–21.

[30] Yamasaki, T., Teel, R.W. and Lau, B.H.S. Effect of allixin, a phytoalexin produced by garlic, on mutagenesis, DNA-binding and metabolism of aflatoxin B1. Cancer Lett. 1991; 59:89–94.

[31] Dorant, E., van den Brandt, P.A., Goldbohm, R.A., Hermus, R.J.J. and Sturmans, F. Garlic and its significance for the prevention of cancer in humans: a critical view. Br. J. Cancer 1993; 67:424–429.

[32] Horie, T., Murayama, T. and Mishima, T. *et al.* Protection of liver microsomal membranes from lipid peroxidation by garlic extract. Planta Med. 1989; 55:506–508.

[33] Cadenas, E. and Sies, H. Low level chemiluminescence of liver microsomal fractions initiated by *tert*-butyl hydroperoxide. Eur. J. Biochem. 1982; 124: 349–356.

[34] Hirao, Y., Sumioka, I. and Nakagami, S. *et al.* Activation of immunoresponder cells by protein fraction from aged garlic extract. Phytother. Res. 1987; 1:161–164.

CHAPTER 8

INTERNATIONAL PHYTOTHERAPEUTIC USES OF GARLIC FOOD FORMS

RAJABATHER KRISHNARAJ

Geriatric Medicine (787)
University of Illinois at Chicago
Chicago, Illinois 60612

ABSTRACT

International therapeutic uses of food forms of garlic tested on sick and healthy human volunteers of varying ethnic and environmental background in various continents representing at least 18 countries are reviewed. The most widely studied effect, viz., the antihypercholesterolemic property of garlic bulb and its extracts is evident in most studies. The antiplatelet aggregative, thrombolytic, hypotensive and hypoglycemic properties of garlic are increasingly being reported. The numerous in vitro *antimicrobial properties of garlic have rarely been extended to* in vivo *testing in humans, except for a few isolated reports on patients with AIDS, tuberculosis and cryptococcal meningitis. While animal studies on chemical induction and initiation-promotion models of cancer strongly support the antineoplastic role of active principles in garlic, no direct clinical trial has yet been reported in humans. Since the immune system is intimately involved in the maintenance of an antineoplastic milieu, one can predict that the anti-cancer effects of garlic would include an interaction with the immune cells. In* vitro *and animal studies show that garlic contains immunopotentiating active principles. If confirmed in human studies, several immunocompromised states and chronic diseases with an abnormal cell-mediated immunity component may be targeted for experimental garlic therapy. Since most of the above pathologies, including some of the diseases involving vascular remodelling (e.g., atherosclerosis) and hypertension, are common during senescence, and garlic appears to have a remittive influence on several of them, it may be worthwhile to explore garlic as an anti-aging nutritional supplement.*

INTRODUCTION

A number of references to the therapeutic effects of garlic can be found in ancient Indian, Chinese and Egyptian literature. Are these claims mere myths or medically useful properties? In order to review the modern scientific evidence on this subject, a detailed computer search was made on several data bases including MEDLINE, IBIS and NAPRALERT. The information presented here is based on scientific articles published up to April 1994, available in the English language, either in original or an abstract form. In this mini review, an attempt will be made to identify significant therapeutic uses of garlic food forms tested in various continents. In general, only clinical trials in humans were considered. Although specific components of garlic have been suggested to have certain therapeutic effects[1], no attempt will be made to imply that individual garlic components are responsible for their therapeutic action or to describe their mechanism of action. In-depth analysis, including the meta-analysis of a specific therapeutic use of garlic can be found in recent reviews.[2,3]

The experimental systems explored for the therapeutic effects of food forms of garlic *in vivo* include humans and at least 11 other vertebrate species (fish, frog, mouse, rat, hamster, guinea pig, rabbit, chick, cat, dog, and sheep). Human volunteers of different race with varying cultural, ethnic and environmental background from at least 18 countries around the world have been subjected to garlic feeding trials. The food forms of garlic tested in sick and healthy humans and their significant therapeutic actions tested in two or more feeding trials in at least two countries are listed in Tables 8.1 and 8.2 respectively. Several study designs (cross over, parallel group, single-blind, double-blind, open-label), have been used on apparently healthy young, middle aged, and elderly male and female volunteers as well as patients with ischemic heart disease, hyperlipidemia, hypertension, diabetes, AIDS, peripheral arterial occlusive disease and general fatigue. Additional therapeutic effects e.g., immunomodulation and unconfirmed single clinical trials have also been reported.

TABLE 8.1
GARLIC FOOD FORMS

FRESH	Whole/Crushed
	(1 clove = 3g fresh garlic)
DRIED	At 50°C (1.35 = 5g fresh garlic)
COOKED	Fried or boiled
POWDERED	Desiccated
	Spray dried
OIL	Steam-distilled
	C_2H_5OH extracted
	Ajoene
OIL-MACERATED	--
AGED EXTRACT	Up to 20 months cold aging
	(Kyolic)

TABLE 8.2
EXPERIMENTAL THERAPEUTIC EFFECTS OF GARLIC IN HUMANS

ACTION	THERAPEUTIC EFFECT	CHANGE
HYPOLIPIDEMIC	Plasma cholesterol	↓
	Plasma triglycerides	↓
	Plasma HDL cholesterol	↑
	Plasma LDL cholesterol	↓
ANTIOXIDANT	LDL peroxidation	↓
ANTITHROMBOTIC	Platelet aggregation	↓
	Plasma fibrinogen level	↓
	Fibrinolytic activity	↑
	Ligand induced platelet aggregation	↑
ANTIHYPERTENSIVE	Blood pressure in high & moderate hypertensives	↓
ANTIMICROBIAL	Mycobacteria, fungi	↓
ANTIHYPOGLYCEMIC	Plasma glucosed	↓
IMMUNOMODULATORY	NK activity (in healthy & HIV-1+)	↑
	Helper/Suppresor T-cell ratio	↑
	Phagocytosis	↑

Plasma Lipid Concentration

Hyperlipoproteinemia and hyperlipidemia are well-known predisposing factors to ischemic heart diseases. One of the earliest reports in English on the lipid-lowering action of garlic appeared in a scientific publication from India [4], a country where garlic had been in use for thousands of years. In systems of traditional medicine, such as Ayurveda and Unani ("Alternative Medicine"), one to ten cloves of garlic may be prescribed as raw, fried or boiled cloves or as a whole extract for various therapeutic effects. Studies on alimentary hyperlipidemia suggest that garlic may offer partial protection after a fatty meal.[5]

The difficulties with some of the earlier controlled clinical trials with special reference to the lipid-lowering actions of garlic have been discussed in recent meta-analyses from the USA[2] and the UK.[3] Using well-designed studies (325 patients from 5 trials in 3 countries), it was concluded that a net decrease in total cholesterol attributable to garlic (equivalent to one-half to one clove per day) was 22.69 mg dl^{-1}, representing an average 9% reduction.[2] One mg dl^{-1} equals 0.026 mmol l^{-1}. In a more recent analysis (952 subjects, 16 trials from

5 countries), significant reductions in total cholesterol (29.62 mg dl^{-1}, 12%), triglyceride (11.92 mg dl^{-1}) and an insignificant change in HDL-cholesterol with garlic therapy beyond the final levels achieved with placebo alone became evident.[3] In one study, the reduction was evident in one month, and persisted up to 10 months (see ref. 3). In general, both dried (600-900 mg/day) and fresh garlic (10-20 g/day) were effective in lowering blood cholesterol levels. Fresh garlic's effects were consistent, but required a high dose (7 cloves/day) and had more side effects. Commercial preparations gave variable results and appear to have less side effects.

Blood Coagulation

Garlic has been reported to decrease plasma fibrinogen levels and ligand-induced platelet aggregation, increase the fibrinolytic and plasminogen activator activities, with or without a change in serum thromboxane B2 levels.[1,6,7] No consistent or significant effect of garlic on coagulation time has been found. The fibrinolytic activity of garlic was reported to be preserved by frying.[5] Garlic in raw or boiled form prevented the increase in plasma cholesterol and fibrinogen and decrease in fibrinolytic activity during alimentary lipidemia.[6] Thus, garlic may have a beneficial effect in thrombotic diseases.

Hypotension and Hypoglycemia

Recent evaluations show a moderate improvement in systolic and diastolic blood pressure in subjects with normal or high blood pressure[7,9], perhaps by decreasing peripheral vascular resistance. Followed by several animal studies that have shown a significant hypoglycemic effect of garlic, data on human patients with or without diabetes mellitus appear to be promising[8,9], and the effect is probably not due to increased insulin release into circulation.[10] The antihypertensive and antihypoglycemic actions of garlic remain the least explored therapeutic properties of garlic.

Infection

The antimicrobial properties of garlic, known since the time of Pasteur, include *in vitro* growth inhibition and/or killing of gram negative, gram positive and acid fast bacteria, protozoa, parasites, as well as inhibitory effects on the clinical isolates of yeasts and fungi and attenuation of intensity of experimental infections in animals. However, no well-controlled human clinical trial (published in English and available to the general scientific community) could be identified except a few isolated reports.[11,13] The *in vivo* antifungal properties may be mild and selective. The antimycobacterial effects and antifungal therapeutic effects of garlic on e.g., *Cryptococci*, *Mycobacteria*, and *Candida* are examples of potential combination chemotherapeutic uses of garlic that may be explored in meningitis, tuberculosis, and vaginitis, respectively. Herpes simplex and influenza viruses are inactivated by garlic *in vitro*.[14] Synthetic stereoisomers of ajoene (4, 5, 9-trithiadodeca-1,6,11-triene-9-oxide) from garlic was claimed to inhibit the replication of HIV-1 virus *in vitro*, suggesting that ajoene, in combination with the current modalities of treatment for AIDS may have a therapeutic potential.[15] Although the *in vivo* antimicrobial effectiveness of garlic would appear to be of limited value, well-designed clinical trials on human patients with infection are yet to be conducted.

Malignancy

Nutritional epidemiological studies in Asia, Europe and USA have suggested the potential anti-cancer properties of (presumably cooked) garlic consumption, i.e., an inverse association with the risk of death from certain types of cancer.[16,18] *In vitro* cytotoxic effects of garlic extracts on several tumor cell lines and *in vivo* studies using chemical induction and initiation-promotion models of animal cancer support this contention.[19] No direct human clinical trial has yet been undertaken on the possible cancer preventive or suppressive role of garlic. However, carcinogenesis is a slow, complex and incompletely understood process. A prospective study in which the dietary characteristics are assessed before cancer (e.g., gut-associated cancer) develops may be a preferable approach to study the possible chemopreventive effect of garlic on cancer development in humans. It is also hoped that future research in humans would rectify a gap in our knowledge on the potential adjuvant therapeutic uses of cooked or commercial garlic preparations in cancer patients undergoing various forms of treatment.

Immunomodulation

Nutrition is one of the critical determinants of immunocompetence and nutritional immunology is a fast-growing subdiscipline. Since the immune system is intimately involved in the maintenance of an antineoplastic *milieu*, it is reasonable to anticipate that the anti-cancer effect of garlic preparations may be partly due to a boosting of immunological defense mechanisms. The immunomodulatory properties of garlic and its active principles (Table 8.3) are currently being explored in several laboratories, including ours. Raw, powdered and liquid extract of garlic, a crude protein extract and F4 are all variably effective in several immune function tests in man and mouse.[20,25] Proteins and complex carbohydrates, which may influence the immune system are beginning to be described, e.g., F4, a 11 Kda protein fraction extracted from aged garlic[23] and mannose-binding lectins isolated from garlic bulbs.[26] Interestingly, Concanavalin A (red kidney bean), a mannose-binding glycoprotein, is one of the standard and

TABLE 8.3
IMMUNOMODULATION OF HUMAN PERIPHERAL BLOOD LYMPHOCYTES
BY GARLIC COMPONENTS

Garlic Form	Immune reactivity	Duration of feeding	Subjects	Effect	Reference
Raw garlic	NK activity	3 wks	Healthy	↑	20
Aged garlic	NK activity	3 wks	Healthy		20
Special preparation (Aged)	NK activity	6-12 wks	AIDS	↑	21
	Helper/Suppresor T-cell ration				
Powdered dragees	Phagocytosis	3 months	Geriatric	↑	22
Protein fraction (F-4)	IL2 Induced proliferation	in vitro	Healthy	↑	23
	%IL2-α+ cells	in vitro		↑	23
	NK activity	in vitro		↑	23
	LAK activity	in vitro		↑	23
	Con A induced	in vitro		↑	23

potent T lymphocyte mitogens used in immunological studies.[27] Bacterial capsules contain mannose. Mannose-binding proteins are known to bind bacteria resulting in a change in their shape, so that they activate the complement cascade and turn on phagocytes, perhaps telling the body which particles must be bound and eliminated. Hypocholesterolemic effects of garlic in human clinical trials are consistent with animal studies. Similarly, if the *in vitro* and *in vivo* immunomodulatory effects of garlic in mice are reproducible in humans, it should open a well-defined area of nutritional immunology for future therapeutic research on garlic. It is speculated that several diseases e.g., autoimmune diseases (rheumatoid arthritis), immune deficiency diseases (e.g., AIDS) and other chronic diseases with an abnormal cell-mediated immunity component (e.g., cancer, diabetes mellitus, tuberculosis) are potential candidates for garlic supplementation therapy. Most importantly, the geriatric population may stand to benefit from various therapeutic effects of garlic. In particular, the T-cell immune deficiency during senescence noted by us[27], presents an opportunity to test the immunomodulatory potentials of garlic. Additionally, a high percentage of ever-increasing world population of the elderly suffers from pathological conditions, such as atherosclerosis, hypercholesterolemia, diabetes mellitus, cancer, hypertension, and frequent microbial infections including pneumonia and urinary tract infections. Since garlic appears to have a remittive influence on most of the above pathologies, it may even qualify to be an anti-aging nutritional supplement of therapeutic importance.

Lessons from Past Trials

The minimal and optimal time needed to produce a desired therapeutic effect needs to be addressed in future clinical trials. The hypolipidemic effect was demonstrable in 4 weeks, but not always[28], although a 12 week feeding trial is more common. Another approach to accelerate our knowledge on garlic therapy would be to assess multiple pharmacological end-points in garlic-fed subjects in a comprehensive manner. Clinically relevant endpoints, such as morbidity could be included. However, the dose, food form, and duration of feeding required to produce multiple effects may not be the same. The effects of garlic are transient. Factors that can influence the therapeutic efficiency, e.g., bioavailability and bioactivity have to be worked out. Side effects other than odor were reported to be relatively rare, but need to be monitored more closely. Blinding has been one of the problems until "odorless" garlic preparations have become available recently. Standardization of the garlic preparations is a major scientific problem. The alliin content (the amino acid precursor to volatile sulfur-containing compounds in garlic), generally adjusted to 1.3% by weight has offered some uniformity, but different active principles appear to be responsible for different therapeutic effects. Also, chemically different garlic constituents could contribute to the same therapeutic effect via biochemically interdependent or independent pathways. Thus, more information on structure-function relationships are crucial to compare various preparations, trials and therapeutic effects. Knowledge on the potential garlic-drug pharmacological interactions, pharmacokinetics and tolerability are lacking at present.

Potential for Future Trials of Garlic Therapy

It is tempting to speculate the potential categories of patients who might benefit from garlic therapy (Table 8.4). It must be emphasized that in all these cases, prior animal experiments, including toxicity studies are needed. In addition, several less explored areas of potential therapeutic uses of garlic exist. For instance, free radical induced damages are common in many pathophysio-logical conditions, e.g., presence of lipid peroxides in atherosclerotic aorta and the increase in lipid peroxides in diabetic plasma. *Allium sativum* has an antioxidant activity and is an hydroxyl radical scavenger. Oxidation of LDL may play a significant role in atherosclerogenesis. Hence it is not impossible that the reported ability of garlic powder to suppress lipid peroxidation could have an interventional effect on the outcome of the coronary heart disease even without a significant reduction in plasma LDL concentration.[28] The latter may be a more sensitive biomarker of cardiovascular risk. In Japan, improvements in the symptoms of stress and general fatigue have been reported in garlic-fed subjects.[29] If confirmed, the detoxification by garlic of lead or mercury poisoning[24,30], perhaps by binding of the metals to sulfur compounds in garlic, may have an important therapeutic application in the field of environmental toxicology. Since the concept of vascular remodelling having a central role in cardiovascular disorders is emerging[31], it may be worthwhile to investigate the effects of garlic on factors involved in vascular diseases such as atherosclerosis and hypertension. As in the case of chemically synthesized drugs, basic research and clinical trials are the only reliable ways of exploring and assessing potential therapeutic uses of garlic and other phytochemical preparations.

TABLE 8.4
POTENTIAL PATIENT POPULATION FOR GARLIC THERAPY IN IMMUNITY,
INFECTION AND CANCER

Therapy	Patient Population
1. Adjunct therapy to immunocompromised patients	Cancer patients undergoing immunotherapy, chemotherapy, post-surgical therapy
2. Combined with chemotherapy in selected microbial infections	e.g., AIDS, tuberculosis
3. Supplemental therapy to boost immunity	Apparently healthy subjects e.g., elderly, depressed, under stress

REFERENCES

[1] Kleijnen, J., Knipschild, P. and Ter Riet, G. Garlic, onions and cardiovascular risk factors. A review of the evidence from human experiments with emphasis on commercially available preparations. Br. J. Clin. Pharmac. 1989; 28:535–544.

[2] Warshafsky, S., Kamer, R.S. and Sivak, S.L. Effect of garlic on total serum cholesterol. Ann. Intern. Med. 1993; 119:599–605.

[3] Silagy, C. and Neil, A. Garlic as a lipid lowering agent – a meta-analysis. J. Royal Coll. Phys. London, 1994; 28:39–45.

[4] Bordia, A. and Bansal, H.C. Essential oil of garlic in prevention of atherosclerosis. Lancet 1973; 2:1591.

[5] Chutani, S.K. and Bordia, A. The effect of fried versus raw garlic on fibrinolytic activity in man. Atherosclerosis 1981; 38:417.

[6] Sharma, K.K., Shrma, S.P. and Arora, R.C. Some observations on the mechanism of fibrinolytic enhancing effect of garlic during alimentary lipaemia. J. Postgrad. Med. 1978; 24:98.

[7] Harenberg, J., Giese, C. and Zimmermann, R. Effect of dried garlic on blood coagulation, fibrinolysis, platelet aggregation and serum cholesterol levels in patients with hyperlipoproteinemia. Atherosclerosis 1988; 74:247–249.

[8] Kiesewetter, H., Jung, F., Pindur, G., Jung, E.M., Mroweitz, C. and Wenzel, E. Effect of garlic on thrombocyte aggregation, microcirculation and other risk factors. I.J. Clin. Pharmacol. Ther. Tox. 1991; 29(4):151– 155.

[9] Aslam, M. and Stockley, I.H. Interaction between curry ingredient (Karela) and drug (chlorpropamide). Lancet 1979,i, 607.

[10] Sitprija, S., Plengvidhya, C., Kangkaya, V. and Bhuvapanich, S. Garlic and diabetes mellitus phase II clinical trial. J. Med. Assoc. Thailand Suppl. 1987; 70(2):223-227.

[11] Caporaso, N., Smith, S.M. and Eng, R.H.K. Antifungal activity in human urine and serum after ingestion of garlic (*Allium sativum*). Antimicro. Agents Chemother. 1983; 23(5):700-702.

[12] Caceres, A., Giron, L.M., Alvarado, S.R. and Torres, M.F. Screening of antimicrobial activity of plants popularly used in Guatemala for the treatment of dermatomucosal diseases. J. Ethnopharmacol. 1987; 20(3):223– 237.

[13] Davis, L.E., Shen, J.K. and Cai, Y. Antifungal activity in human cerebrospinal fluid and plasma after intravenous administration of *Allium sativum*. Antimicro. Agents Chemother. 1990; 34(4):651-653.

[14] Tsai, Y., Cole, L.L., Davis, L.E., Lockwood, S.J., Simmons, V. and Wild, G.C. Antiviral properties of garlic: *in vitro* effects on influenza B, *Herpes simplex* and *coxsackie* viruses. Planta Med. 1985; 5:460-461.

[15] Tatarintsev, A.V., Vrzheshch, P.V., Yershov, D.E., et al. Ajoene blockade of integrin-dependent processes in the HIV-infected cell systems. Vestnik Rossiiskoi Akademii Meditsinskikh-nauk 1992; 11:6-10.

[16] You, W.C., Blot, W.J., Chang, Y.S. et al. *Alliium* vegetables and reduced risk of stomach cancer. J. Natl. Cancer Inst. 1989; 81:162-164.

[17] Van den Brandt, P.A., Goldbohm, R.A., Van t Veer, P. et al. A large scale prospective cohort study on diet and cancer in The Netherlands. J. Clin. Epidemiol. 1990; 43:285-295.

[18] Steinmetz, K.A., Kushi, L.H., Bostick, R.M., Folsom, A.R. and Potter, J.D. Vegetables, fruit, and colon cancer in the Iowa Women's Health Study. Amer. J. Epidemiology 1994; 139:1-15.

[19] Liu, J. *et al.* Inhibition of 7,12-dimethylbenz[a]anthracene-induced mammary tumors and DNA adducts by garlic powder. Carcinogenesis 1992; 13:1847–1851.

[20] Kandil, O., Abdullab, T.H., Elkadi, A. and Carter, J. Garlic and the immune systems in humans: Its effect on natural killer cells. Fed. Proc. 1987; 46:441 (abstract).

[21] Abdullah, T.H., Kirkpatrick, D.V. and Carter, J. Enhancement of natural killer cell activity in AIDS with garlic. J. Oncol. 1989; 21:52–53.

[22] Brosche, T. and Platt, D. On the immunomodulatory action of garlic (*Allium sativum* L.). Medizinische-welt 1993; 44:309–313.

[23] Morioka, N., SZE, L., Morton, D. and Irie, R. A protein fraction from aged garlic extract enhances cytotoxicity and proliferation of human lymphocytes mediated by interleukin-2 and concanavalin A. Cancer Immunol. Immunother. 1993; 37:316–322.

[24] Lau, B.H.S. Detoxifying, radioprotective and phagocyte-enhancing effects of garlic. Int. Clin. Nutr. Rev. 1989; 9:27–31.

[25] Hirao, Y., Sumioka, I., Nakagami, S. *et al.* Activation of immunoresponder cells by the protein fraction from aged garlic extract. Phytotherapy Res. 1987; 1(4):161–164.

[26] Damme, Smeets, K., Torrekens, S., Van Leuven, F. and Goldstein, I., Peumans. The closely related homomeric and heterodimeric mannose-binding lectins from garlic are encoded by one-domain and two-domain lectin genes, respectively. Eur. J. Biochem. 1992; 206:413–420.

[27] Krishnaraj, R. Immunosenescence: The activity, subset profile and modulation of human natural killer cells. AGING: Immunology and Infectious Dis. 1990; 2:127–134.

[28] Phelps, S. and Harris, W.S. Garlic supplementation and lipoprotein oxidation susceptibility. Lipids 1993; 28:475–477.

[29] Kawanishi, Y. *et al.* Clinical experience of Kyoleopin: Clinical study on general fatigue associated with cold. Treatment and New Medicine 1985; 22: 3012–3024.

[30] Edward, M.G. Pharmacological and clinical studies of garlic. Biol. Abstract 1967; 66:74823E.

[31] Gibbons, G.H. and Dzau, V.J. Mechanism of disease: The emerging concept of vascular remodelling. N. Eng. J. Med. 1994; 330:1431–1438.

CHAPTER 9

LIPID SOLUBLE PHYTOCHEMICAL CONSTITUENTS IN GARLIC FOOD FORMS

DAVID S. WEINBERG

Southern Research Institute
2000 Ninth Avenue South
Birmingham, AL 35205

ABSTRACT

We reviewed studies that had been published on the determination of biologically active organosulfur phytochemicals in garlic food products. Such phytochemicals may be produced when garlic is crushed and processed. The array of organosulfur phytochemicals present in a specific garlic product at a specific time is highly dependent on the history of the product.

Garlic products that are antibacterial or antifungal may contain a high concentration of allicin or related alk(en)yl thiosulfinates. Garlic products that are antithrombic may contain a high concentration of (Z)-ajoene, (E)-ajoene, 2-vinyl-[4H]-1,3-dithiin, or 3-vinyl-[4H]-1,2-dithiin. Garlic products that have antioxidant or antitumor properties may contain a high concentration of diallyl disulfide or other alk(en)yl polysulfides.

Analyses of samples of garlic are more informative when thoroughly validated analytical methods are used and the history of the samples is very well-documented.

TECHNICAL REVIEW

Introduction

Allium, a genus that belongs to the family Liliaceae, contains more than 600 different species.[1] Alliaceous garlic species are important as food plants and include garlic (*Allium sativum* L.), great headed garlic, (*Allium ampeloprasum* L.), and wild garlic (*Allium ursinum* L.). Garlic has a long history as a folk medicine.[2] Currently there is considerable interest in the biological properties of garlic that are related to infectious disease, cardiovascular disease, and cancer.[1]

Garlic exhibits little odor in the natural state, but when it is crushed it develops an intense odor.[3] Chemical reactions of precursors and enzymes produce thermally unstable organosulfur phytochemicals that spontaneously transform into other phytochemicals.[4] Therefore chemical reactions are critically important in determining the precise array of organosulfur phytochemicals that will be present in any particular garlic sample at any particular time.[4] Phytochemicals detected depend on the identity of the garlic species, the location where the garlic was grown, the storage of the garlic after harvesting, the processing of the garlic, the isolation of the phytochemicals from garlic, the storage of the isolated phytochemicals, and the analysis of the isolated phytochemicals.[4]

Alliin, a Precursor to Biologically Active, Lipid-Soluble Phytochemicals

Alliin [(+)-S-2-propenyl-L-cysteine S-oxide] is a white, crystalline, odorless solid with a melting point of 163-165°C. It is highly soluble in water and is insoluble in most organic solvents.[3] When garlic is crushed, the enzyme alliinase reacts with alliin to give 2-propenyl 2-propenylthiosulfinate (allicin). Other phytochemicals related to alliin present in garlic include S-(E)-1-propenyl-L-cysteine S-oxide and S-methyl-L-cysteine S-oxide.[4] These phytochemicals also react with alliinase in crushed garlic to give rise to the corresponding biologically active thiosulfinate.[4]

Allicin, an Antibacterial Phytochemical and Precursor to Other Biologically Active Phytochemicals

Allicin is the predominant olfactory, gustatory, and antibacterial principle of freshly cut garlic.[5] It is a thermally unstable phytochemical. Cavallito and Bailey extracted 4 kg of ground garlic cloves with aqueous alcohol at ambient temperature, concentrated the extract, and distilled the extract at low temperature to obtain about 6 grams of allicin.[6] In 1947, Small et al. synthesized allicin from diallyl disulfide by treatment with hydrogen peroxide.[7] Cavallito et al. found that aqueous, nonaqueous, or dry preparations of allicin upon standing at ambient temperature for two days lost all of their antimicrobial activity and produced a viscous liquid that was no longer water soluble and cannot be distilled.[8] Brodnitz et al. also found allicin to be highly thermally unstable.[9] They found that nearly complete decomposition of synthetic allicin occurred when it was held at 20°C for 20 h. Laakso et al. observed that the degradation of allicin in vegetable oil was much faster than in water.[10] Hughs and Lawson reported that garlic clove homogenates stored at 22°C showed an 86% decrease occurred in 1 month while that stored at 4°C showed no decrease in allicin content in 1 month and only 5% in 6 months.[11]

Jansen et al. developed an improved high-performance liquid-chromatographic method to determine allicin in crushed garlic and other garlic food forms.[12] Lawson et al. used a variation of the method to determine the concentration of allicin and other alk(en)yl thiosulfinates in crushed garlic bulbs and found 2-propenyl 2-propenethiosulfinate (3.72 mg/g), 2-propenyl 1-

propenethiosulfinate (0.35 mg/g), 2-propenyl methanethiosulfinate (0.23 mg/g), 1-propenyl 2-propenethiosulfinate (0.34 mg/g), 1-propenyl methanethiosulfinate (0.14 mg/g), methyl 2-propenethiosulfinate (0.50 mg/g), and methyl methanethiosulfinate (0.11 mg/g).[13] Other authors analyzed other garlic samples for allicin and other alk(en)yl thiosulfinates.[5,14] Block found that the major thiosulfinate in garlic was allicin.[5] In addition, he found that the major unsymmetrical thiosulfinates were methyl 2-propenethiosulfinate and 2-propenyl methanethiosulfinate (2:1 ratio), and that the ratio of methyl/allyl thiosulfinates depended on whether the garlic was grown in a warm or cold climate.[5] He also found that the concentration of the minor unsymmetrical thiosulfinates that possess a 1-propenyl group varied with age and storage conditions of the garlic.[5]

Ajoenes and Dithiins, Antithrombic Phytochemicals Present in Garlic

Block *et al.* isolated from garlic (E,Z)-4,5,9-trithiadodeca-1,6,11-triene-9 oxide (0.065 g, 0.011%), which they named (E,Z)-ajoene.[15] They also isolated 2-vinyl-4H-[1,3]-dithiin (0.09 g, 0.015%) and 3-vinyl-4H-[1,2]-dithiin (0.02 g, 0.0033%).[15] They found that all of these phytochemicals were effective in inhibiting platelet aggregation. They discovered a simple synthesis of ajoenes. They dissolved 5.8 g of synthetic allicin in 58 mL of 40% aqueous acetone and heated the solution at 64°C for 4 h. After chromatographic purification they obtained 1 g (17% yield) of 90:10 E/Z ratio of ajoene.[15]

Lawson *et al.* determined the concentration of ajoene and vinyldithiins in fresh picked or store purchased garlic or in commercial garlic products.[16] They detected ajoene (0.015 to 0.118 mg/mL) and vinyldithiins (0.070 to 0.689 mg/mL) in oil-macerated garlic. They detected no ajoene or vinyldithiins in crushed garlic, in garlic powder tablets, in garlic powder suspended in a gel, or in steam-distilled garlic oil.[16] Iberl *et al.* analyzed centrifuged supernatants of garlic products.[14] They crushed garlic bulbs (2.5 g), left the crushed bulbs at ambient temperature for 30 min, and then suspended the crushed bulbs in linseed oil (10 mL), or suspended the garlic bulbs in the oil and then crushed them. They found that the allicin content (maximum about 5 mg/g garlic) rapidly decreased and the ajoene content (maximum about 1 mg/g garlic) and vinyldithiin content (maximum about 1.8 mg/g garlic) in the oil rapidly increased at ambient temperature over 24 h.[14]

Diallyl Disulfide and Related Phytochemicals, Antioxidant and Antitumor Compounds Present in Garlic

Garlic oil contains diallyl sulfide, which inhibits colon cancer in mice exposed to 1,2- dimethylhydrazine.[17] Other constituents of garlic, allyl methyl sulfide, allyl methyl disulfide, and diallyl sulfide, all inhibited benzo[a]pyrene-induced neoplasms of the forestomach and lung of female A/J mice when administered 48 to 96 h prior to carcinogen challenge.[18] Wertheim and later Semmler crushed and steam distilled garlic cloves and obtained "oil of garlic." The major phytochemical found to be present was diallyl disulfide.[19,20,21] Although diallyl disulfide and other volatile, nonpolar sulfides and polysulfides are created upon the steam distillation of garlic and are major components of the garlic oil so obtained, it is not really proper to consider those components as part of the "essential oil" of garlic because they are formed during the processing of garlic and are not present as such in unprocessed garlic.[4]

Lawson et al. used high-performance liquid chromatography to determine alk(en)yl polysulfides in crushed, steam-distilled garlic and obtained the following results: diallyl sulfide (0.087 mg/g), diallyl disulfide (1.13 mg/g), diallyl trisulfide (0.810 mg/g), diallyl tetrasulfide (0.368 mg/g), diallyl pentasulfide (0.094 mg/g), and diallyl hexasulfide (0.018 mg/g). They also determined the corresponding allyl methyl polysulfides and dimethyl polysulfides.[16] Weinberg et al. extracted an experimental, aged, odorless garlic product with n-pentane and analyzed the extract using low temperature gas chromatography/mass spectrometry.[22] They found dimethyl sulfide (0.000607 mg/g), dimethyl disulfide (0.000181 mg/g), diallyl sulfide (0.002016 mg/g), diallyl disulfide (0.000784 mg/g), diallyl trisulfide (0.000695 mg/g), allyl methyl disulfide (0.001643 mg/g), allyl methyl disulfide, (0.000411 mg/g), allyl methyl trisulfide (0.000795 mg/g), and ethyl 2-propenesulfinate (0.011363 mg/g).[21] The latter compound was probably formed upon reaction of allicin with ethyl alcohol employed in the original preparation of the odorless garlic.[22]

REFERENCES

1 Block, E., Naganathan, S., Putman, D. and Zhao, S.-H. Organosulfur chemistry of garlic and onion: recent results. Pure & Appl. Chem. 1993; 65:625–632.

2 Block, E. Antithrombotic agent of garlic: A lesson from 5000 years of folk medicine. In Folk Medicine, (Steiner, R.P., ed.) American Chemical Society, Washington, DC, 1986:111–124.

3 Stoll, A. and Seebeck, E. Chemical investigation on allicin, the specific principle of garlic. Advan. Enzymol. 1951; 11:377–400.

4 Block, E. The organo-sulfur chemistry of the genus Allium — implications for the organic chemistry of sulfur. Angew Chem. Int. Ed. 1992; 31:1135–1178.

5 Block, E., Naganathan, S., Putman, D. and Zhao, S.-H., Allium chemistry: HPLC analysis of thiosulfinates from onion, garlic, wild garlic (ramsoms), leek, scallions, shallot, elephant (great-headed) garlic, chive, and chinese

chive. Uniquely high allyl to methyl ratios in some garlic samples. J. Agric. Food Chem. 1992; 40:2418–2430.

[6] Cavallito, C.J. and Bailey, J.H. Allicin, the antibacterial principle of *Allium sativum*. I. Isolation, physical properties and bacterial action. J. Am. Chem. Soc. 1944; 66:1950–1951.

[7] Small, L.D., Bailey, J.H. and Cavallito, C.J. Alkyl thiosulfinates. J. Am. Chem. Soc. 1947; 67;1710–1713.

[8] Cavallito, C.J., Buck, J.S. and Suter, C.M. Allicin, the antibacterial principle of *Allium sativum*. II. Determination of the chemical structure. J Am. Chem. Soc. 1944; 66:952.

[9] Brodnitz, M.H., Pascale, J.V. and Van Derslice, L. Flavor components of garlic extract. J. Agric. Food Chem. 1971; 19:273–275.

[10] Laakso, I., Seppanen-Laakso, T., Hiltunen, R., Muller, B., Jansen, H. and Knobloch, K. Volatile garlic odor components: gas phases and adsorbed exhaled air analysed by headspace gas chromatography-mass spectrometry. Planta Medica 1989; 55:257–261.

[11] Hughes, B.G. and Lawson, L.D. Antimicrobial effects of *Allium ampeloprasum* L. (elephant garlic), and *Allium cepa* (onion), garlic compounds, and commercial garlic supplement products. Phytotherapy Res. 1991; 5:154–158.

[12] Jansen, H., Muller, B. and Knobloch, K. Allicin characterization and its determination by HPLC, Planta Medica 1987:559–562.

[13] Lawson, L.D., Wood, S.G. and Hughes, B.G. HPLC analysis of allicin and other thiosulfinates. Planta Medica 1991; 57:263–270.

[14] Iberl, B., Winkler, G. and Knobloch, K. Products of allicin transformation: ajoenes and dithiins, characterization and their determination by HPLC. Planta Medica 1990; 56:202–211.

[15] Block, E., Ahmad, S., Catalfamo, J.L., Jain, M.K. and Apitz-Castro, R. Antithrombotic organosulfur compounds from garlic: structural, mechanistic, and synthetic studies. J. Am. Chem. Soc. 1986; 108:7045–7055.

[16] Lawson, L.D., Wang, Z.-Y.J. and Hughes, B.G. Identification and HPLC quantitation of the sulfides and dialk(en)yl thiosulfinates in commercial garlic products. Planta Medica 1990; 56:363–370.

[17] Wargovitch, M.J. Diallyl sulfide, a flavor component of garlic (*Allium sativum*), inhibits dimethylhydrazine-induced colon cancer. Carcinogenisis 1987; 8:487–489.

[18] Sparnins, V.L., Barany, G. and Wattenberg, L.W. Effects of organosulfur compounds from garlic and onions on benzo[a]pyrene-induced neoplasia and glutathione S-transferase activity in mouse. Carcinogenesis 1988; 9:131–134.

[19]   Wertheim, T. Ann. 1844;51;289 cited in Stoll, A. and Seebeck, E. Chemical investigations on allicin, the specific principle of garlic. Advan. Enzymol. 1951; 11:377–400.

[20]   Wertheim, T. Ann. 1845;55:297, cited in Stoll, A. and Seebeck, E. Chemical investigations on allicin, the specific principle of garlic. Advan. Enzymol. 1951; 11:377–400.

[21]   Semmler, F.W. Arch. Pharm, 1892;230:434, cited in Stoll, A. and Seebeck, E. Chemical investigations on allicin, the specific principle of garlic. Advan. Enzymol. 1951; 11:377–400.

[22]   Weinberg, D.S., Manier, M.L., Richardson, M.D. and Haibach, F.G. Identification and quantification of organosulfur compliance markers in a garlic extract. J. Agric. Food Chem. 1993; 41:37–41.

CHAPTER 10

CHEMICAL METHODS DEVELOPMENT FOR QUANTITATION OF PHYTOCHEMICALS IN AQUEOUS GARLIC EXTRACT

RICHARD S. GEARY and MICHAEL A. MILLER

Department of Applied Chemistry and Chemical Engineering
Southwest Research Institute
San Antonio, Texas

ABSTRACT

Methods were developed to analyze the volatile and nonvolatile constituents in an aged aqueous garlic extract (AAGE) using gas chromatography (GC) with flame ionization detection (FID) and HPLC with electrochemical and UV/VIS detection. Constituents of AAGE were structurally characterized using GC/MS and FAB/MS. Five sulfur-containing allylic compounds were identified in the volatile fraction of AAGE, and four of these were successfully quantified. Two sulfur-containing allylic compounds were identified in the aqueous fraction of AAGE, and one of these was successfully quantified. The degradation rate of S-allyl-L-cysteine, a constituent of the aqueous fraction of AAGE, was inversely related to temperature storage conditions over the twelve-week sampling period. The concentration of volatile AAGE constituents did not change significantly over the twelve-week sampling period at the 4- and 25°C temperature storage conditions, but increased significantly at 45°C over the same period.

INTRODUCTION

A history of the chemistry of garlic is informative and directive for the current study. One of the earliest chemical studies was made in 1844 by the German chemist Theodor Wertheim. Wertheim employed steam distillation to produce steam which contained small amounts of garlic oil. Redistillation of the oil yielded some odoriferous volatile substances. He proposed the name allyl for the hydrocarbon group in the oil (C_3H_5). In 1892 another German investigator, F.W. Semmler, applied steam distillation to cloves of garlic, producing one or two grams of an odoriferous oil per kilogram of garlic. In turn the oil yielded diallyl disulfide accompanied by lesser amounts of diallyl trisulfide and diallyl tetrasulfide.

The next key discovery in the chemistry of garlic was made in 1944 by Chester J. Cavallito and his colleagues at the Sterling-Winthrop Chemical Company in Rensselaer, New York. They established that extraction methods less vigorous than steam distillation yield rather different substances. Cavallito applied ethyl alcohol to four kilograms of garlic at room temperature and eventually produced six grams of an oil whose formula was $C_6H_{10}S_2O$. The oil proved to be both antibacterial and antifungal. Chemically, Cavallito's oil is the oxide of diallyl disulfide, the principal substance Semmler had isolated by steam distillation half a century earlier. The name of the chemical is diallyl thiosulfinate and was named allicin by Cavallito. It is a chemically unstable, colorless liquid that accounts for the odor of garlic — much more so than the diallyl disulfides. It was later shown that allicin develops in garlic when an enzyme initiates its formation from an odorless precursor molecule, which Stoll and Seebeck identified as (+)-S-allyl-L-cysteine sulfoxide. Evidently, the cutting or crushing of garlic enables the enzyme, called alliinase, to come in contact with the precursor of allicin. Stoll and Seebeck named the precursor alliin which accounts for about 0.24 percent of the weight of a typical garlic bulb.

Early chemistry done by Eric Block and John O'Connor proved important in elucidating the structure and mode of formation of the garlic antithrombotic factor.[1] The compound was determined to be 4,5,9-trithiadodeca-1,6,11-triene 9-oxide and was named ajoene. Eric Block's work on the self-condensation of methyl methanethiosulfinate suggested that ajoene might form by self-condensation from allicin. This was verified by simply heating allicin with a mixture of water and an organic solvent such as acetone.

The medicinal properties of garlic have long been recognized. Antiseptic and antithrombotic activities have been documented and additional studies have reported reduction of serum cholesterol and retarding hyperglycemia attributed to garlic.[2-5] Tumor-inhibitory effects of garlic and garlic components have been demonstrated in several different experimental systems.[6-9]. In addition, epidemiological reports concerning the preventive effect of garlic on gastric cancer in humans provide data supporting the experimental studies.[10]

The objectives of this study were to identify and quantitate at least ten major sulfur-containing allylic compounds in a garlic extract supplied to our laboratories by the National Cancer Institute (NCI) and to determine shelf-life stability of the compounds in the extract over at least a three-month period. The selected sulfur-containing compounds are dietary phytochemicals that have been shown to be effective for their anticarcinogenic potential by epidemiological and test system evidence.

METHODS AND MATERIALS

The gas chromatographic analyses were conducted on a system equipped with a flame ionization detector and a 50 m × 0.22 mm fused silica capillary column coated with polyethylene glycol. The volatile components of the AAGE (Sample Code No.: RRGAR400001) were structurally characterized on a GC/MS quadrupole system equipped with a DB-5 chromatographic column. Additional structural characterization was accomplished using a Finnigan 4800 quadrupole system equipped with the same capillary column used on the GC/FID system (50 m × 0.22 mm fused silica column coated with polyethylene glycol).

The volatile constituents of AAGE were successfully extracted from the aqueous matrix with diethyl ether. A spinning band distillation apparatus was used with low heating to concentrate the ether extracts.

AAGE (Sample Code No.: RRGAR400001) was analyzed by reverse-phase high performance liquid chromatography (RP-HPLC) using both UV detection (at 254 nm and 205 nm), and electrochemical detection (ECD) at dual mercury/gold electrodes (reduction = − 1 V; oxidation = +0.15 V vs. Ag/AgCl). This chromatography was compared to standards and duplicate spikes of the AAGE to confirm coelution of standard and sample peaks.

The electrochemical HPLC system was comprised of an LDC Model III dual piston pump, an SSI (Model LP-21) low-pulse dampener, a refrigerated Waters® WISP autoinjector, a Supelcosil® LC-18-DB 5 μm analytical HPLC column (length: 25 cm; i.d.: 4.6 mm), and a BAS amperometric detector with a dual mercury-gold transducer cell. The mobile phase for separation of very polar garlic compounds consisted of 5% acetonitrile and 95% 0.1 M KH_2PO_4/0.159 M KH_2PO_4 aqueous buffer (pH 7.0). Isocratic conditions were maintained on this system due to the detector requirements. Because of the variability of electrochemical detectors, it is imperative that the samples be internally standardized. Dipropanolsulfide (thiodipropanol) was selected as the internal standard for the polar amino acid sulfur-containing allylic compounds.

Six unopened vials containing AAGE (Code No.: RRGAR400001) were randomly assigned to three different storage temperature treatment groups (two vials per temperature). Prior to placing the vials on test, an aliquot was separated and extracted for GC/FID analysis and a second aliquot was separated for direct injection on HPLC (both ECD and UV detection).

After removal of sample aliquot from the test vials, the vials were purged with nitrogen, tightly capped, and placed back into the test conditions. Temperatures were monitored and documented for each of the test conditions twice each week. Sample aliquots were analyzed by GC/FID, HPLC-UV, and HPLC-ECD at 0, 1, 2, 3, 6, 8, and 12 weeks.

RESULTS

Quadrupole GC/MS analysis with DB-5 chromatographic separation, based primarily on interpretation of mass fragmentation patterns, suggest the presence of diallyl sulfide, diallyl disulfide, allyl methyl disulfide, allyl propyl disulfide, bis(l-propenyl) sulfide, allyl vinyl sulfide, and 2,6-bis(1,1-dimethyl-ethyl)-4-methyl-phenol.

Additional analysis facilitated by a Finnigan 4800 quadrupole system equipped with a fused silica capillary column (polyethylene coated) identical to the column utilized on our GC/FID system suggests the presence of diallyl sulfide, diallyl disulfide, allyl methyldisulfide, diallyl trisulfide, and 2,6-bis(1,1 -dimethylethyl)-4-methyl-phenol. The most prominent peak in the profile has been identified as acetic acid by its characteristic fragmentation pattern.

A total of five sulfur-containing allylic compounds was identified in the volatile fraction of the AAGE. Three of these (diallyl sulfide, diallyl disulfide, and diallyl trisulfide) were successfully quantified using standard curves of purified standards developed over an appropriate concentration range for the observed level of the compound of interest in AAGE.

Allyl methyl sulfide was apparent in trace amounts in early analysis of AAGE. Later analysis of the AAGE indicated that this peak was no longer measurable. This observation suggests that allyl methyl sulfide is unstable in AAGE and was therefore not quantifiable. Table 10.1 contains a summary of the quantitation of these compounds in AAGE.

TABLE 10.1
SUMMARY OF QUANTITATIVE DATA FOR THE AAGE
(HPLC and GC/FID)

Extract Compound	Concentration in ($\mu q/mL$)
Diallyl sulfide	6.91 ± 2.37*
Diallyl disulfide	52.9 ± 9.40*
Diallyl trisulfide	18.3 ± 2.91*
Allyl methyl sulfide	NQ
Allyl methyl disulfide	4.33 ± 1.25*
S-allyl-L-cysteine	1890 ± 305[†]
2, 6-Bis (1,1-dimethylethyl) 4-methyl-phenol	24[‡]
Acetic acid	491[‡]

* Quantitated over 2 months with storage at 4°C (n=6).
† n=3
‡ Semiquantitative (no standard; based on mass-equivalent-response assumption for the FID to internal standard and sample compound).
NQ = Not Quantifiable.

S-allyl-L-cysteine was isolated by EC detection as compared to the pure standard. This compound eluted at approximately five minutes under the initial conditions of the gradient. Other peaks which were identified in AAGE by coelution of standards include diallyl sulfide, diallyl disulfide, diallyl trisulfide, allyl methyl sulfide, and allyl methyl disulfide. Compounds of interest which were not found in AAGE include propyl allyl disulfide, allicin, allyl methanethiosulfinate, dimethyl disulfide, and allylmercaptan (2-propenethiol).

Quantitation of the less polar constituents was not conducted by HPLC because the GC/FID system appeared to provide more reproducible chromatography with good sensitivity. Quantitation of S-allyl-L-cysteine was, however, pursued by HPLC-ECD methods. Quantitation was accomplished by construction of a five-point standard curve (0.1, 0.25, 0.5, 1.0, and 2.0 mg/mL) using a pure external standard of S-allyl-L-cysteine. Standards were prepared in methanol and injected directly on the HPLC column. Twenty microliters of the garlic extract (AAGE) with no extraction or sample treatment was directly injected on the HPLC column.

Following analysis of stored extract at various temperatures, stability profiles (concentration-time curves) for allylic sulfur containing constituents of AAGE were developed. Where appropriate, the data were fitted to mono- and bi-exponential expressions representing pseudo-first-order, nonreversible kinetics and reversible consecutive kinetics, respectively. A comparison of the pseudo rate constant is provided in Table 10.2.

TABLE 10.2
COMPARISON OF FIRST-ORDER RATE CONSTANTS
FOR AAGE CONSTITUENTS

Constituent	Isotherm (°C)	*Alpha*	*Beta*
S-allyl-L-cysteine*	4	1.43	0.0622
	25	1.24	-
	45	0.881	0.0077
Cysteine analog[†]	4	0.475	-
	25	0.286	-
	45	0.270	-
Diallyl sulfide[‡]	45	0.497	-
Diallyl disulfide*	45	2.09	0.0400
Diallyl trisulfide*	45	3.00	0.0600

* Reversible, consecutive reaction model
[†] Simple first-order nonreversible reaction model.
[‡] Simple first-order nonreversible formation model.

Isothermal concentration-time curves for the nonpolar AAGE constituents indicate little or no degradation at 4- and 25 °C storage conditions over an eight-week period. However, the 45 °C isotherm for diallyl sulfide increased to steady-state concentrations in a first-order fashion. A similar behavior was observed for the polysulfides, but with an additional degradation phase after reaching a peak concentration as is typically observed for reversible consecutive kinetics.

S-allyl-L-cysteine exhibited biphasic decay, analogous to complex first-order, reversible consecutive kinetics. As indicated in Table 10.2, the initial rate of degradation, *alpha*, was inversely related to temperature storage conditions. This suggests that a reversible "feeding" mechanism may play a significant role in its apparent rate of degradation. The source for such a mechanism may be attributed to less stable cysteine and allylic analogs that degrade more rapidly with increasing temperature and that yield reactive precursors of S-allyl-L-cysteine. The second polar AAGE constituent, an undetermined allylic cysteine analog, demonstrated pseudo first-order kinetics with a degradation rate that was independent of temperature storage conditions.

DISCUSSION

Although a large number of compounds were separated and character-ized, only a few of the compounds identified in the AAGE extract were sulfur-containing allylic compounds. Indeed, some of the compounds identified are clearly not of plant origin (such as the pyridines and substituted phenols), which may be compounds leached from packaging or during processing or added to the extract for other reasons (such as an antioxidant for preservation). Five of the six compounds quantified were nonpolar AAGE constituents of mono- and polysulfide allylic chemical structure and were present in very small quantities. However, a polar sulfur-containing compound, S-allyl-L-cysteine, is present in substantial quantity as determined by HPLC-ECD (> 1 mg/mL). An additional polar sulfur-containing compound, which has not been completely characterized, but is believed to be an allylic cysteine analog, appears to be present in substantial quantities as well.

In spite of the relatively large amount of the two cysteine allylic analogs, these compounds are nevertheless unstable at all storage conditions (4-, 25-, 45 °C) and S-allyl-L-cysteine decays in a biphasic manner over time. The initial degradation rate of S-allyl-L-cysteine appears to be inversely related with temperature storage conditions. The mechanism of degradation is complex and has not been fully elucidated. However, one may speculate that because of the biphasic profile, at least one reversible "feeding" reaction may be involved. For example, S-allyl-L-cysteine, which is a chemically reduced form of alliin, may oxidize to the sulfoxide in the presence of water and thus be susceptible to enzymatic cleavage by allinase. Clearly, this postulate depends on the assumption that allinase is present in AAGE and it is probable that allinase does not survive the aging process. More complete information on AAGE preparation is necessary to elucidate the mechanism of degradation.

Although the nonpolar sulfur-containing allylic compounds are stable at 4- and 25 °C over a twelve-week period, the concentration of each of these compounds increases to a maximum concentration over time at 45 °C. For the case of the polysulfides, there appears to be a competing effect between the rate of formation and the rate of degradation, thus yielding a peak concentration followed by first-order degradation. The apparent lack of a degradation pathway for diallyl sulfide may be attributed to its stability as compared to the less stable polysulfides. The effect of increasing concentration over time is dependent on the availability of a precursor compound(s), probably the unstable polar allylic cysteine analogs. The data are consistent with thermodynamically-controlled formation of the polysulfides.

It is important to differentiate between thermodynamically and kinetically controlled formation of the polysulfides. The former mechanism would result in long-term stability at or below room temperature, whereas the latter would yield temperature-dependent shelf-life. The definitive mechanism may be elucidated by conducting stability studies with temperature progression (i.e., 25-, 30-, 35-, 40 °C). A highly correlative relationship between temperature and rate of formation or degradation is indicative of a kinetically-controlled reaction. A thermodynamically-controlled reaction should result in an activation response at a temperature corresponding to the activation energy of precursor reactions.

ACKNOWLEDGMENT

The authors would like to express their gratitude to Dr. Eric Block, SUNYA, for his professional consultation on this project and for his capable technical skills in providing sulfur-containing allylic standards synthesized in his laboratory. This work was supported by NCI Contract No. NO1-CN-05283-O1.

REFERENCES

[1] Block, E. The chemistry of garlic and onions. Scientific American 1985; 252:114-119.

[2] Boullin, D.J. Garlic, a platelet inhibitor. Lancet 1981;776-777.

3   Chang, M.J.W. and Johnson, M.A. Effect of garlic on carbohydrate metabolism and lipid synthesis in rats. J. Nutr. 1980; 110:931-936.

4   Ernst, E., Weihmayr, T.H. and Matrai, A. Garlic and blood lipids. Br. Med. J. 1985;291:139.

5   Sodimu, O., Joseph, P.K. and Augusti, K.T. Certain biochemical effects of garlic oil on rats maintained on high fat-high cholesterol diet. Experientia 1984; 40:78-80.

6   Belman, S. Onion and garlic oils inhibit tumor promotion. Carcinogenesis 1983; 4:1063-1065.

7   Criss, W.E., Fakunle, J., Knight, E., Adkins, J., Morris, H.P. and Dhillon, G. Inhibition of tumor growth with low dietary protein and with dietary garlic extracts. Fed. Proc. 1982; 41:281.

8   Wargovich, M.J. Diallyl sulfide, a flavor component of garlic (allium sativum), inhibits dimethylhydrazine-induced colon cancer. Carcinogenesis 1987; 8:487-489.

9   Nishino, H., Iwashima, A., Itakura, Y., Makuura, H. and Fuwa, T. Antitumor-promoting activity of garlic extract. Oncology 1989; 46:277-280.

10   Horwitz, N. Garlic as a plant du jour: chinese study finds it could prevent G.I. cancer. Medical Tribune, August 12, 1981.

CHAPTER 11

DIETARY TOLERANCE/ABSORPTION/METABOLISM OF PHYTOCHEMICALS IN GARLIC

YUKIHIRO KODERA

Department of Chemistry
University of California, Irvine
Irvine, CA 92717

ABSTRACT

The metabolism and pharmacokinetics of constituents in garlic are described in this paper. Alliin, which is enzymatically converted to allicin, was not transformed into allicin in the digestive tract when alliin alone was administered orally. Under simulated digestive conditions, little allicin was released from a garlic powder which contained both alliin and alliinase. When allicin was mixed with blood, it disappeared very rapidly, and the formation of allylmercaptan and diallyl disulfide were observed. Allicin also converted the hemoglobin in red blood cells to methemoglobin. Furthermore, no allicin could be detected in the effluent when allicin was perfused into an isolated liver. These data indicate that allicin does not seem to be a biologically active compound inside the body. The pharmacokinetics of other organosulfur compounds, such as alliin, diallyl sulfides, vinyl dithiins and S-allyl cysteine, derived from garlic are also discussed in this paper.

INTRODUCTION

Garlic is rich in a variety of organosulfur compounds. However, garlic cloves contain a limited number of sulfur compounds. Processing of garlic is necessary for the formation of a variety of unique organosulfur compounds. Allicin, which is an unstable enzymatic transformation product generated from alliin when garlic cloves are crushed, has been shown to be the transient compound which decomposes into a variety of organosulfur compounds such as diallyl sulfides, vinyl dithiins, and ajoene. Analyses of garlic preparations have shown that their components and composition vary depending on both the processing method of the raw garlic and the type of preparations. This variance could be related to the complete or partial inactivation of the enzymes present in the garlic during processing.

A number of studies related to the *in vitro* and *in vivo* biological effects of organosulfur compounds derived from garlic have been reported as shown in Table 11.1.[1-3] Little data are available concerning the absorption, metabolism, and distribution of these compounds in laboratory animals and humans, because of the difficultly in analyzing labile organosulfur compounds *in vivo*. For example, dried garlic powders, which contains alliin and alliinase, produce allicin upon contact with water. This characteristic of dried garlic powder is referred to as "allicin-producing potential." However, there is little information on the release of allicin after oral administration of these powders.

TABLE 11.1
PRINCIPAL BIOLOGICAL ACTIVITY OF SOME KEY COMPOUNDS OF GARLIC

Compound	Biological Activities	
	In Vitro	*In Vivo*
Alliin	Antiplatelet Antihepatotoxic	
Allicin	Antimicrobial Antiplatelet	Hypoglycemic
Diallyl disulfide	Chemopreventive Antiplatelet	Chemopreventive Hypoglycemic
Vinyl dithiins	Antiplatelet	
Ajoene	Antiplatelet Antimycotic	Antiplatelet
S-Allylcysteine	Chemopreventive Antihepatotoxic	Chemopreventive Antihepatotoxic
Allixin	Chemopreventive Antimutagenic	Chemopreventive

In this paper, we elucidate whether or not alliin and garlic powder containing alliin and alliinase can release allicin in the digestive tract, and determine whether or not allicin can reach target organs to demonstrate a biological effect. Furthermore, the pharmacokinetics of organosulfur compounds, such as alliin, diallyl sulfides, vinyl dithiins and S-allyl cysteine, derived from garlic are also discussed.

Metabolism of Alliin

Alliin, a precursor found in garlic cloves, is enzymatically converted into allicin, a transient compound which is then transformed into a variety of organosulfur compounds. As shown in Table 11.2, all of the garlic powders and their products tested did not contain any allicin. Since they contained alliin and alliinase, they were able to release variable amounts of allicin in water.[4] However, this observation can not be assumed to be indicative of allicin production after ingestion of these garlic powders when considering the conditions in the digestive tract. After incubating garlic powder containing alliin and alliinase for 1 h in simulated intestinal fluids (SIF), 62% of the amount of allicin produced in water was observed. In contrast, after 1 h of incubation of the same garlic powder in simulated gastric fluids (SGF), only 4% of the allicin could be found. These observations may be due to the inactivation of alliinase by acid and pepsin. Under the sequential combination of SIF and SGF digestive conditions, only about 1% of the allicin was observed.

TABLE 11.2

ALLICIN CONTENT AND ALLICIN PRODUCING POTENTIAL IN WATER OF VARIOUS TYPES OF GARLIC PREPARATIONS*

Product	Allicin Content ($\mu g/g$)	Allicin[1] Potential ($\mu g/g$)
Sugar-coated tablet A	N.D.[2]	1.46[3]
Sugar-coated tablet B	N.D.	0.53[3]
Film-coated capsule A	N.D.	0.56
Film-coated capsule B	N.D.	0.72
Film-coated tablet	N.D.	0.28
Hard capsule	N.D.	0.32
Soft capsule	N.D.	0.64
Garlic oil soft capsule	N.D.	N.D.

* Adapt from reference 4.
[1] The allicin potential was determined after incubating the garlic products in water for 30 min at room temperature.
[2] N.D.: Not detected (Detection limit was 1 $\mu g/g$ product).
[3] Mean value.

We found that more than 95% of alliin remained intact after being incubated for 20 min in the homogenates of the stomach, duodenum, colon, liver and kidney of rats, and no allicin and diallyl disulfide were formed (Table 11.3). Moreover, 10 min after oral administration of alliin (10 mg/mouse), alliin was found in the stomach (7.2%), intestine (22.4%), and liver (2.5%), but allicin and its degradation products including diallyl disulfide, vinyl dithiins, and allyl-SS conjugated compounds, which may arise from a reaction between allicin and amino acids or proteins containing the SH groups, were not detected. These data suggest that allicin is not released in the digestive tract after oral administration of alliin alone.

TABLE 11.3
STABILITY OF ALLIIN IN HOMOGENATES OF RATS ORGANS FOR
20 MINUTES AT 37°C

Compound	Content (%)				
	Stomach	Duodenum	Colon	Liver	Kidney
Alliin	99.8	95.0	97.8	94.4	96.2
Metabolites					
Allicin	N.D.*	N.D.	N.D.	N.D.	N.D.
Diallyl disulfide	N.D.	N.D.	N.D.	N.D.	N.D.

*N.D.: Not detected (Detection limit was 0.2%).

Siegers *et al.* reported that after oral administration of alliin to rats, only a small amount of alliin was detected in the plasma and its bioavailability was 16.5% within 4 h.[5] They also reported that alliin was detected in the perfusate after passage through an isolated liver, indicating a small first pass effect. However, no allicin was detected when alliin was perfused into the liver.[6]

Metabolism of Allicin

Allicin linearly decreased at gastric pH (1.2) and at intestinal pH (7.5), with about 80% and 60%, respectively remaining after incubation for 20 h at 37C. There are no reports which have studied the absorption of allicin in the digestive tract, probably, because of its instability and high reactivity. We investigated the stability and metabolism of allicin in blood and found that allicin completely disappeared from whole blood within a few minutes accompanied by the formation of allylmercaptan and diallyl disulfide. We also found that allicin was more reactive to the blood cell fraction than to the plasma fraction (Table 11.4). In addition, we noted a change in the color of the blood from a red to a brown color immediately after the addition of allicin. Spectrophotometric analysis of the brown colored blood showed the appearance of a maximum band at 630 nm, and a decrease in the absorbance at 541 nm and 577 nm bands which are characteristic of hemoglobin, indicating the formation of methemoglobin which can not bind oxygen. This was, probably, caused by rapid oxidation of the iron in hemoglobin through an interaction with allicin.

TABLE 11.4
ALLICIN STABILITY IN WHOLE BLOOD, BLOOD CELL FRACTION AND
PLASMA FRACTION

Incubation time (min)	Residual Content (%)		
	Whole Blood	Blood Cell Fraction	Plasma Fraction
0	100	100	100
3	-	N.D.*	-
5	N.D.	-	76
12	N.D.	-	-
30	N.D.	N.D.	-
35			60
60			48
120			28

*N.D.: Not detected (Detection limit was 0.04 μg/ml).

Egen-Schwind et al. reported a remarkable first liver pass effect of allicin.[6] A small amount of allicin was found in the perfusate only when allicin was perfused into an isolated liver at a high concentration which caused severe liver cell damage. At a low concentration, which did not cause cell injury, allicin could not be detected in the perfusate. They also identified diallyl disulfide and allylmercaptan as metabolites of allicin.

Pharmacokinetics of Organosulfur Volatiles

Diallyl disulfide (DADS) is one of the major organosulfur volatiles in garlic preparations, such as garlic essential oil and garlic oil macerates. Pushpendran et al. reported the metabolism of [^{35}S]-labeled DADS after intraperitoneal injection.[7] The uptake of [^{35}S]-labeled DADS by the liver of mice was highest 90 min after administration of DADS, and 70% of the radioactivity was observed in the liver cytosol of which 80% was metabolized to sulfate. The metabolism of DADS using an isolated rat liver was also studied by Egen-

Schwind and his colleagues.[6] They found that a small part of the infused DADS was detected in the perfusion medium after the liver passage and, interestingly, a small amount of allylmercaptan was also identified in the perfusate.

On the other hand, the pharmacokinetics of vinyl dithiins found in oily preparations of garlic have been reported after oral administration to rats.[8] 1,3-vinyldithiin seemed to be less lipophilic and was rapidly eliminated from the serum, kidney and fat tissue, whereas 1,2 vinyldithiin was more lipophilic and showed a tendency to accumulate in the fat tissue.

Pharmacokinetics of S-allylcysteine and S-allylmercaptocysteine

S-allylcysteine (SAC), one of the water-soluble transformation products from garlic, has been shown to protect hepatocytes in mice,[3] and to have a chemopreventive effect on 1,2-dimethyhydrazine-induced colon tumors in rats[9] and 7,12-dimethylbenz[a]anthracene-induced mammary tumor in rats.[10] Thus, SAC appears to be one of the chemically and biologically remarkable compounds derived from garlic. Recently, Nagae *et al.* reported the pharmacokinetic behavior of SAC after oral administration to rats, mice and dogs.[11] As shown in Table 11.5, SAC was rapidly and easily absorbed in the gastrointestinal tract and distributed mainly in the plasma, liver, and kidney. In rats, SAC is excreted into the urine mainly in the N-acetyl form. N-Acetyl-S-allylcysteine has been identified in the urine of rats after the consumption of garlic.[12] On the other hand, mice excrete both SAC and its N-acetyl form into the urine, suggesting a slightly different metabolism depending on the species. Jones *et al.* reported that S-propylcysteine sulfoxide was identified as one of the major metabolites from the urine of rats after oral administration of S-propylcysteine.[13] However, alliin (S-allylcysteine sulfoxide) was not detected in the urine after oral administration of SAC.

TABLE 11.5

PHARMACOKINETIC PARAMETERS AND EXCRETION AFTER ORAL ADMINISTRATION OF A-ALLYLCYSTEINE (SAC) IN RATS, MICE AND DOGS

Animal Species	Dose (mg/kg)	Tmax (hr)	Cmax (µg/ml)	T 1/2 (hr)	Bio-availa-bility (%)	Metabolites in 24 h Urine (%)	
						SAC	N-Ac-SAC[1]
Rats	50	1.0	36.9	2.33	98.2	0.4	38.1
Mice	50	0.25	11.9	0.77	103.0	16.5	7.2
Dogs	50	0.25	23.8	10.34	87.2	N.D.[2]	N.D.

[1] N-Acetyl-S-allylcysteine.
[2] N.D.: Not determined.

Since SAC is one of the compounds verified to be both biologically active and bioavailable among the organosulfur compounds derived from garlic, it may be prudent to quantify this important compound in garlic preparations. As shown in Table 11.6, a significant formation of allylmercaptan and diallyl polysulfides, including diallyl mono-, di-, tri-, and tetrasulfides was observed after incubating S-allylmercaptocysteine (SAMC) in the rat liver homogenate. This finding is of interest when considering the relationship between the biological activities of SAMC and its metabolism. Horie *et al.* have reported that diallyl polysulfides (S = 3 to 7) demonstrate a protective effect from lipid peroxidation in rat liver microsomes.[14] SAMC has been shown to have antioxidant activity *in vitro*[15] and a hepato-protective effect both *in vitro*[3,16] and *in vivo*.[3] These results suggest that the biotransformation of SAMC to diallyl polysulfides in the liver might be responsible for antihepatotoxic action of garlic, which protects the liver from free radicals generated from carbon tetrachloride. Interestingly, among the garlic preparations tested, this unique compound, SAMC, has been isolated and identified only from the aged garlic extract.

TABLE 11.6

BIOTRANSFORMATION OF S-ALLYLMERCAPTOCYSTEINE (SAMC) IN RAT LIVER HOMOGENATE AT 37°C

Incubation time (min)	Yield of Transformed Compounds[1]			
	DADS	DATS	DATES	DAPS
10	8.3	1.9	0.2	N.D.[2]
20	12.4	5.0	0.5	0.1
30	11.2	5.9	0.9	0.1
60	7.8	6.4	1.6	0.3

[1] DADS: Diallyl disulfide, DATS: Diallyl trisulfide. DATES: Diallyl tetrasulfide, DAPS: Diallyl pentasulfide.

[2] N.D.: Not detected (Detection limit was 0.01%).

Metabolites after the Consumption of Garlic in Humans

Few scientific studies have been conducted in humans to elucidate and quantify metabolites which arise after the consumption of garlic and its preparations. Minami *et al.* reported that after the ingestion of grated garlic, allylmercaptan was found to be the major garlic odor compound released in human breath, and diallyl disulfide was a secondary compound released according to GC-MS analysis.[17] Jandke *et al.* identified N-acetyl-S-allylcysteine in human urine, suggesting that it was a metabolite of γ-glutamyl-S-allylcysteine found in garlic.[12] Lawson *et al.* unsuccessfully attempted to identify allicin in the human blood and organs/tissues after the ingestion of a garlic preparation which contained a significant amount of allicin.[18] As of yet no one has been able to confirm the bioavailability of allicin.

SUMMARY

Though allicin was considered to be a key compound in garlic in the past, recent scientific findings, including the pharmacokinetics and metabolism of organosulfur compounds in garlic, have revealed that allicin is not biologically active inside of the body, and the release of allicin from garlic preparation containing alliin and alliinase is very questionable, as alliinase is inactivated under digestive conditions. In addition, these studies have also revealed that oil-soluble organosulfur compounds such as DADS are converted to other compounds including allylmercaptan, whereas vinyldithiins are not metabolized, and that water-soluble organosulfur compounds, such as SAC and SAMC, could play an important role in the medicinal properties of garlic. Though allylmercaptan was identified as a metabolite of the organosulfur compounds, its yield was only a small part of those original compounds and their major metabolic pathway needs to be elucidated by further studies.

ACKNOWLEDGMENTS

I would like to thank Professor Dr. F. Freeman of the University of California, Irvine, for his valuable suggestions. I would also thank Mr. Shinji Nagae at Wakunaga Pharmaceutical Co., Ltd. for his kind permission to report on pharmacokinetics studies of organosulfur compounds before publication.

REFERENCES

[1] Kleijnen, J., Knipschild, P. and Ter Riet, G. Garlic, onion and cardiovascular risk factors; a review of the evidence from human experiments with emphasis on commercially available preparation. Br. J. Clin. Pharmac. 1989; 28:535–544.

[2] Dorant, E., Van den Brandt, P.A., Goldbohm, R.A., Hermus, R.J.J. and Sturmans, F. Garlic and its significance for the prevention of cancer in humans: a critical view. Br. J. Cancer 1993; 67:424–429.

[3] Nakagawa, S., Kasuga, S. and Matsuura, H. Prevention of liver damage by aged garlic extract and its components in mice. Phytother. Res. 1989; 3:50–53.

4 Freeman, F. and Kodera, Y. The stability of allicin in solvents, blood, and simulated physiological fluids. J. Agric. Food Chem. 1995; 43(9):2332-2338.

5 Guo, Z., Miller, D., Pentz, R., Kress, G. and Siegers, C.P. Bioavailability of sulfur containing ingredients of garlic in the rat. Planta Med. 1990; 56:692.

6 Egen-Schwind, C., Eckard, R. and Kemper, F.H. Metabolism of garlic constituents in the isolated perfused rat liver. Planta Med. 1992; 58:301-305.

7 Pushpendran, C.K., Devasagayam, T.P.A., Chintalwar, G.J., Banerji, A. and Eapen, J. The metabolic fate of [^{35}S]-diallyl disulfide in mice. Experientia 1980; 36:1000-1001.

8 Egen-Schwind, C., Eckard, R., Jekat, F.W. and Winterhoff, H. Pharmacokinetics of vinyldithiins, transformation products of allicin. Planta Med. 1992; 58:8-13.

9 Sumiyoshi, H. and Wargovich, M.J. Chemoprevention of 1,2-dimethylhydrazine induced colon cancer in mice by naturally occurring organosulfur compounds. Cancer Res. 1990; 50:5084-5087.

10 Amagase, H. and Milner, J.A. Impact of various sources of garlic and their constituents on 7,12-Dimethylbenz[a]anthracene binding to mammary cell DNA. Carcinogenesis 1993; 14:1627-1631.

11 Nagae, S., Ushijima, M., Hatono, S. et al. Pharmacokinetics of the garlic compound S-allylcysteine. Planta Med. 1994; 60(3):214-217.

12 Jandke, J. and Spiteller, G. Unusual conjugates in biological profiles originating from consumption of onions and garlic. J. Chromatogr. 1987; 421:1-8.

13 Jones, A.R. and Walsh, D.A. The fate of S-propylcysteine in the rat. Xenobiotica 1980; 10:827-834.

14 Horie, T. et al. Identified diallyl polysulfides from an aged garlic extract which protects the membranes from lipid peroxidation. Planta Med. 1992; 58:468-469.

15 Imai, J., Ide, N., Nagae, S., Moriguchi, T., Matsuura, H. and Itakura, Y. Antioxidant and radical scavenging effects of aged garlic extract and its constituents. Planta Med. 1994; 60(5):417-420.

16 Hikino, H., Tohkin, M., Kiso, Y., Samiki, T., Sishimura, S. and Takeyama, K. Antihepatotoxic actions of Allium sativum bulbs. Planta Med. 1986; 52:163-168.

17 Minami, T., Boku, T., Inada, K., Morita, M. and Okazaki, Y. Odor components of human breath after the ingestion of grated raw garlic. J. Food Sci. 1989; 54:763-765.

18 Lawson, L.D., Ransom, D.K. and Hughes, B.G. Inhibition of whole blood platelet aggregation by compounds in garlic clove extracts and commercial garlic products. Thromb. Res. 1992; 65:141-156.

CHAPTER 12

ANTIOXIDANT ACTIVITIES OF AGED GARLIC EXTRACTS AND CANCER CHEMOTHERAPY

S. TSUYOSHI OHNISHI and RYUSUKE KOJINRA

Philadelphia Biomedical Research Institute
100 Ross and Royal Roads
King of Prussia, PA 19406

ABSTRACT

In vitro *antioxidant activities of aged garlic extracts were evaluated by measuring their ability in: (a) quenching stable chemical radicals produced by 2,2-diphenyl-1-picrylhydrazyl (DPPH) in ethanol solutions; (b) inhibiting superoxide-mediated autoxidation of pyrogallol; and (c) inhibiting peroxidation of linoleic acid micelles induced by a water soluble radical generator, 2,2'-azo-bis(2-amidinopropane)HCl (AAPH).* In vivo *protective effects of aged garlic extracts against the cardiotoxicity of doxorubicin (DOX) were evaluated in the mouse. DOX (1.5 mg/kg body weight) was administered intraperitoneally 3 times a week for 40 days. Aged garlic extracts, either WG-1 (a preserved stock solution prepared by WAKUNAGA Pharm. Co., Ltd.; administered 0.05 ml/20 g body weight intraperitoneally) or KYOLIC (tablets made from the same stock solution by WAKUNAGA; administered 0.01-0.02 g/20 g body weight orally) was given 6 times weekly. The electrocardiogram measurement especially, the QRS width, indicated that aged garlic extracts protected the heart against the cardiotoxicity of DOX. Lipid peroxidation (as measured from the level of thiobarbituric acid reactive substances, TBARS) in the heart homogenates caused by DOX was also inhibited by the extracts. The decreases in the survival rate and body weight caused by the DOX administration were also prevented by the simultaneous administration of garlic extracts.*

INTRODUCTION

Garlic (*Allium sativum*) contains various biochemically active substances.[1,2] The thioallyl compounds seem to be especially interesting.[3-6] It was reported that these compounds have antioxidant activity.[7-11] In this chapter, we will discuss the possibility of applying the antioxidant activity of garlic for the improvement of cancer chemotherapy.

Doxorubicin (DOX), also known as adriamycin, is an anthracycline antibiotic which has been widely used as an effective antineoplastic drug. However, its therapeutic usefulness is limited by its cardiotoxicity. It is well known that DOX causes heart failure and cardiac arrhythmia, especially ventricular extrasystole.[12-25] Therefore, a drug which could prevent these side effects of DOX has been desired in the treatment of patients with neoplastic disorders. The cardiotoxicity of DOX may be related to oxidative stress[26,27] caused by the semiquinone radicals[28] (Fig. 12.1). Antioxidants[29] and iron chelating agents[30] were reported to counteract against this toxicity.

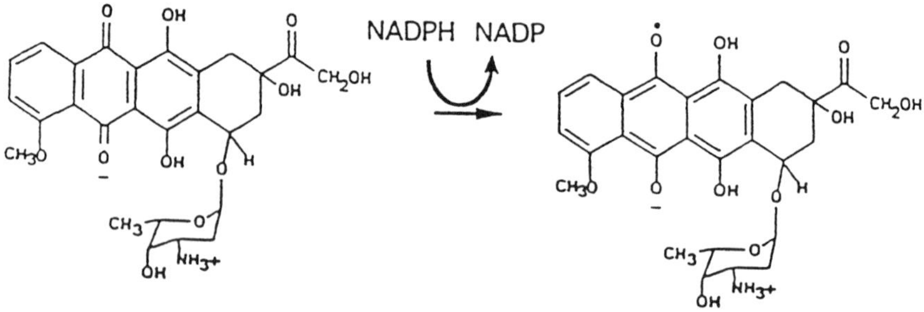

FIG. 12.1. DOXORUBICIN AND ITS SEMIQUINONE RADICAL

Therefore, we undertook this study to test whether aged garlic extracts could protect the heart against the side effects of DOX. First, we measured the antioxidant activities of aged garlic extracts *in vitro*. We then evaluated the degree of cardiotoxicity in mice by measuring body weight and survival rate, by electrocardiograph (ECG) monitoring, and by measuring the level of lipid peroxidation of the heart homogenates using the TBARS assay.

MATERIALS AND METHODS

WG-1 (Aged garlic extracts), KYOLIC garlic tablets, allicin (Diallyldisulfide-S-oxide) and S-Allyl cysteine (SAC) were provided by WAKUNAGA Pharm. Co. Ltd., Mission Viejo, CA. DPPH (2,2-diphenyl-1-picrylhydrazyl), DOX (doxorubicin) and linoleic acid were purchased from Sigma Chemicals (St. Louis, MO). AAPH (2,2'-azo-bis(2-amidinopropane)HCl) was obtained from Polysciences, Inc. (Warington, PA). For the *in vitro* study, a KYOLIC tablet was ground and extracted in 2.5 ml ice-cold water for 10 minutes. The suspension was centrifuged and the supernatant was used.

For the *in vivo* study, Balb/C mice (18-20 g) were purchased from Charles River (Wilmington, MA). DOX (1.5 mg/kg/day; 3 times per week) was given via i.p. WG-1 (0.05 ml/20 g body weight/day; 6 times per week) was given via i.p.; KYOLIC tablets were cut into small pieces, the weight measured, and 0.01 or 0.02 g per 20 g body weight was orally given once a day 6 times per week. Electrocardiography (ECG) (Model 1511B, Hewlett Packard) was done twice a week under isoflurane anesthesia using an electrocardiograph at a chart speed of 50 mm/sec. The level of TBARS was measured by the method of Ohkawa.[31] The lipid peroxidation level was expressed in terms of nmol malonyldialdehyde (MDA)/ 100 mg protein.

RESULTS AND DISCUSSION

In Vitro Study

Allicin and S-allylcysteine are used as the reference controls.

(a) **DPPH-Experiments.** Figure 12.2 shows radical structure of DPPH. The changes in the radical concentration in an ethanol solution can be measured by absorption at 480 nm.[32] By adding the same amounts of SAC, KYOLIC or Allicin, we measured the rate of color decrease at 480 nm. The changes in the rate of absorbance was in the order of Allicin > SAC > KYOLIC (Fig. 12.5; top panel).

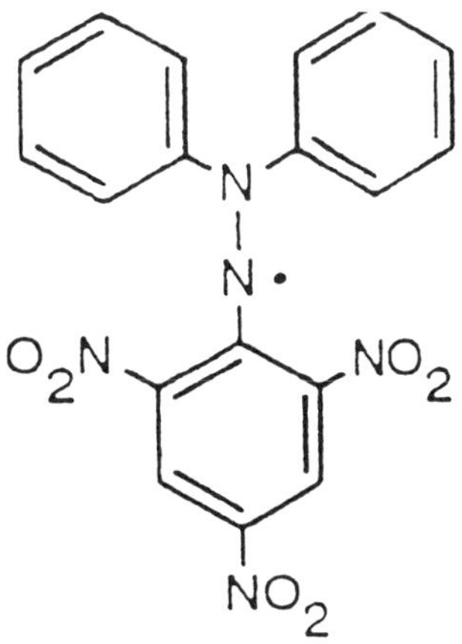

FIG. 12.2. DPPH RADICAL

(b) Pyrogallol-Experiments. As shown in Fig. 12.3, the auto-oxidation of pyrogallol is mediated by superoxide radicals.[33-35] We found that the autoxidation was inhibited by the addition of SAC, KYOLIC or Allicin. This suggests that they could quench superoxide radicals. The strongest inhibitory activity was seen with SAC. KYOLIC had similar activity to that of Allicin (Fig. 12.5; middle panel).

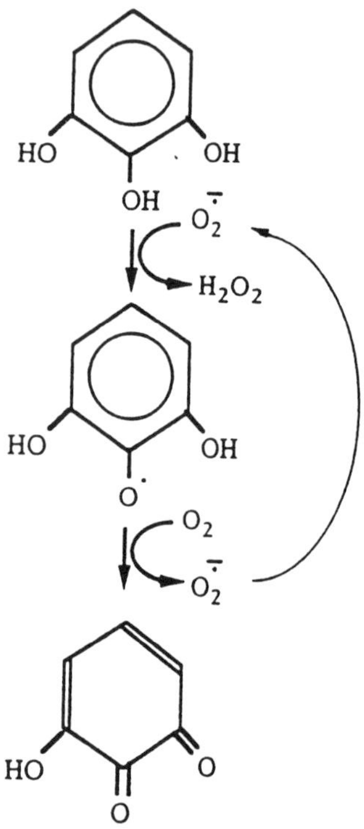

FIG. 12.3. MECHANISM OF AUTO-OXIDATION OF PYROGALLOL

(c) Micellar-suspension Experiments. When AAPH was added to a linoleic acid micelle suspension, the oxygen concentration decreased, showing that peroxidation was induced according to the scheme shown in Fig. 12.4.[36] With the addition of SAC, KYOLIC or Allicin, the rate of oxygen decrease was reduced, demonstrating that these compounds had antioxidant activity. (Fig. 12.5; bottom panel). Allicin had the strongest antioxidant activity. It is interesting to note that KYOLIC was stronger than SAC in the inhibition of AAPH-induced lipid peroxidation.

initiation: $A - N = N - A \xrightarrow{Kd} [A \cdot N_2 \cdot A]$

$$\begin{cases} \rightarrow (1-e)A - A & (1) \\ \rightarrow 2eA^{\cdot} & (2) \end{cases}$$

$$A^{\cdot} + O_2 \rightarrow AO_2^{\cdot} \qquad (3)$$

$$AO_2^{\cdot} + LH \xrightarrow{O_2} AOOH + LO_2^{\cdot} \qquad (4)$$

propagation: $LO_2^{\cdot} + LH \xrightarrow{Kp} LOOH + L^{\cdot} \qquad (5)$

$$L^{\cdot} + O_2 \rightarrow LO_2^{\cdot} \qquad (6)$$

termination: $2LO_2^{\cdot} \xrightarrow{Kt} \text{nonradical products} \qquad (7)$

FIG. 12.4. MECHANISM OF RADICAL FORMATION AND LIPID PEROXIDATION
BY AZO-BIS COMPOUNDS[36]

The order of activity in quenching radicals produced by DPPH was
Allicin $> >$ SAC $>$ KYOLIC. In the quenching of superoxide-mediated auto-
oxidation of pyrogallol, the order was SAC $>$ Allicin $>$ KYOLIC. In the
inhibition of lipid peroxidation induced by AAPH, the order of antioxidant
activities was Allicin $> >$ KYOLIC $>$ SAC. This suggests that the efficacy of
these compounds in quenching radicals is different for different radicals. In these
reactions, perhaps the most biologically relevant reaction may be lipid
peroxidation. It was found that KYOLIC was more effective than a pure
compound, SAC, in inhibiting lipid peroxidation.

In Vivo. After 40 days of experiment, the body weight of mice in the
DOX group decreased from the pre-treatment level by 1.1 - 1.5 g, but the
simultaneous administration of WG-1 prevented the weight decrease. The final
body weight of the WG-1+DOX group increased by 1.5 g.[38] At this level of
WG-1, the administration of WG-1 alone slightly inhibited normal growth rate.[37]
With DOX administration, the survival rate was only 3/10, but
simultaneous administration of WG-1 completely protected the mice as indicated
by the complete survival of 10/10.[37]

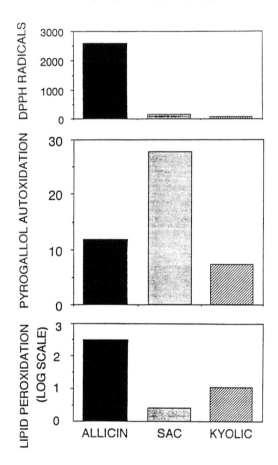

INHIBITION
(ARBITRARY SCALE)

FIG. 12.5. (A) THE DEGREE OF QUENCHING OF DPPH RADICALS BY SAC, KYOLIC
AND ALLICIN AT THE SAME DOSE (0.27 MG/ML).
The values were obtained from the measurements of absorption at 480 nm. Concentrations: 0.1
mM DPPH (in ethanol), 25°C. In 300 ml cuvette (6 mm i.d.) Abbreviations are S: SAC;
K: KYOLIC; and A: Allicin.
(B) THE DOSE-DEPENDENT INHIBITION OF SUPEROXIDE-MEDIATED
AUTO-OXIDATION OF PYROGALLOL BY SAC, KYOLIC AND ALLICIN
The rate of auto-oxidation was followed from the absorption changes at 480 nm.
(C) THE DOSE-DEPENDENT INHIBITION OF LINOLEIC ACID PEROXIDATION
INDUCED BY 20 MM AAPH
The rate of peroxidation was measured by the oxygen uptake using a Clarke electrode.
Concentrations: 100 mM linoleic acid, 2.7% BRIJ-35, 5 mM Na-phosphate buffer (pH 7.4), 1
mM EDTA and 1 mM DTPA. 25°C. The volume of this chamber was 1.2 ml.

The QRS width (see Fig. 12.6) is known to be a sensitive indicator for the DOX-induced cardiotoxicity.[12] Therefore, we used this value to assess the degree of cardiac injury. As shown in Fig. 12.7, the QRS width in the DOX group (middle panel) was much extended compared to that of the normal group (top panel). The simultaneous administration of KYOLIC inhibited this change (bottom panel). As compared to the control group, the level of TBARS increased by 76% in the DOX group after 40 days, but the increase was only 17% in the group with the simultaneous administration of WG-1 or DOX.[37]

In these simultaneous administration experiments, the effects of KYOLIC tablets (0.01 g/20 g body weight; oral) were slightly smaller than those of WG-1 (i.p.).[38] This dose of KYOLIC tablets (0.5 g/kg/day) corresponds to 25-30 g KYOLIC tablets/adult/day, which is more than 10 times higher than what is normally taken as a food supplement, but not unreasonable for use with a patient receiving DOX treatment.

In summary, we confirmed that aged garlic extracts have antioxidant activity. We also found that they can protect the mouse heart against the cardiotoxicity of doxorubicin. Although the daily dose required to demonstrate such protective actions in humans is tenfold, it may still be possible to take such a dose. If we could identify an effective component in the extract, then such a therapeutic intervention could become more practical. We have preliminary data to suggest that the administration of 50mg/kg/day of SAC (i.p.) protected the mice from DOX-induced cardiotoxicity (Mima and Ohnishi; unpublished results). Therefore, SAC is at least one of the effective components in aged garlic extracts. Using other animal models (ischemia-reperfusion injury in rats),

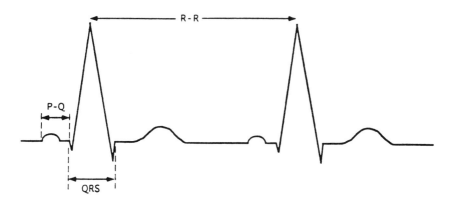

FIG. 12.6. ECG OF A NORMAL MOUSE
The intervals of P-Q, QRS and R-R are shown.

we have demonstrated that water-soluble SAC reduced ischemia-reperfusion injury of the rat brain, but that water-insoluble compounds, such as allyl sulfide and allyl disulfide, had no effects.[38] SAC may not be the only active components. It is possible that there may be different effective components in garlic extracts as the results of Fig. 12.5 suggest. Further investigation is needed to identify effective components in aged garlic extracts.

CONTROL

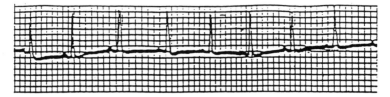

DOX (1.5 mg/kg/day; 3 times /week)

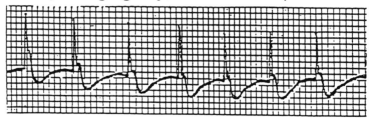

DOX (1.5 mg/kg/day; 3 times /week) + KYOLIC (0.5 g/kg/day; 6 times/week)

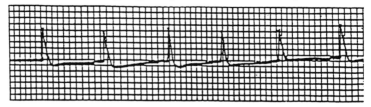

FIG. 12.7. EXAMPLES OF ECG AFTER 40 EXPERIMENTAL DAYS
(Top) control, (middle) DOX alone, and (bottom) DOX + KYOLIC.

ACKNOWLEDGEMENT

This study was supported in part by an NIH grant NS30186 to STO.

REFERENCES

1   Block, E., Ahmad, S., Jain, M.K., Crecely, R.W., Apitz-Castro, R., and Cruz, M.R. (E Z)-Ajoene: A potent antithrombotic agent from garlic. J. Am. Chem. Soc. 1984; 106:8295–8296.
2   Sodimu, O., Joseph, P.K. and Augusti, K.T. Certain biochemical effect of garlic oil on rats maintained on high fat-high cholesterol diet. Experientia 1984; 40:78–80.
3   Apitz-Castro, R., Cabrera, S., Cruz, M.R., Ledezma, E. and Jain, M.K. Effects of garlic extract and of three pure components isolated from it on human platelet aggregation, arachidonate metabolism, release reaction and platelet ultrastructure. Thromb. Res.1983; 32:155–169.
4   Welch, C., Wuarin, L. and Sidell, N. Antiproliferative effect of the garlic compound S-allyl cysteine on human neuroblastoma cells in vitro. Cancer Letters 1992; 63:211–219.
5   Takeyama, H., Hoon, D.S.B., Saxton, R.E., Morton, D.L. and Irie, R.F. Growth inhibition and modulation of cell markers of melanoma by s-allyl cysteine. Oncology 1993; 50:63–69.
6   Augusti, K.T. and Mathew, P.T. Lipid lowering effect of allicin (diallyl disulphide-oxide) on long-term feeding to normal rats. Experientia 1973; 30 (5):468–470.
7   Phelps, S. and Harris, W.S. Garlic supplementation and lipoprotein oxidation susceptibility. Lipids 1993; 28(5):475–477.
8   Hikono, H., Tohkin, M., Kiso, Y., Namiei, T., Nishimura, S. and Takeyama, K. Antihepatoxic actions of Allium sativum bulbs. Planta Med. 1986; 52:163–168.
9   Nakagawa, S., Kasuga, S. and Matsuura, H. Prevention of liver damage by aged garlic extract and its components in mice. Phyto. Res. 1988; 3:50–53.
10 Horie, T., Awazu, S., Itakura, Y. and Fuwa, T. Identified diallyl polysulfides from an aged garlic extract which protects the membranes from lipid peroxidation. Planta Medica 1992; 58(5):468–469.
11 Kourounakis, P.N. and Rekka, E.A. Effect on active oxygen species of alliin and allium sativum (garlic) powder. Res. Commun. Chem. Pathol. Pharmacol. 1991; 74(2):249–252.
12 Zbinden, G. and Brändle, E. Toxocologic screening of daunorubicin (NSC-82151), adriamycin (NSC-123127), and their derivatives in rats. Cancer Chem. Rep. Part 1 1975; 59:707–715.
13 Henderson, I.C. and Frei, E. Adriamycin and the heart. N. Engl. J. Med. 1979; 300:310–312.
14 Olson, H.M., Young, D.M., Dieur, D.J., Leroy, A.F. and Reagan, R.L Electrolyte and morpholoic alterations of myocardium in adriamycin-treated rabbits. Am. J. Pathol. 1974; 77:439–454.

[15] Ferrans, V.J. Overview of cardiac pathology in relation to anthracycline cardiotoxicity. Cancer Treat. Rep. 1978; 62:955–961.

[16] Myers, C.E, McGoire, W.P., Liss, R.H., Ifrim, I., Grotzinger, K. and Young, R.C. Adriamycin: The role of lipid peroxidation in cardiac toxicity and tumor response. Science 1977; 197:165–167.

[17] Valvere, V.I.U., Shkhvatsabaia, L.V. and Niu-Tian-de, G.B. Comparative evaluation of the cardiotoxic effects of antineoplastic antibiotics adriamycin and pharmorubicin. Kardiologiia 1989; 29:64–67.

[18] Kauel, E., Koschel, G., Gatzemeyer, U. and Sarewski, E. A phase II study of pirarubicin in malignant pleural mesothelioma. Cancer 1990; 66:651–654.

[19] Ringenberg, Q.S., Propert, K.J., Muss, H.B., Weiss, R.B., Schilsky, R.L., Modeas, C., Perry, M.C., Norton, L. and Green, M. Clinical cardiotoxicity of esorubicin (4'-deoxydoxorubicin,DxDx). Prospective studies with serial gated heart scans and reports of selected cases. A cancer and Leukemia Group B report. Invest. New Drugs 1990; 8:221–226.

[20] Villani, F., Galimberti, M. and Comazzi, R. Early cardiac toxicity of 4'-iodo-4'deoxydoxorubicin. Euro. J. Cancer 1991; 27:1601–1604.

[21] Raabe, N.K. and Storstein, L. Cardiac arrhythmias in patients with small cell lung cancer and cardiac disease before, during and after doxorubicin administration. An evaluation of acute cardiotoxicity by continuous 24-hour Holter monitoring. Acta Oncol. 1991; 30:843–846.

[22] Danesi, R., Bernardini, N., Agen, C., Costa, M., Macchiarini, P., Torre, P.D. and Tocca, M.D. Cardiotoxicity and cytotoxicity of the anthracycline analogue 4'-deoxy-4'-iodo- doxorubicin. Toxicology 1991; 70:243–253.

[23] Doherty, J.D. and Cobbe, S.M. Electrophysiological changes in a model of chronic cardiac failure. Cardiovasc. Res. 1990; 24:309–316.

[24] Steinherz, L. and Steinherz, P. Delayed cardiac toxicity from anthracycline therapy. Pediatrician 1991; 18:49–52.

[25] Mauldin, G.E., Fox, P.R., Patnaik, A.K. and Bond, B.R. Doxorubicin-induced cardiotoxicosis. Clinical features in 32 dogs. J. Vet. Intern. Med. 1992; 6:82–88.

[26] Thayer, W.S. Adriamycin-stimulated superoxide formation in submitochondrial particles. Chem. Biol. Interact. 1977; 19:265–278.

[27] Thayer, W.S. Evaluation of tissue indicators of oxidative stress in rats treated chronically with adriamycin. Biochem. Pharmacol. 1988; 37:2189–2193.

[28] Kalyanaraman, B., Morehouse, K.M. and Mason, R.P. An electron paramagnetic resonance study of the interactions between the adriamycin semiquinone, hydrogen peroxide, iron-chelators, and radical scavengers. Archiv. Biochem. Biophy. 1991; 286:164–170.

[29] Milei, J., Boveris, A., Llesuy, S., Molina, H.A., Storino, R., Ortega, D. and Milei, S.E. Amelioration of adriamycin-induced cardiotoxicity in rabbits by prenylamine and vitamins A and E. Am. Heart J. 1986; 111:95–102.

30 Konig, J., Palmer, P., Franks, C.R., Mulder, D.E., Speyer, J.L., Green, M.D. and Hellmann, K. Cardioxane ICRF-187, towards anticancer drug specificity through selective toxicity reduction. Cancer Treat. Reviews 1991; 18:1–9.

31 Ohkawa, H., Ohishi, N. and Yagi, K. Assay for lipid peroxides in animal tissues by thiobarbituric acid reaction. Analytical Biochemistry 1979; 95:351–358.

32 Bloios, M.S. Antioxidant determinations by the use of a stable free radical. Nature 1958; 181:1199–1120.

33 Marklund, S. and Marklund, G. Involvement of the superoxide anion in the auto-oxidation of pyrogallol and a convenient assay for superoxide dismutase. Eur. J. Biochem. 1974; 47:469–474.

34 Marklund, S.L. Spectrophotometric study of spontaneous disproportionation of superoxide anion radical and sensitive direct assay for superoxide dismutase. J. Biol. Chem. 1976; 251:7504–7507.

35 Marklund, S.L. In Handbook of Methods for Oxygen Radical Research, (Greeneald, R.A., ed.) CRC Press, Boca Raton, FL. 1985:243–247.

36 Yamamoto, Y., Niki, E., Eguchi, J., Kamiya, Y. and Shimasaki, H. Oxidation of biological membranes and its inhibition: Free radical chain oxidation of erythrocyte ghost membranes by oxygen. Biochim. Biophys. Acta. 1985; 819:29–36.

37 Kojima, R. and Ohnishi, S.T. Protective effects of an aged garlic extract on doxorubicin-induced cardiotoxicity in the mouse. Nutrition and Cancer. 1994; 22:163–173.

38 Numagami, Y., Sato, S. and Ohnishi, S.T. Attenuation of rat ischemic brain damage by an aged garlic extract: a possible protecting mechanism as antioxidants. Neurochemistry International 1996; 29:135–143.

CHAPTER 13

IMPROVEMENT OF AGE-RELATED DETERIORATION OF LEARNING BEHAVIORS AND IMMUNE RESPONSES BY AGED GARLIC EXTRACT

YONGXIANG ZHANG, TORU MORIGUCHI, HIROSHI SAITO and
NOBUYOSHI NISHIYAMA

Department of Chemical Pharmacology
Faculty of Pharmaceutical Sciences
The University of Tokyo
Tokyo 113, Japan

ABSTRACT

The antiaging effects of aged garlic extract (AGE) were studied on age-related deficiencies of learning and memory and immune response using two animal models, thymectomized ddY mice and senescence accelerated mouse-prone/8 (SAMP8). Chronic oral administration of AGE significantly ameliorated both thymectomy-reduced antibody production response and thymectomy-induced impairment of learning behaviors in passive avoidance performances, and in a spatial memory task in both thymectomized ddY mice and SAMP8. The levels of hypothalamic norepinephrine, 3,4-dihydroxyphenylacetic acid and homovanilic acid, and the hypothalamic choline acetyltransferase activity were significantly increased in thymectomized ddY mice, and AGE restored them to the control levels. Chronic ingestion of AGE significantly potentiated lymphocyte proliferation induced by concanavalin A or lipopolysaccharides in both SAMP8 and SAMR1, a senescence-resistant substrain of SAM. These results suggested that AGE ameliorated age-related deterioration of learning and memory ability and immune response.

INTRODUCTION

Garlic has been utilized as a flavoring agent in almost all nations for centuries, during which its medicinal benefits have been discovered. Accumulated data in both western and traditional Chinese medicine recorded that garlic was used as a remedy with antibacterial, expectorant, anthelmintic, diuretics, diaphoretics, etc. properties.[1] Modern pharmacological studies reveal that garlic or its extract possesses antitumor activity[2,3] and it potentiates immune function, such as promoting the activity of macrophages and natural killer cells, and augmenting lymphocyte proliferation induced by mitogens or interleukin-2 *in vitro*.[4-6] Thus, the antitumor activities of garlic are apparently achieved by potentiating or modulating host immunity.[5,7,8] Recent studies from our laboratory show that the chronic ingestion of aged garlic extract (AGE) ameliorated the learning and memory ability in the senescence-accelerated mouse (SAM) (unpublished observation), suggesting that AGE could also affect the central nervous system.

Age-related cognitive and immune deficiencies are two predominant features during the aging process, which is supposed to be largely due to the gradual imbalance of the neuroendocrine immunomodulation (NIM) network.[9-11] The thymus has been found to mediate the reciprocal interactions between neuroendocrine and immune systems[12-15], suggesting that age-related deterioration of central cognitive and immune functions could be functionally linked by the thymus. Taking the immunopotentiating and learning-ameliorative effects of AGE into consideration, it is reasonable to postulate that AGE may affect or modulate the NIM network to exert anti-aging effects.

With these findings and the idea, we studied the effects of AGE on learning and memory abilities in both thymectomized ddY mice and thymecto- mized SAM. It has been established that thymectomy in mice reduces the immune function[16-20] and causes neuroendocrine imbalance[15], which is supposed to accelerate the systematic aging process in less than 1 year after thymectomy.[16] In addition, the effects of AGE on brain monoamine levels and choline acetyltransferase (ChAT) activity were determined in thymectomized ddY mice. The effect of AGE on immune responses was also evaluated in naive SAM, a genetic mouse model of accelerated aging.[21]

Animal Models and AGE Treatment

Male mice were thymectomized by a sterile procedure 4 weeks after birth. Sham-operated and the age-matched intact mice served as controls. AGE was formulated by dipping sliced raw garlic (*Allium sativum*) in aqueous ethanol in a tank at room temperature for more than 10 months. In this study, AGE (2% w/w) was prepared in CE-2 mouse food and fed to thymectomized mice after the operation till all the experiments were completed. All the evaluations were performed 10 months after thymectomy in ddY mice and 5 months after thymectomy in SAM-prone/8 (SAMP8//HS) mice. For the evaluation of the immune response in SAM, naive mice were fed AGE-containing food for 10 months from the age of 2 months. Mice in the control groups were fed CE-2 control food.

AGE Treatment Increased the Body Weight in Thymectomized Mice

Mice were weighed regularly after thymectomy. The growth rate of thymectomized animal showed a significant decline from the age of 8 months in ddY mice and from the age of 2 months in SAMP8. AGE treatment showed a significant growth-promoting effect in thymectomized ddY mice and prevented the early decline of body weight gain in thymectomized SAMP8 (Fig 13.1).

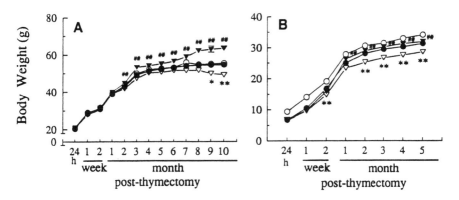

FIG. 13.1. EFFECTS OF AGE ON THE GAIN OF BODY WEIGHT SUPPRESSED BY
THYMECTOMY
Mice were weighed regularly after thymectomy. A: ddY mice, ○ : control (CON), ● : sham-operated (SHAM), ▽: thymectomized (ATx), ▼: AGE-treated thymectomized (AGE + ATx). B: SAM, ○: SAMR1 control (RC), ●: SAMP8 control (PC), ▽ : thymectomized SAMP8 (Px), ▼: AGE-treated Px (AGE+Px). *: $p < 0.05$, **: $p < 0.01$ vs SHAM in A, vs PC in B; ##: $P < 0.01$ vs ATx in A, vs Px in B, in Tukey's test; n=10.

AGE Treatment Ameliorated the Learning Behaviors Impaired by Thymectomy

The learning and memory abilities were evaluated with passive avoidance performances in step through (ST) and step down (SD) tests, active avoidance performances in shuttle box (SB) and lever press (LP) tests, and the spatial memory water maze task of Morris.

ST and SD tests were conducted for 10 consecutive days including one learning trial and nine test trials according to the methods developed in our laboratory.[22] In the ST test, learning performance was measured by observing the latency for a mouse to enter the dark chamber equipped with electric shock. In SD test, performance was observed by recording the number of events ("errors") of the mouse stepping down to the electric grid floor. The latency or the number of errors in the first testing trial reflects the memory registration ability in the ST and SD tests. The latency in the ST test tended to decrease in both thymectomized ddY mice and SAMP8. AGE treatment tended to prolong them, but did not reach statistical significance in either test (data not shown). In the SD test, thymectomy in ddY mice increased the number of errors and also the total number of errors, which indicated a lower memory retention ability. The number of both kinds of errors tended to increase in thymectomized SAMP8. AGE treatment decreased the number of errors and the total number of errors (Fig. 13.2) in both strains of thymectomized mice.

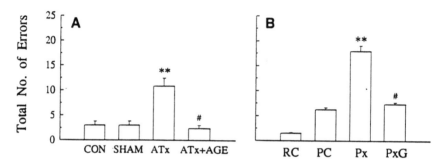

FIG. 13.2. EFFECTS OF AGE ON MEMORY RETENTION IMPAIRED BY THYMECTOMY
IN STEP DOWN TEST

Total number of errors indicated the average number of errors, i.e., stepping down the platform, in each testing trial in ddY mice (A) and SAM (B). For explanation of the abbreviations, see FIG 13.1. ** $p < 0.01$ vs SHAM in A, vs PC in B; # $p < 0.05$ vs ATx in A and vs Px in B, in Mann-Whitney's U test; mean $\pm$ SEM, n=10.

SB and LP tests were performed for 7 and 10 consecutive days with 60 trials per day, respectively.[23] The chambers of both tests were equipped with buzzer and light as a conditioned stimulation (CS) and electric shock as an unconditioned stimulation (US). The programmed schedule of one trial was 60 sec, in which 40 sec was set for interval, 10 sec for CS and 10 sec for US. A mouse must move from one side of the chamber to the other in SB test, or press a lever on the wall of the chamber in LP test to avoid the electric shock. The periods of CS or interval were recorded as conditioned avoidance response (CAR) or spontaneous response (SR), respectively. Thymectomy significantly delayed the memory registration in the SB test. The CAR of thymectomized mice remained at the lower level from trial day 2 in ddy mice and from trials day 2, 3 and 5 in SAMP8. But removal of thymus did not affect CAR in the LP test in both strains. AGE treatment showed no ameliorative effect on CAR in the SB test, but tended to increase CAR in the LP tests in the two thymectomy models (data not shown).

Mice were subjected to a 7-day trial in a water maze test to evaluate their spatial memory ability. A mouse remembered the clues arranged around the water pool to reach the invisible platform, which was located in a constant position of region 1 (4 regions in total) of the pool. In the trials 1-6, each mouse was allowed 4 trials (90 sec/trial) per day. The total time spent for a mouse to reach the platform ("escaping latency"), the swimming (not floating) distance, the swimming time spent in finding the goal, and the average swimming speed, were recorded automatically. In the last trial (day 7), the platform was taken out and a mouse was subjected to a single trial of 90 sec. The summation of time when a mouse stayed in region 1, the ratio of swimming distance and swimming time in region 1 over the total 4 regions, and the number of coming across the platform-located position were recorded.

The escaping latency, the swimming distance and the swimming time were significantly prolonged in thymectomized ddY mice compared with sham-operated mice in trials 2, 3 and 5, but the swimming speed was not altered by thymectomy. AGE treatment shortened the escaping latency significantly, and decreased the swimming distance and the swimming time in the testing trials of day 2-6 without affecting the swimming speed (Fig. 13.3). In the last trial, the staying time in region 1 and the ratios of the swimming time and the swimming distance in region 1 tended to decrease in thymectomized ddY mice. The number crossing the platform position was significantly decreased in thymectomized ddY mice compared with sham operated mice. AGE treatment showed significant ameliorative effects on all of the four parameters observed in the last trial (data not shown).

In thymectomized SAMP8, the escaping latency in the trials of day 5 and 6 were significantly prolonged, and the swimming distance and the time increased correspondingly. AGE-treatment showed an improved tendency on these parameters, but the differences were not statistically significant. Unlike the results of thymectomized ddY mice, the swimming speed of thymectomized SAMP8 were significantly slower than in SAMP8 controls. AGE treatment raised the swimming speed to the control level (Fig. 13.3). In the last trial, thymectomy tended to impair the four parameters observed and AGE tended to improve them, but the differences were not statistically significant (data not shown).

AGE Treatment Did Not Alter Motor Activity

A mouse was put into a round testing cage of a tilting-type ambulo-meter, and the amount of motor activity (MA) was measured for 30 min every

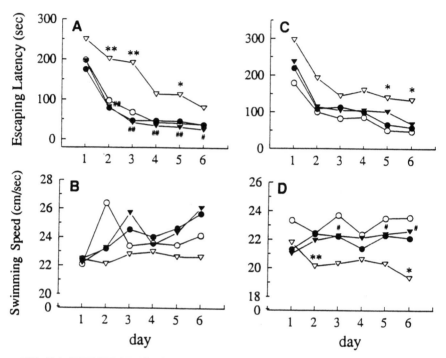

FIG. 13.3. EFFECTS OF AGE ON SPATIAL MEMORY PERFORMANCE IMPAIRED BY THYMECTOMY

In the trials through day 1 to day 6, each mouse subjected to 4 trials per day started from randomly arranged 4-positions. Escaping latency indicates the total time for a mouse to reach the platform in the 4 trials in ddY mice (A) or SAM (C). The average swimming speed of mice in the 4 trials in ddY (B) or SAM (D). For explanation of the abbreviations, see FIG 13.1. *: $p < 0.05$, **: $p < 0.01$ vs SHAM in A and B, vs PC in C and D; # P < 0.05, ## P < 0.01 vs ATx in A and B, vs Px in D, in Tukey's or Steel-Dwass's test; $n = 10$.

day before the trials of passive avoidance performance. Neither thymectomy nor AGE treatment affected MA in both strains of thymectomized mice (data not shown).

AGE Treatment Restored Thymectomy-induced Elevations of Hypothalamic Monoamine Contents and ChAT Activity in ddY Mice

The contents of monoamines, including norepinephrine (NE), dopamine (DA), 3,4-dihydroxyphenylacetic acid (DOPAC), homovanilic acid (HVA), serotonin (5-HT) and 5-hydroxyindole acetic acid (5-HIAA) of the cortex, hippocampus and hypothalamus were assayed by high pressure liquid chromatography combined with an electrochemical detector. Brain ChAT activity was radio assayed using ^{14}C-Acetyl CoA. The hypothalamic contents of NE, DOPAC and HVA in thymectomized mice were significantly elevated. DA also tended to increase. AGE treatment restored them to the control level (Table 13.1). Neither thymectomy nor AGE altered the monoamine contents of the cortex and hippocampus, except that 5-HIAA of the cortex was decreased in AGE-treated group (data not shown). Similar to the monoamines results, thymectomy increased the hypothalamic ChAT activity. AGE prevented this elevation and restored the enzyme activity to control level (Table 13.1). The ChAT activity of the cortex and hippocampus were not influenced by thymectomy or AGE treatment (data not shown).

AGE Treatment Potentiated the Immune Responses

Antibody production ability (plaque-forming cells)(PFC) in response to the *in vivo* challenge of sheep red blood cells[24]; and lymphocyte proliferation (LT) ability induced by mitogens[5,6]; were examined to evaluate immune response. Thymectomy in ddY mice significantly reduced the number of PFC formed by splenocytes (Fig. 13.4 A). Six-months-old SAMP8 also showed a significantly lower antibody production ability compared with the age-matched SAMR1, and thymectomy further reduced the number of PFC (Fig. 13.4 B). Chronic administration of AGE significantly increased the number of PFC in thymectomized ddY mice and tended to increase it in thymectomized SAMP8 (Fig. 13.4). In naive SAM, long-term AGE treatment significantly increased the PFC number in both SAMR1 and SAMP8 compared to that of their age-matched controls (Fig. 13.5 A). In addition, the lymphocyte proliferation responses to the mitogens, such as Con A and LPS, were dramatically suppressed in SAMP8 compared to that of the age-matched SAMR1. AGE treatment significantly augmented the proliferative abilities induced by both mitogens in both substrains, while it showed no effect on basal proliferation (Fig. 13.5 B).

DISCUSSION

Thymectomy in both ddY mice and SAMP8 resulted in an accelerated manifestation of learning and memory deficiency. AGE treatment significantly ameliorated memory registration and retention in passive avoidance performances and tended to improve learning behaviors in LP test. It significantly improved thymectomy-induced impairment of spatial memory performance in ddY mice and showed an obvious ameliorative tendency in SAMP8. These results suggest that AGE ameliorated thymus-related deteriorations of learning and memory. Neither thymectomy nor AGE treatment affected motor activity, indicating that the changes in learning behaviors that were observed were not attributable to

TABLE 13.1.
EFFECTS OF AGE ON HYPOTHALAMIC MONOAMINE CONTENTS AND ChAT ACTIVITY IN THYMECTOMIZED ddY MICE

Group	NE	DA	DOPAC	HVA	5-HT	5-HIAA	ChAT Activity (nmol/mg/protein)
			Monoamines (ng/mg tissue)				
CON	16.67±0.44	5.61±0.71	1.50±0.07	2.23±0.09	18.67±0.65	7.84±0.45	0.26±0.01
SHAM	16.34±0.26	5.25±0.50	1.56±0.13	2.17±0.20	17.12±0.95	7.12±0.44	0.25±0.02
ATx	19.11±0.76**	5.87±0.55	2.00±0.21*	2.62±0.16*	20.07±1.03	8.96±0.74**	0.32±0.03**
ATx±AGE	16.20±0.73##	4.92±0.37	1.42±0.12##	1.92±0.09##	17.94±0.56	6.97±0.37	0.26±0.02##

The hypothalamic contents of monoamines and ChAT activity were determined 10 months after thymectomy. Data were represented with mean±SEM of 8 individual samples. *: $p < 0.05$, **: $p < 0.01$ vs SHAM, ##: $p < 0.01$ vs ATx in Duncan's test.

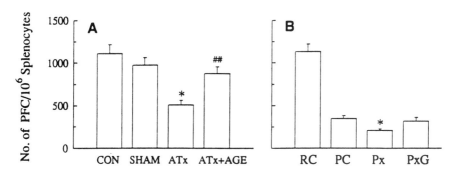

FIG. 13.4. EFFECTS OF AGE ON ANTIBODY PRODUCTION RESPONSE REDUCED
BY THYMECTOMY
PFC assay was performed 10 and 6 months after thymectomy in ddY mice (A) and SAMP8 (B),
respectively. For explanation of the abbreviations, see FIG. 13.1. *: p < 0.05 vs SHAM in A,
vs PC in B; ##: p < 0.01 vs ATx in A in Duncan's test; mean ± SEM, n=5-8.

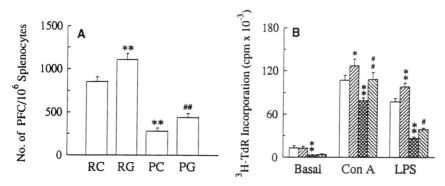

FIG. 13.5. EFFECT OF AGE ON IMMUNE RESPONSE IN NAIVE SAM
Both PFC and LT assays were performed at the age of 12 months after chronic AGE treatment
for 10 months. A: PFC assay. B: LT assay, splenocytes were cultured without mitogen (Basal),
with Con A (2 μg/ml) or LPS (20 μg/ml). □: SAMR1 control (RC), ▨: AGE-treated
SAMRI (RG), ▩: SAMP8 control (PC), ▥: AGE-treated SAMP8 (PG). *: p < 0.05, **:
p < 0.01 vs RC, #: p < 0.05, ##: p < 0.01 vs PC in Duncan's test; mean ± SEM, n=5-6.

factors which alter motor activity. In addition, AGE treatment potentiated
thymectomy-reduced antibody production and enhanced lymphocyte proliferation
induced by Con A, or in LPS in both SAMR1 and SAMP8. These results
suggested that AGE may potentiate the T and B cell functions of aged animals.

Taken together, the results obtained indicate that chronic oral administration of AGE improves the deficits in both central cognitive and immune functions caused by thymus dysfunction or senescence. AGE treatment prevented thymectomy-induced suppression of the body weight, functions of the body.

The imbalance of NIM network is supposed to accelerate systematic aging, in which the involution of thymus plays an important role.[9-11] Thymectomy in mice at either neonatal or weaning and early adult period not only dramatically reduces the immune response[16-20], but also cause neuroendocrine imbalances, such as reduced plasma level of thyroid hormones and disordered plasma level of adrenocorticotrophic hormone or corticosterone etc., which also occur during the natural aging process.[15] In addition, the progressive reduction of submandibular glands responsiveness to beta-adrenoceptor stimulation and the presence of tetraploid cells in the liver, two typical aging parameters in the rodent, occurs much earlier in thymectomized mice than in normal litter mates.[25,26] Thus, thymectomy-induced learning and memory deficiency may be one of the signs of systematic aging primarily caused by thymus dysfunction. Therefore, the ameliorative effects of AGE on thymectomy-induced learning and memory deficiency may be attributable, at least partially, to improved immune function or the restoration of the thymus-mediated balance of NIM network disordered by thymectomy.

The aging pattern of SAM is an accelerated senescence rather than a premature aging[27] and the characteristic phenotype of SAMP8 has been demonstrated to be age-related learning and memory deficiency.[27,28] Our results demonstrate that the antibody production ability is dramatically reduced at the age of 6 months. Mitogen-induced lymphocyte proliferation abilities also significantly declined at the age of 12 months. Powers et al., however, reported that the significant reduction of splenocyte proliferation induced by phytohemagglutinin was obvious only at the age of 15 months in SAMP8.[29] AGE treatment significantly potentiated the immune response of SAM. These results indicated that the immunomodulative effect of AGE is not restricted in normal mice[4-6], but is also applicable to senile animals. These results support the possibility that AGE modulates NIM network to cause ameliorative effects on learning behaviors in intact SAM (unpublished observations).

The hypothalamus has been assured to be the center of NIM network.[30-32] Hypothalamic NE and DA modulate the release of hypophysiotropic peptide hormones, such as gonadotropin-releasing hormone, thyrotropin-releasing hormone and growth hormone-releasing hormone.[33] These modulations could influence immune function by changing the levels of their corresponding peripheral hormones, since these hormones have proved to be immunomodulators.[15] Our results show that the elevated hypothalamic NE, DOPAC and HVA, and ChAT activity in thymectomized ddY mice were restored to normal levels by AGE treatment. These results suggest that AGE influences the hypothalamic function by modulating both adrenergic and cholinergic transmissions, which may consequently affect the NIM network.

Although the present study did not evaluate the effects of AGE on other age-related deteriorations of body functions, it suggests that AGE possesses antiaging effect in regard to its ameliorative effects on age-related deficiencies of learning and memory abilities and immune responses in thymectomized mice and SAM. The operative mechanisms of AGE remains unclear. Improvement of immune function and restoration of NIM network may play a basic role in exerting such effects. Recent studies from our laboratory reveal that AGE has an neurotrophic effect on cultured neurons from rat embryo (Saito *et al.* Chapter 13), which may partially account for its CNS beneficial effects. Further studies on AGE and its ingredients are now being conducted in our laboratory.

REFERENCES

[1] Takasugi, N., Kotoo, K., Fuwa, T. and Saito, H. Effect of garlic on mice exposed to various stresses. Oyo Yakuri 1984; 28:991–1022.

[2] Lau, B.H.S., Tadi, P.P. and Tosk, J.M. *Allium Sativum* (garlic) and cancer prevention. Nutr. Res. 1990; 10:937–948.

[3] Liu, J., Lin, R.I. and Milner, J.A. Inhibition of 7,12-dimethylbenz[a]anthracene-induced mammary tumors and DNA adducts by garlic powder. Carcinogenesis 1992; 13:1847–1851.

[4] Hirao, Y., Sumioka, I., Nakagami, S. *et al.* Activation of immunoresponder cells by the protein fraction from aged garlic extract. Phytother. Res. 1987; 1:161–164.

[5] Lau, B.H.S., Yamasaki, T. and Gridley, D.S. Garlic compounds modulate macrophage and lymphocyte function. Mol. Biother. 1991; 3:103–107.

[6] Morioka, N., Sze, L.L., Morton, D.L. and Irie, R.F. A protein fraction from aged garlic extract enhances cytotoxicity and proliferation of human lymphocytes mediated by interleukin-2 and concanavalin A. Cancer Immununol. Immununother. 1993; 37:316–322.

[7] Aboul-Enein, A. Inhibition of tumor growth with possible immunity by Egyptian garlic extracts. Die Nahrung 1986; 30:161–169.

[8] Lau, B.H.S. Detoxifying, radioprotective and phagocyte-enhancing effects of garlic. Int. Clin. Nutr. Rev. 1989; 9:27–31.

[9] Fabris, N. and Piantanelli, L. Thymus-neuroendocrine interactions during development and aging. *In* Hormones and Aging, (Adelman, R.C. and Roth, G.S., eds.) CRC Press, Boca Raton, Fla., 1982:167–181.

[10] Fabris, N. Pathways of neuroendocrine-immune interactions and their impact with aging processes. *In* Immunoregulation in Aging, (Facchini, A.,

Haaijman, J.J. and Labo, G., eds.) Rijswijk, The Netherlands, Eurage 1986:117–130.

[11] Goya, R.G. The immune-neuroendocrine homeostatic network and aging. Gerontology 1991; 37:208–213.

[12] Deschaux, P., Massengo, B. and Fontanges, R. Endocrine interaction of the thymus with the hypophysis, adrenals and testes: effects for two thymic extracts. Thymus 1979; 1:95–108.

[13] Hall, N.R., McGillis, J.P., Spangelo, B.L. and Goldstein, A.L. Evidence that thymosins and other biologic response modifiers can function as neuroactive immunotransmitters. J. Immunol. 1985; 135:8065–8115.

[14] Folch, H., Eller, G., Mena, M. and Esquivel, P. Neuroendocrine regulation of thymus hormones: hypothalamic dependence of "facteur thymique serique" level. Cell Immunol. 1986; 102:211–216.

[15] Fabris, N., Mocchegiani, E., Muzzioli, M. and Provinciali, M. Neuroendocrine-thymus interaction: perspectives for intervention in aging. Ann. NY Acad. Sci. 1988; 521:72–87.

[16] Dutartre, P. and Pascal, M. Thymectomy at weaning. An accelerated aging model for the mouse immune system. Mech. Ageing. Dev. 1991; 59:275–289.

[17] Kappler, J.W., Hunter, P.C., Jacobs, D. and Lord, E. Functional heterogeneity among the T derived lymphocytes of the mouse. I. Analysis by adult thymectomy. J. Immunol. 1974; 113:27–38.

[18] Hirokawa, K. and Utsuyama, M. Combined grafting of bone marrow and thymus, and sequential multiple thymus graftings in various strains of mice. The effect on immune functions and life span. Mech. Ageing Dev. 1989; 49:49–60.

[19] Mitani, M., Mori, K., Himeno, K. and Nomoto, K. Anti-tumor cytostatic mechanism and delayed-type hypersensitivity against a syngeneic murine tumor, comparison between neonatally thymectomized mice and congenitally athymic nude mice. J. Immunol. 1989; 142:2148–2154.

[20] Utsuyama, M., Kasai, M., Kurashima, C. and Hirokawa, K. Age influence on the thymic capacity to promote differentiation of T cells: induction of different composition of T cell subsets by aging thymus. Mech. Ageing Dev. 1991; 58:267–277.

[21] Takeda, S. Development of senescence accelerated mouse (SAM). Tr. Soc. Pathol. Jpn. 1990; 79:39–48.

[22] Ishihara, A., Saito, H., Ohata, H. and Nishiyama, N. Basal forebrain lesioned mice exhibit deterioration in memory acquisition process in step through passive avoidance test. Jpn. J. Pharmacol. 1991; 57:329–336.

[23] Nishiyama, N., Wang, Y.L. and Saito, H. Effects of Biota (Bsi-Zi-Ren), a traditional Chinese medicine, on learning performances in mice. Shoyakugaku Zasshi 1992; 46:62–70.

[24] Cunnigham, A.J. and Szenberg, A. Further improvements in the plaque technique for detecting single antibody-forming cells. Immunol. 1968; 14: 599–600.

[25] Piantanelli, L., Basso, A., Muzzioli, M. and Fabris, N. Thymus-dependent reversibility of physiological and isoproterenol evoked age-related parameters in athymic (nude) and old normal mice. Mech. Ageing Dev. 1978; 7: 171–182.

[26] Pieri, C., Giuli, C., Dei Moro, M. and Piantanelli, L. Electron microscopic morphometric analysis of mouse liver. II. Effect of aging and thymus transplantation in old animals. Mech. Ageing Dev. 1980; 13:275–280.

[27] Yagi, H., Kato, S., Akiguchi, I. and Takeda T. Age-related deterioration of ability of acquisition in memory and learning in senescence accelerated mouse: SAM-P/8 as an animal model of disturbances in recent memory. Brain Res. 1988; 474:86–93.

[28] Miyamoto, M., Kyota, Y., Yamazaki, N., et al. Age-related changes in learning and memory in the senescence-accelerated mouse (SAM). Physiol. Behavior 1986; 38:399–406.

[29] Powers, D.C., Moley, J.E. and Flood, J.F. Age-related changes in LFA-1 expression, cell adhesion, and PHA-induced proliferation by lymphocytes from senescence accelerated mouse (SAM)-P/8 and SAM-R/1 substrains. Cell Immunol 1992; 141:444–456.

[30] Jancovic, B.D. and Isakovic, K. Neuro-endocrine correlates of immune response I. Effects of brain lesions on antibody production, Arthus reactivity and delayed-type hypersensitivity in rat. Int. Arch. Allergy 1973; 45: 360–372.

[31] Cross, R.J., Markesbery, W.R., Brooks, W.H. and Roszman, T.L. Hypothalamic-immune interactions. I. The acute effect of anterior hypothalamic lesion on the immune response. Brain Res. 1980; 196:79–87.

[32] Katayama, M., Kobayashi, S., Kuramoto, N. and Yokoyama, M. Effects of hypothalamic lesion on lymphocyte subsets in mice. Ann. NY Acad. Sci. 1987; 496:366–376.

[33] Meites, J. Role of hypothalamic catecholamines in aging processes. Acta Endocrinologica (Copenh) 1991; 125:98–103.

CHAPTER 14

ANTIOXIDANTS IN GARLIC. II. PROTECTION OF HEART MITOCHONDRIA BY GARLIC EXTRACT AND DIALLYL POLYSULFIDE FROM THE DOXORUBICIN-INDUCED LIPID PEROXIDATION

SHOJI AWAZU

Department of Biopharmaceutics
Tokyo University of Pharmacy and Science
1432-1 Horinouchi, Hachioji
Tokyo 192-03, Japan

and

TOSHIHARU HORIE

Department of Biopharmaceutics
Faculty of Pharmaceutical Sciences
Chiba University
1-33 Yayoi-cho, Inage-ku, Chiba-shi
Chiba 263, Japan

ABSTRACT

We previously reported that garlic extract protected cytomembranes from lipid peroxidation and that the major components in the garlic extract which had the antioxidant activity were diallyl polysulfides. In the present report, the effect of diallyl pentasulfide on the doxorubicin-induced lipid peroxidation in beef heart mitochondria was studied, using diallyl pentasulfide. It completely inhibited the production of thiobarbituric acid-reactive substances (TBA-RS) and chemiluminescence in a heart mitochondrial suspension incubated with doxorubicin and NADH. Diallyl pentasulfide decreased the oxygen consumption in the beef heart submitochondrial particles. Garlic extract, also inhibited the doxorubicin-induced lipid peroxidation of submitochondrial particles, but affected the oxygen consumption less than diallyl pentasulfide did. Garlic extract may have some components which protect the membranes from the diallyl polysulfides.

INTRODUCTION

Lipid peroxidation is closely associated with the etiology of disease states, such as atherosclerosis and diabetes, drug toxicity and ageing.[1,2] This is induced by free radicals, such as reactive oxygens generated in biological systems, which are initiated not only by metabolism, but also by non-enzymatic reactions such as ultraviolet light and x-ray irradiation.

Garlic has been used as a food, condiment, and a medicine since ancient times. Various effects of garlic on biological systems have been reported: tumor inhibiting effects[3,4], antibacterial activity[5], decreases in serum and liver cholesterol levels[6,7], and inhibition of platelet aggregation.[8-10] We have reported that aged garlic extract protected biological membranes from lipid peroxidation both *in vitro* and *in vivo*.[11,12] Further, we identified that the chemical components of this garlic extract, which had the potent protective effect were diallyl polysulfides, i.e., diallyl trisulfide, diallyl tetrasulfide, diallyl pentasulfide, diallyl hexasulfide and diallyl heptasulfide.[13] These compounds may be formed in food-processed forms of garlic that are widely available to consumers.

Doxorubicin is a widely used antineoplastic agent with valid therapeutic activity. This drug is known to induce cardiotoxicity. This toxicity is considered to be mediated by reactive free radicals produced by doxorubicin and is characterized by changes in the morphology and function in mitochondria.[14, 15] Therefore, we investigated the potential of diallyl polysulfides in protecting the mitochondria from the reactive oxygens generated during doxorubicin redox cycling. Such a potential protective effect could have significant application in cancer therapy.

We have detected the chemiluminescence produced during doxorubicin-induced lipid peroxidation in rat heart mitochondrial suspension and have identified its origins by chemiluminescence spectroscopy and by the use of radical scavengers.[16] Chemiluminescence is a useful and sensitive tool to monitor oxidative stress[17] Thus in this chapter, we show the effect of the garlic extract and diallyl pentasulfide on the doxorubicin-induced lipid peroxidation in bovine heart mitochondria, measured by chemiluminescence.

MATERIALS AND METHODS

Materials

The garlic extract and diallyl pentasulfide were supplied by Wakunaga Pharmaceutical Co., Ltd. (Hiroshima, Japan). Doxorubicin was supplied by Daiichi Pharmaceutical Co., Ltd. (Tokyo, Japan). All other reagents were of the highest grade available.

Preparation of Beef Heart Mitochondria and Submitochondrial Particles

The mitochondria and submitochondrial particles were prepared according to the method of Lee and Ernster.[18] The protein concentrations of the mitochondrial and submitochondrial particles were determined according to the method of Lowry et al.[19] with bovine serum albumin as the standard.

Reaction of Mitochondria with Doxorubicin and Diallyl Pentasulfide

The heart mitochondria were preliminarily treated with diethyl maleate as described elsewhere[16], a KCl(150 mM)-Tris (50 mM) - HCl buffer (pH 7.4) through which 95% O_2-5% CO_2 gas had been bubbled for more than 15 min was used. The heart mitochondrial suspension (0.5 mg protein/ml) containing 25 μM $FeCl_3$ was incubated at 37°C with 50 μM doxorubicin and/or 0.1 mM diallyl pentasulfide. The reaction was terminated by adding 1 mM EDTA to the mitochondrial suspension.

Assay of TBA-RS

TBA-RS formed in the heart mitochondrial suspension incubated 60 min with doxorubicin and/or diallyl pentasulfide was determined according to Buege and Aust[20] and expressed as nanomoles of malondialdehyde (MDA) equivalents per milligram of protein. In brief, 0.5 ml of KCl-Tris-HCl buffer and 2 ml of TBA stock solution consisting of 15% (w/v) trichloroacetic acid, 0.375% (w/v) 2-thiobarbituric acid and 0.25 N HCl were added to 0.5 ml of mitochondrial suspension. The mixture was boiled for 15 min, cooled and centrifuged at 1,000 g for 15 min. The precipitate was removed. The absorbance of the supernatant at 535 nm was determined with 1,1,3,3-tetraethoxypropane as standard.

Chemiluminescence. Chemiluminescence was measured as described elsewhere[21], using a single photoelectron counting system, CLD-100 and CLC-10 (Tohoku Electronic Industries Co. Ltd., Sendai, Japan), connected to a personal computer PC-9801 NS (NEC Corp., Tokyo, Japan) for integration. Two milliliters of the mitochondrial suspension (0.5 mg protein/ml) containing 50 μM doxorubicin and 25 μM $FeCl_3$ with and without diallyl pentasulfide were placed in a stainless steel dish (diameter: 50 mm, height: 10 mm) for 5 min where the temperature was kept at 37°C. Then the reaction was started by adding 2.5 mM NADH solution to the mixture. The chemiluminescence emitting from the mitochondrial suspension was measured continuously as counts per minute. The chemiluminescence intensity was shown by subtracting the background counts from the observed counts of the reaction mixtures according to Nakano et al.[22]

Oxygen consumption. The oxygen consumption in heart submitochondrial particle suspension (0.1 M potassium phosphate buffer, pH 7.4) in the absence and presence of doxorubicin and/or diallyl pentasulfide or garlic extract was measured at 37°C using a Clark-type oxygen electrode. NADH (10 mM) was used as a substrate to determine the NADH-stimulated respiration.

RESULTS AND DISCUSSION

Incubation of the heart mitochondrial suspension (0.5 mg protein/ml) containing 2.5 mM NADH and 25 μM $FeCl_3$ with 50 μM doxorubicin for 60 min markedly produced TBA-RS (42.5 nmole MDA equivalents/mg protein). Adding diallyl pentasulfide (0.1 mM) to the mitochondrial suspension with doxorubicin, the production of TBA-RS was inhibited (14.3 nmole MDA equivalents/mg protein), which approximated the control level (10.4 nmole MDA equivalents/mg protein). Diallyl pentasulfide itself had no effect on the production of TBA-RS here, since TBA-RS value in the mitochondrial suspension in the presence of diallyl pentasulfide alone was 6.6 nmole MDA equivalents/mg protein. As we reported previously[13], diallyl polysulfides protect liver microsomes from the ascorbic acid/$FeSO_4$-induced lipid peroxidation. In this study, diallyl pentasulfide was found to prevent doxorubicin-induced lipid peroxidation in heart mitochondria as well.

Chemiluminescence is sensitive to oxidative stress.[17] We have previously reported[16] that the presence of doxorubicin in heart mitochondrial suspension produces chemiluminescence. Chemiluminescence was shown to originate from a singlet oxygen and excited carbonyls. The molecular mechanism of the chemiluminescence generation from heart mitochondria on the doxorubicin-induced lipid peroxidation can be represented as in Fig. 14.1. Briefly, superoxide anions are produced during doxorubicin redox cycling in the presence of NADH and oxygen.[23] They react with mitochondrial lipids (LH) and produce lipid peroxyl radicals (LOO·). The lipid peroxyl radicals (LOO·) produce singlet oxygen ($^1O_2$), excited carbonyls (L=0*) and hydroxyl compounds (LOH). The singlet oxygen ($^1O_2$) and excited carbonyls (L=0*) emit chemiluminescence upon relaxation to the ground state, triplet oxygen ($^3O_2$) and carbonyls (L=0), respectively.

The effect of diallyl pentasulfide on chemiluminescence from the mitochondrial suspension with doxorubicin was investigated (Fig. 14.2). As previously reported[16], chemiluminescence is markedly produced in a mitochondrial suspension containing 2.5 mM NADH, 25 μM $FeCl_3$ with 50 μM doxorubicin. In this experiment it increased with the progress of reaction, reached a maximum intensity around 30 min and then decreased. Interestingly, the chemiluminescence was completely inhibited by the presence of diallyl pentasulfide. Considering the molecular mechanism described above (Fig. 14.1), this depression of the chemiluminescence indicates that both the production of singlet oxygen and excited carbonyls were inhibited by diallyl pentasulfide. The amounts of TBA-RS produced in the mitochondrial suspension at the end of the chemiluminescence measurement (60 min) was 39.0 nmole MDA equivalents/mg protein for the mitochondria with doxorubicin and 9.5 nmole MDA equivalents /mg protein for the mitochondria with doxorubicin and diallyl pentasulfide.

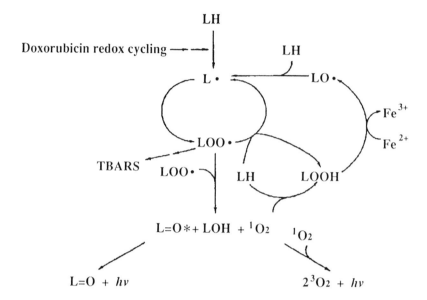

FIG. 14.1. MOLECULAR MECHANISM OF THE CHEMILUMINESCENCE FROM HEART
MITOCHONDRIA UPON DOXORUBICIN-INDUCED LIPID PEROXIDATION
LH, lipids; LOOH, lipid hydroperoxide; LOO·, lipid peroxyl radicals; LOH, lipid hydroxyl
compounds; L=O*, excited carbonyls; $^1O_2$, singlet oxygen; $^3O_2$, triplet oxygen.

Oxygen consumption in heart submitochondrial particles was measured
at 37°C with a Clark-type oxygen electrode. The suspension of heart submitochondrial particles was incubated for 60 min with or without doxorubicin (50
μM) and/or diallyl pentasulfide (0.1mM). Oxygen consumption in the submitochondrial particles incubated for 60 min was affected by the presence of
doxorubicin and/or diallyl pentasulfide. The percentage of oxygen consumption
of the submitochondrial particles with doxorubicin and/or diallyl pentasulfide as
compared to the control submitochondrial particles was 41.2% for the
submitochondrial particles with doxorubicin, 30.3% for doxorubicin and diallyl
pentasulfide, and 30.8% for diallyl pentasulfide. These results indicate that the
presence of diallyl pentasulfide decreases oxygen consumption markedly.

As we reported elsewhere[11], the garlic extract at a concentration of 40
mg/ml inhibited the ascorbic acid/$FeSO_4$-induced lipid peroxidation in liver
microsomes completely. Garlic extract has an antioxidant effect without toxicity.
Thus the effect of garlic extract on the oxygen consumption in the submitochondrial particles was examined as well. The percentages of oxygen consumption

of the submitochondrial particles with doxorubicin and/or garlic extract as compared to the control submitochondrial particles were 41.2% for the submitochondrial particles with doxorubicin, 70.5% for doxorubicin and garlic extract, and 71.9% for garlic extract. Doxorubicin decreased the oxygen consumption, but the presence of garlic extract in submitochondrial particles improved the doxorubicin-induced decrease of oxygen consumption. Further, the oxygen consumption of submitochondrial particles with garlic extract alone was more than twice that with diallyl pentasulfide alone. These results suggest that there may be some specific components of garlic extract, which could protect the mitochondria from the depression of mitochondrial respiration induced by doxorubicin and/or diallyl polysulfides.

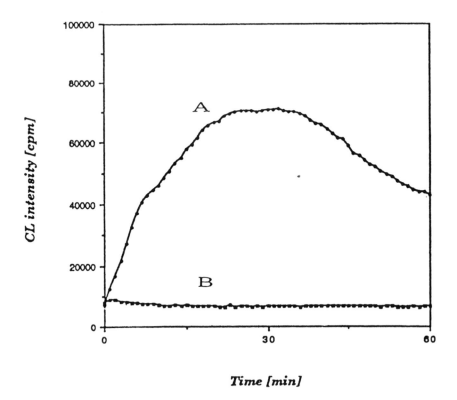

Time [min]

FIG. 14.2. EFFECT OF DIALLYL PENTASULFIDE ON THE CHEMILUMINESCENCE (CL) FROM THE BOVINE HEART MITOCHONDRIAL SUSPENSION (0.5 MG PROTEIN/ML) CONTAINING 50 μM DOXORUBICIN, 2.5 mM NADH AND 25 μM FeCl$_3$
(A) without diallyl pentasulfide, (B) with diallyl pentasulfide (0.1 mM).

In conclusion, diallyl pentasulfide protected heart mitochondria from doxorubicin-induced lipid peroxidation, but decreased the oxygen consumption of submitochondrial particles. The garlic extract protected the mitochondria from lipid peroxidation and less effect on oxygen consumption, suggesting that some components of the garlic extract may protect heart mitochondria from polysulfides.

REFERENCES

[1] Sies, H. *et al.* Oxidative Stress, Academic Press, London, 1985.

[2] Yagi, K. *et al.* Lipid Peroxides in Biology and Medicine, Academic Press, New York, London, 1982.

[3] Wargovich, M.J. Diallyl sulfide, a flavor component of garlic (allium sativum), inhibits dimethylhydrazin-induced colon cancer. Carcinogenesis 1987; 8:487–489.

[4] Nishino, H., Iwashima, A., Itakura, Y., Matsuura, H. and Fuwa, T. Antitumor-promoting activity of garlic extracts. Oncology 1989; 46:277–280.

[5] Cavallito, C.J. and Bailey, J.H. Allicin, the antibacterial principle of Allium sativum. I. Isolation, physical properties and antibacterial action. J. Am. Chem. Soc. 1944; 66:1950–1954.

[6] Kamarma, V.S. and Chandrasekhara, N. Hypocholesteremic activity of different fractions of garlic. Indian J. Med. Res. 1984; 79:580–583.

[7] Qureshi, A.A., Abuirmeileh, N., Din, Z.Z., Elson, C.E. and Burger, W.C. Inhibition of cholesterol and fatty acid biosynthesis in liver enzymes and chicken hepatocytes by polar fractions of garlic. Lipids 1983; 18:343–348.

[8] Ariga, T., Oshiba, S. and Tamada, T. Platelet aggregation inhibitor in garlic. Lancet 1981; I:150–151.

[9] Apitz-Castro, R., Cabrera, S., Cruz, M.R., Ledezma, E. and Jain, M.K. Effects of garlic extract and of three pure components isolated from it on human platelet aggregation, arachidonate metabolism, release reaction and platelet ultrastructure. Thrombosis Res. 1983; 32:155–169.

[10] Block, E., Ahmad, S., Jain, M.K., Crecely, R.W., Apitz-Castro, R. and Cruz, M.R. (E,Z)-Ajoene: A potent antithrombotic agent from garlic. J. Am. Chem. Soc. 1984; 106:8295–8296.

[11] Horie, T., Murayama, T., Mishima, T., Itoh, F., Minamide, Y., Fuwa, T. and Awazu, S. Protection of liver microsomal membranes from lipid peroxidation by garlic extract. Planta Medica 1989; 55:506–508.

[12] Horie, T. Protection of cytomembranes from peroxidation by aged garlic extract. First World Congress on the Health Significance of Garlic and Garlic Constituents 1990; p.23 (abstr).

[13] Horie, T., Awazu, S., Itakura, Y. and Fuwa, T. Identified diallyl polysulfides from aged garlic extract which protects the membranes from lipid peroxidation. Planta Medica 1992; 58:468–469.

14   Bachmann, E., Weber, E. and Zbinden, G. Effects of seven anthracycline antibiotics on electrocardiogram and mitochondrial function of rat hearts. Agents Actions 1975; 5:383-393.

15   Porta, E.A., John, N.S., Matsumura, L., Nakasone, B. and Sablau, H. Acute adriamycin cardiotoxicity in rats. Res. Commun. Chem. Pathol. Pharmacol. 1983; 41:125-137.

16   Kitada, M., Horie, T. and Awazu, S. Chemiluminescence associated with doxorubicin-induced lipid peroxidation in rat heart mitochondria. Biochem. Pharmacol. 1994; 48:93-99.

17   Cadenas, E., Boveris, A. and Chance, B. Low-level chemiluminescence of biological systems. Free Radicals in Biology (Pryor, W.A., ed.) Vol 6, pp. 211-242. Academic Press, New York, 1984.

18   Lee, C. and Ernster, L. Energy-coupling in nonphosphorylating submito-chondrial particles. Methods Enzymol. 1967; 10:543-548.

19   Lowry, O.H., Rosebrough, N.J., Farr, A.L. and Randall, R.J. Protein measurement with the Folin phenol reagent. J. Biol. Chem. 1951; 193,265-275.

20   Buege, J.A. and Aust, S.D. Microsomal lipid peroxidation. Methods Enzymol. 1978; 52:302-310.

21   Yokoyama, H., Horie, T. and Awazu, S. Lipid peroxidation in rat liver microsome during naproxen metabolism. Biochem. Pharmacol. 1993; 45: 1721-1724.

22   Nakano, M., Noguchi, T., Sugioka, K., Fukuyama, H. and Sato, M. Spectroscopic evidence for the generation of singlet oxygen in the reduced nicotinamide adenine dinucleotide phosphate-dependent microsomal lipid peroxidation. J. Biol. Chem. 1975; 250:2404-2406.

23   Davies, K.J.A. and Doroshow, J.H. Redox cycling of anthracyclines by cardiac mitochondria: I. Anthracycline radical formation by NADH dehydrogenase. J. Biol. Chem. 1986; 261:3060-3067.

CHAPTER 15

ROLE OF GARLIC IN DISEASE PREVENTION — PRECLINICAL MODELS

SHUNSO HATONO and MICHAEL J. WARGOVICH

Section of Gastrointestinal Oncology
and Digestive Disease
University of Texas
M.D. Anderson Cancer Center
Houston, Texas

ABSTRACT

In the present study, the chemopreventive activity of the main components of garlic, S-allylcysteine (SAC), diallyl sulfide (DAS) and their derivatives were investigated using the dimethylhydrazine or azoxymethane-induced colonic aberrant crypt assay in rats. Of the DAS-related compounds tested, DAS, allylmethyl sulfide and allylpropyl sulfide significantly decreased the number of aberrant crypt foci by 48.7%, 47.9% and 20.9%, respectively. Among the newly synthesized SAC-related compounds tested, orally administered SAC and S-propylcysteine significantly decreased the number of aberrant crypt foci by 25.4% and 20.3%, respectively. Structure-activity analysis revealed that three carbon atoms, especially an allyl group on the side chain are essential for the expression of chemopreventive activity. However, substituting the position or composition of the unsaturated group diminished the chemopreventive activity. Compounds with a disulfide bond in their structure had no chemopreventive effect. In another study, the effect of administration of SAC during the initiation or the promotion stage, was compared. The administration of 0.4 and 0.8 maximum tolerated dose (MTD) of SAC incorporated in the experimental diet significantly decreased the number of aberrant crypt foci when given during initiation, but had no effect during promotion. This result suggests that SAC affects the initiation of carcinogenesis but not growth or differentiation.

Rats given S-ethylmercaptocysteine, S-propylmercaptocysteine and S-propagylcysteine, exhibited increased numbers of aberrant crypt foci by 37.0%, 34.7% and 60.9%, respectively. However, the dose levels at which these compounds were tested caused decreased food intake and resultant decreased body weight. Dietary restriction in the absence of any chemopreventive compounds also resulted in increased number of aberrant crypt foci.

The latter results suggest that the observed increase in the number of aberrant crypt foci in this experiment was not a specific effect of the organosulfur compounds tested, but related to a deficiency in food intake.

INTRODUCTION

The medical properties of garlic (*Allium sativum*) have been known since ancient times.[1,2] Recent epidemiological studies conducted in China and Italy revealed that gastric cancer incidence was remarkably lowered in districts with high garlic consumption compared with the other areas.[3-7] The various pharmacological activities of garlic, such as antimicrobial, antithrombic, antihypertensive, antihyperglycemic and antihyperlipidemic activities, have been extensively studied and validated.[1,8-10] The relationship between cancer, garlic and its constituents have attracted the interest of scientists. Many studies on garlic and the organosulfur compounds derived from this herb have been reported.[11] However, the structure-activity relationship of these compounds to their chemopreventive activity is not precisely elucidated. DAS and SAC, which are two of the most important components of garlic, were reported to have chemopreventive activities in the dimethylhydrazine-induced colon and nitrosomethylbenzylamine-induced esophageal cancer in rodents.[12-14] Aberrant crypts are putative precursor lesions from which adenomas and carcinomas in the colon develop and have been observed in humans, as well as in rodents given colonic carcinogens.[15,16] Aberrant crypts, which are characterized by increased size, thicker epithelial lining and increased pericryptal zone, can be observed shortly after the injection of colonic carcinogens, such as dimethyl-hydrazine and azoxymethane.[17] The inhibitory effects of various agents on aberrant crypt induction are used as a measure of chemopreventive activity.

The aberrant crypt assay has been shown to accurately predict 73-80% of long-term tumorigenesis tests in the colon.[18] In the present study, chemopreventive activities of SAC, DAS and their derivatives differing in the length of carbon chain, sulfur number and composition of unsaturated group were tested for their potency to prevent carcinogenesis using the rat aberrant crypt assay.

MATERIALS AND METHODS

Animals and Diets

Six-week-old male Fischer-344 rats weighing 116-135g were obtained from Harlan Sprague-Dawley (Indianapolis, IN). AIN-76A semipurified diet (Dyets, Bethlehem, PA) was given *ad libitum* throughout the experimental period. The rats were acclimated to the diet and environment for a period of 1 week after arrival. The rats were housed three to a cage and had constant access to water. The temperature (20-22°C), humidity (45-55%), and lighting (12 h day-night cycle) were maintained.

Carcinogen

Dimethylhydrazine (DMH, Aldrich, Milwaukee, WI) dissolved in 0.1% EDTA, pH 6.8, 25 mg/kg body weight as a base was given intraperitoneally once a week for two weeks. Azoxymethane (AOM) (Ash Stevens, Inc. Detroit, Ml) was dissolved in saline and injected intraperitoneally weekly at a dose of 15 mg/kg body weight once a week for two weeks.

Test Compounds

The chemical structures of test compounds are shown in Fig. 15.1. S-methylcysteine (SMC), S-ethylcysteine (SEC), S-propylcysteine (SPC), S-allylcysteine (SAC), S-vinylcysteine (SVC), cis- and $trans$-S-propenylcysteine

S-allylcysteine (SAC) derivatives

$R-CH_2-CH(NH_2)-COOH$

R= -S-CH$_3$
  S-methylcysteine (SMC)
R= -S-C$_3$H$_7$
  S-n-propylcysteine (SPC)
R= -S-S-CH$_3$
  S-methylmercaptocysteine (SMMC)
R= -S-CH=CH-CH$_3$
  cis- and $trans$-S-propenylcysteine
  (cis- and $trans$-SPeC)
R= -S-S-C$_3$H$_7$
  S-n-propylmercaptocysteine (SPMC)
R= -S-S-CH$_2$-CH=CH$_2$
  S-allylmercaptocysteine (SAMC)

R= -S-C$_2$H$_5$
  S-ethylcysteine (SEC)
R= -S-CH$_2$-CH=CH$_2$
  S-allylcysteine (SAC)
R= -S-S-C$_2$H$_5$
  S-ethylmercaptocysteine (SEMC)
R= -S-CH=CH$_2$
  S-vinylcysteine (SVC)
R= -S-CH$_2$-C $\equiv$ CH
  S-propagylcysteine (SPgC)

Diallyl sulfide (DAS) derivatives

$CH_2=CH-CH_2-S-CH_2-CH=CH_2$
  diallyl sulfide (DAS)

$CH_3-S-CH_2-CH=CH_2$
  allylmethyl sulfide (AMS)

$C_2H_5-S-CH_2-CH=CH_2$
  allylethyl sulfide (AES)

$CH_2=CH-CH_2-S-S-CH_2-CH=CH_2$
  diallyl disulfide (DADS)

$C_3H_7-S-C_3H_7$
  di-n-propyl sulfide (DPS)

$C_3H_7-S-CH_2-CH=CH_2$
  allyl-n-propyl sulfide (APS)

$CH_3-S-C_3H_7$
  methyl-n-propyl sulfide (MPS)

$C_3H_7-S-S-C_3H_7$
  di-n-propyl disulfide (DPDS)

FIG. 15.1. CHEMICAL STRUCTURES OF TEST COMPOUNDS

(*cis* and *trans*-SPeC), S-propagylcysteine (SPgC), S-methylmercaptocysteine (SMMC), S-ethylmercaptocysteine (SEMC) and S-propylmercaptocysteine (SPMC) were synthesized in Department of Chemistry, University of CA, Irvine. The structures of these compounds were confirmed using IR, UV, NMR and mass spectrometry. Allylmethyl sulfide (AMS), di-*n*-propyl sulfide (DPS), diallyl disulfide (DADS) and di-*n*-propyl disulfide (DPDS) were purchased from Aldrich (Milwaukee, WI). Allyl-*n*-propyl sulfide (APS) and methyl-*n*- propyl sulfide (MPS) were purchased from Pfaltz & Bauer, Inc. (Waterbury, CT).

Experimental Schedule

(1)   Structure-activity Relationship Study

Each SAC derivative was dissolved or suspended in distilled water or polyethylene glycol 400 (Sigma, St. Louis, MO) and DAS derivatives were dissolved in corn oil (CPC International, Englewood Cliffs, NJ). These compounds were administered orally once a day for three days before each DMH injection at a dose of 1 mmol/kg/day (n = 10). DMH was injected as described above. The rats were sacrificed 2 weeks after the last injection of DMH (Fig. 15.2).

(2)   Stage of Chemopreventive Activity

The dietary level of SAC, 0.125 g/kg diet and 0.2 g/kg diet corresponding to 0.4 maximum-tolerated dose (MTD) and 0.8 MTD, respectively were used.

(a)   Chemopreventive effect on initiation: Rats were fed a diet containing SAC one week prior to AOM injection and throughout the experiment (n = 10). One week after the test diet was initiated, the rats received AOM as described above. The rats were sacrificed 2 weeks after the last AOM injection (Fig. 15.3).

(b)   Chemopreventive effect on promotion: Two weeks after the last injection of AOM, the rats were placed on a SAC- containing diet for 4 weeks (n = 10). The rats were sacrificed 6 weeks after the last injection of AOM (Fig. 15.3).

(3)   Effect of Dietary Restriction on DMH-induced Aberrant Crypts

Three experimental groups were used. Group 1 had diet *ad libitum*. Group 2 had limited access to the diet at the rate of 50% of their usual daily intake and group 3 was fasted. These dietary conditions were maintained for 48 h prior to each DMH injection. *Ad libitum* feeding was re-introduced 6 h after DMH injection (n = 10).

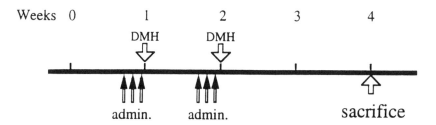

FIG. 15.2. EXPERIMENTAL SCHEDULE OF STRUCTURE-ACTIVITY
RELATIONSHIP STUDY
admin.: oral administration of test compound (1 mmol/kg/day); DMH: intraperitoneal injection of DMH (25mg/kg as a base), 3 h after last administration of test compound.

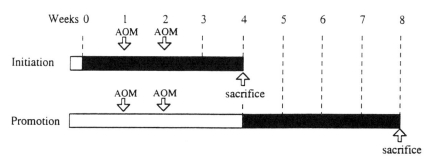

AOM; intraperitoneal injection of AOM (15mg/kg)

■ ; SAC 0.4 or 0.8 MTD (0.125 or 0.25 g/kg diet)

FIG. 15.3. EXPERIMENTAL SCHEDULE OF CHEMOPREVENTIVE STAGE

Aberrant Crypt Assay

At the times described above, rats were sacrificed by CO_2 asphyxiation. Colons were removed immediately, slit longitudinally, flattened and fixed in 70% EtOH for a minimum of 24 h. The number and multiplicity of aberrant crypt foci along the entire length of the colon were determined using a dissecting microscope after staining with 0.3% methylene blue.

Statistics

Data are presented as the mean $\pm$ SE. An unpaired two-tailed Student's t test was used to detect statistically significant differences. When the variance among groups were non-uniform, the Mann-Whitney U test was used instead of a t test. Differences with a p value < 0.05 were considered significant.

Results

(1)   Structure-activity Relationship Study

The results of structure-activity relationship study are shown in Fig. 15.4 and Fig. 15.5. The whole colons of rats were examined using a dissecting microscope after staining with methylene blue. No aberrant

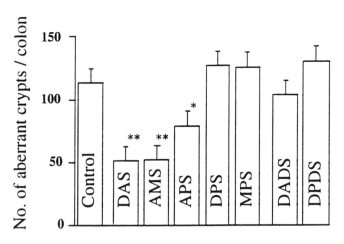

FIG. 15.4. CHEMOPREVENTIVE ACTIVITY OF DAS DERIVATIVES IN ABERRANT CRYPT ASSAY
Test compounds were administered orally at a dose of 1 mmol/kg/day.
*: $p < 0.05$, **: $p < 0.001$.

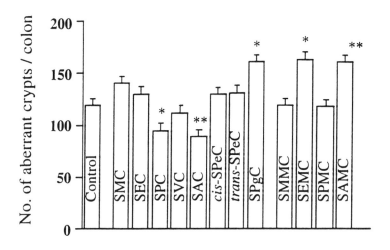

FIG. 15.5. CHEMOPREVENTIVE ACTIVITY OF SAC DERIVATIVES IN ABERRANT
CRYPT ASSAY
Test compounds were administered orally at a dose of 1 mmol/kg/day.
*: $p < 0.05$, **: $p < 0.01$.

crypt was found in colons of rats not given the carcinogen. Control rats
which received the DMH only, but not test compounds developed a
mean number of 113 aberrant crypts. Among the DAS derivatives
tested, the number of aberrant crypt foci was significantly decreased by
54.6%, 53.9% and 30.0% in the group given DAS, AMS and APS,
respectively. Significantly decreased number of aberrant crypts foci
were found in the group given SPC (20.3%) and SAC (25.4%) among
the SAC derivatives. Administration of SMC, SEC, SVC, *cis* and
trans-SPeC, and SPMC exhibited no significant chemopreventive
activity.

Interestingly, the number of ACF was significantly increased by
35.2%, 37.0% and 34.7% in the group given SPgC, SEMC and
SAMC, respectively. Simultaneously, significantly decreased body
weights were observed in these groups. The relationship between the
body weight and the number of aberrant crypt foci are shown in Fig.
15.6. Linear regression analysis showed that there was significant
correlation between the number of aberrant crypt foci and body weight
($r = 0.876$, $p < 0.05$).

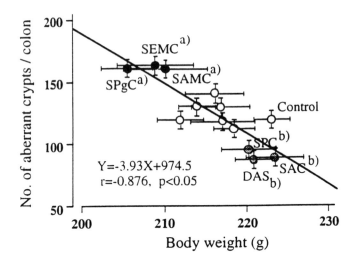

FIG. 15.6. RELATIONSHIP BETWEEN THE NUMBER OF ABERRANT CRYPTS
AND BODY WEIGHT
a) significantly different from control in the number of aberrant crypts and body weight.
b) significantly different from the control group in the number of aberrant crypts.

(2)   Stage of Chemopreventive Activity

The effect of dietary administration of SAC on initiation and promotion
of colon carcinogenesis by AOM is shown in Fig. 15.7. A mean of 85
aberrant crypt foci were found in the colons of control rats given SAC
during initiation. The number of aberrant crypt foci was significantly
decreased by 32.9% and 54.1% in the group given 0.4 and 0.8 MTD
of SAC. However, the number of aberrant crypt foci was not changed
in the group when SAC was given during promotion.

(3)   Effect of Dietary Restriction on DMH-induced Aberrant Crypts

The results of the dietary restriction study are shown in Fig. 15.8. The
number of aberrant crypt foci was increased by 35.2% and 87.1% in
the restricted and fasted group, respectively. An inverse relationship
between dietary intake and number of DMH-induced aberrant crypt foci
was found.

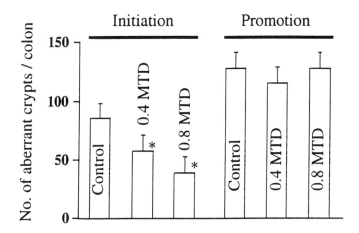

FIG. 15.7. COMPARISON OF ADMINISTRATION PERIOD OF SAC
SAC was administered orally in two different periods (see text). *:p < 0.05

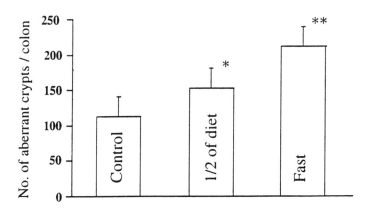

FIG. 15.8. EFFECT OF DIETARY RESTRICTION ON DMH-INDUCED ABERRANT
CRYPTS, *:p < 0.05, **:p < 0.001

DISCUSSION

In the present study, several derivatives of SAC, differing in the length of side chain, and position or composition of the unsaturated group, were synthesized. Activities of these compounds were compared with DAS and its derivatives using the rat colonic aberrant crypt assay. The structure-activity relationships of test compounds are summarized in Tables 15.1 and 15.2. DAS and SAC showed chemopreventive activity as reported previously.[12-14] We found that SPC, AMS, APS also have chemopreventive activity in this assay. All of the compounds which showed chemopreventive activity have three carbon atoms on their side chain. Particularly, an allyl group on the side chain is essential for expression of the activity. However, this activity was diminished by substituting the side chain with shorter chains or changing the position and composition of the unsaturated group. No compounds with a disulfide bond in their structure showed chemopreventive activity, even in the group given diallyl disulfide which has two allyl groups bound to a disulfide group. The lack of chemopreventive activity of compounds containing a disulfide bond may be due to the differences in physiochemical or biochemical characteristics such as solubility, reactivity, pharmacodynamics or metabolic fate. Some of these factors may regulate the expression of the chemopreventive activities of organosulfur compounds.

TABLE 15.1
SUMMARY OF CHEMOPREVENTIVE ACTIVITY OF SAC DERIVATIVES

R	Cy-S-R	Cy-S-S-R
-CH$_3$	–	–
-C$_2$H$_5$	–	x
-C$_3$H$_7$	+	–
-CH$_2$-CH=CH$_2$	+	x
-CH=CH$_2$	–	n.t.
cis -CH=CH-CH$_3$	–	n.t.
trans -CH=CH-CH$_3$	–	n.t.
-CH$_2$-C≡CH	x	n.t.

Cy = -CH$_2$-CH(NH$_2$)-COOH
+, positive; –, negative; x, toxic; n.t., not tested

Significantly increased numbers of aberrant crypt foci were observed in the groups given SEMC, SPMC and SPgC. After administration of these compounds, the rats were found to suffer weight loss. Two explanations for these observations can be made. First, the fact that the administration of these compounds caused increased aberrant crypt foci suggests they may act as promoters. Secondly, a deficiency of food intake may have had a promoting effect on the carcinogenic process. To test the second hypothesis, a food intake regulation experiment was carried out. As shown in Fig. 15.8, a fasting or dietary restriction resulted in increased numbers of aberrant crypt foci. The degree of food intake was negatively correlated with the induction of aberrant crypt foci. These results suggest that the observed increase of aberrant crypt foci in the groups given SEMC, SPMC and SPgC was not related to these organosulfur compounds, but may be a function of insufficient food intake.

TABLE 15.2
SUMMARY OF CHEMOPREVENTIVE ACTIVITY OF DAS DERIVATIVES

	R1-S-R2			R1-S-S-R2	
	R2			R2	
R1	Me	Pr	Al	Pr	Al
$-CH_3$ (Me)	n.t.	–	+	n.t.	n.t.
$-C_3H_7$ (Pr)	–	–	+	–	n.t.
$-CH_2-CH=CH_2$ (Al)	+	+	+	n.t.	–

+, positive; –, negative; n.t., not tested

SAC was effective only when given during initiation. Our preliminary study showed that the levels of the detoxifying enzyme (GST) were significantly increased by the administration of SAC. This increase was observed not only in the liver, but also in the mucosa of the small intestine and colon (data not shown). Of interest is the fact that several SAC and DAS analogues were effective in suppressing DMH- or AOM-induced aberrant crypts. Our previous work indicates that elevation of liver and intestinal GST activity is, at least, in part responsible for chemopreventive activity. Whether the positively chemopreventive analogues tested in this study work by the same mechanism is the subject of our current research.

ACKNOWLEDGMENTS

I would like to express my gratitude to Mr. Arnold Jimenez, University of Texas MD Anderson Cancer Center, Houston, Texas. I would like to express my gratitude to Dr. Fillmore Freeman and Mr. Yukihiro Kodera, UCLA Irvine, Irvine, CA, for synthesizing SAC analogues.

REFERENCES

1   Block, E. The chemistry of garlic and onions. Sci. Amer. 1985; 252: 114–119.

2   Fenwick, G.R. and Hanley, A.B. The genus *Allium*. Part 1. Critical Reviews in Food Science 1985; 22:199–271.

3   Holwitz, N. Garlic as a plant du jour: Chinese study finds it could prevent GI cancer. Med. Tribune, Aug. 12, 1981.

4   You, W.C., Blot, W.J., Chang, Y.S., Qi, A., Henderson, B.E., Fraumeni, J.F. and Wang, T.G. *Allium* vegetables and reduced risk of stomach cancer. J. Natl. Cancer Inst. 1989; 8:162–164.

5   Mei, X., Wang, M.L. and Pan, X.Y. Garlic and gastric cancer. 1. The influence of garlic on the level of nitrate and nitrite in gastric juice. Acta Nutri. Sini. 1982; 4:53–56.

6   Mei, X., Wang, M.L. and Han, N. Garlic and gastric cancer. II. The inhibitory effect of garlic on the growth of nitrate-reducing bacteria and on the production of nitrite. Acta Nutri. Sini. 1985; 7:173–176.

7   Buiatti, E., Palli, D., Decarli, A., Amadori, D., Ayellini, C., Bianchi, S., Biserni, R., Cipriani, F., Cocco, P., Giacosa, A., Mrubini, E., Puntori, R., Vindigi, C., Fraumeni, J. Jr. and Blot, W. A case control study of gastric cancer and diet in Italy. Int. J. Cancer 1989; 44:611–616.

8   Kengler, B.S. Garlic (*Allium sativum*) and onion (*Allium cepa*): A review of their relationship to cardiovascular disease. Preventive Medicine 1987; 16:670–685.

9   Adtumbi, M.A. and Lau, B.H.S. *Allium sativum* (garlic) – a natural antibiotic. Medical Hypothesis 1983; 12:227–239.

10 Abdullah, T.H., Kandili, O., Elkadi, A. and Carter, J. Garlic revisited: Therapeutic for the major diseases of our times. J. National Medical Assoc. 1988; 80:439–445.

11 Sumiyoshi, H. and Wargovich, M.J. Garlic (*Allium sativum*): A review of its relationship to cancer. Asia Pacific J. Pharmacol. 1989; 4:133–140.

12 Wargovich, M.J. and Goldberg, M.T. Diallyl sulfide: A naturally occurring thioether that inhibits carcinogen-induced damage to colon epithelial cells in vivo. Mutation Res. 1985; 143:127–129.

13 Wargovich, M.J. Diallyl sulfide, a flavor component of garlic (*Allium sativum*), inhibits dimethylhydrazine induced colon cancer. Carcinogenesis 1987; 8:487–489.

14 Wargovich, M.J., Woods, C.E., Eng, V.W.S., Stephens, L.C. and Gray, K. Chemoprevention of nitrosomethylamine-induced esophageal cancer in rats by thioether, diallyl sulfide. Cancer Res. 1988; 48:6872–6875.

15 Pretlow, T.P., Barrow, B.J., Ashton, W.S., O'Riordan, M.A., Pretlow, T.G., Juricisek, J.A. and Stellato, T.A. Aberrant crypts: putative preneoplastic foci in human colon mucosa. Cancer Res. 1991; 51: 1564–1567.

16 Roncucci, L., Stamp, D., Medline, A., Cullen, J.B. and Bruce, W.R. Identification and quantification of aberrant crypt foci and microadenomas in the human colon. Human Pathol. 1991; 22:287–294.

17 Bird, R.P., Mclellan, A.E. and Bruce, W.R. Aberrant crypts, putative precancerous lesions, in the study of the role of diet in the aetiology of colon cancer. Cancer Surv. 1989; 8:189–200.

18 Steele, V.E. and Kelloff, G.J. Evaluation of a rat colon crypt assay to identify chemopreventive agents compared to rodent colon tumor assay system. Proc. Am. Assoc. Cancer Res. 1993; 34:552.

CHAPTER 16

EFFECTS OF AGED GARLIC EXTRACT ON RAT BRAIN NEURONS

HIROSHI SAITO, TORU MORIGUCHI, YONGXIANG ZHANG,
HIROSHI KATSUKI and NOBUYOSHI NISHIYAMA

Department of Chemical Pharmacology
Faculty of Pharmaceutical Sciences
The University of Tokyo
7-3-1 Hongo, Bunkyo-Ku
Tokyo 1 13, Japan

ABSTRACT

The pharmacological effects of garlic have been widely studied, but the central nervous system was omitted from the research. We investigated the effects of aged garlic extract (AGE) on the survival and morphology of primary cultured brain neurons, and also on the induction of long-term potentiation (LTP) in rat hippocampus. AGE supported the survival of brain neurons in culture concentration-dependently. A high molecular weight protein fraction of AGE also promoted neuronal survival. However, the fructan fraction of AGE was devoid of the neuronal survival effect. In addition, AGE promoted the axonal branching of hippocampal neurons. LTP induced by tetanic stimulation, however, was not affected by oral administration of AGE. This is the first report providing direct evidence that AGE and/or its protein fraction promote neuronal survival and axon branching of rat brain neurons in culture. Fractionation of AGE may lead to a discovery of novel substances regulating synaptogenesis in the brain.

INTRODUCTION

Garlic (*Allium sativum*) has a long history of use, not only as food but also as nutrient and tonic in folk medicine. In recent years, the effects of garlic have attracted great attention in the field of pharmacy and medicine. Aged garlic extract (AGE), extracted for more than 10 months, has been shown to be less irritating than raw garlic.[1,2] Pharmacological studies of AGE and its components have revealed: its anti-stress effect[3]; the protective activity against carbon tetrachloride-induced liver damage[4]; anti-tumor promoting activity against phorbor 12-myristate 13 acetate-induced skin tumor[5]; and a chemopreventive effect in 1,2-dimethylhydrazine-induced colon cancer.[6] We recently found that

AGE prolonged the life span and ameliorated learning and memory performances in the senescence-accelerated mouse (Moriguchi *et al.*, unpublished observation). Moreover Zhang *et al.* showed that AGE improved learning and immune disorders in thymectomized mice as reported in this book. They also noted that AGE restored the altered monoamine levels and choline acetyltransferase activity in the hypothalamus of thymectomized mice. These results raise the possibility that AGE might have direct effect on the central nervous system. In the present study we investigated the effects of AGE on the survival and morphology of primary cultured brain neurons and also on the long-term potentiation (LTP) of evoked potential in rat hippocampus. The LTP in the hippocampus is a form of activity-dependent synaptic plasticity that may underlie learning and memory.

MATERIALS AND METHODS

Test Substances

AGE and its fractions were provided by Wakunaga Pharmaceutical Co. Ltd. (Hiroshima, Japan). The extract was formulated by the following procedures: briefly, sliced raw garlic was dipped into aqueous ethanol in a tank and extracted at room temperature for more than 10 months to give AGE. The quality of the extract was regularly examined chemically and microbiologically over the period of extraction. Fractionation of AGE was performed by the method of Hirao *et al.*[7]

Evaluation of Neuronal Survival

Procedures for neuronal cell culture are described elsewhere.[8] Briefly, the desired brain region was isolated from 18-day-old Wistar rat embryos and dissociated by incubation with 0.25% trypsin and 0.01% DNase I, followed by gentle pipetting. The brain regions tested in the present study were the cerebral cortex, septum, hippocampus and cerebellum. The cells were plated on polylysine-coated plastic 4-8-well plates (1 cm^2/well) at a density of 4 × 10^4 cells/cm^2 with modified Eagle's medium (MEM) supplemented with 10% fetal bovine serum. Twenty-four hours after culture in serum-containing medium, the medium was switched to serum-free medium, and then the test compounds were added. The medium for serum-free culture was 1:1(v/v) mixture of Dulbecco's modified Eagle's medium and Ham's nutrient mixture F-12 supplemented with transferrin, insulin and progesterone. The cultures were fixed with 4% paraformaldehyde three days after the drug application, and the number of surviving neurons were visualized by Nissl staining with cresyl violet.

Observation of Neuronal Morphology

The dissociated hippocampal neurons from 18-day-old rat embryos were plated on polylysine-coated 35 mm plastic dishes at a density of 2×10^3 cells/cm^2 with MEM supplemented with 1 mM pyruvate and 10% fetal bovine serum. This low-cell-density culture made it possible to minimize neuronal or glial-neuronal interactions and to clearly observe individual cell morphology. Twenty-four hours after plating, the cell medium was changed to serum-free MEM supplemented with 1 mM pyruvate, 5 μg/ml human transferrin, 5 μg/ml bovine insulin, 20 nM progesterone and 100 μM putrescine. Another 24 h later, the neurons bearing processes were randomly selected and photographed using an inverted microscope with phase-contrast optics. The test compounds were added to the cultured medium immediately after the recording. The same neurons were photographed 24 and 48 h after addition. If within 48 h, the selected neurons had died or its processes had crossed with neurites derived from other neuronal cell bodies, the data were omitted from the result. Measurement of morphological parameters was made by tracing the photographs on a digitizing tablet. To quantify the effects of the tested samples, the following parameters were measured: (1) the axon length, (2) the number of branch points per axon, and (3) the number of processes sprouted from the neuronal soma.

Recording of Long-Term Potentiation LTP in Rat Dentate Gyrus

The experiment was performed according to the method employed in our laboratory.[9,10] Male Wistar rats, 8-10 weeks old, were anesthetized by i.p. injection of a mixture of urethane and α-chloralose (1 g/kg and 25 mg/kg, respectively) and fixed in a stereotaxic frame. A bipolar stainless-steel electrode with a tip separation of 0.8 mm was inserted into the left entorhinal cortex (8.1 mm posterior to bregma, 4.4 mm lateral to midline and approximately 3.0 mm below the dura) to stimulate the perforant path. A monopolar recording electrode was placed in the granule cell layer of the ipsilateral dentate gyrus (3.5 mm posterior to bregma, 2.0 mm lateral to midline and about 3.0 mm below the dura). The depths of both stimulating and recording electrodes were adjusted to obtain a desired response. A single test stimulus (0.08 ms duration) was applied at a constant interval of 30 s and the evoked field potential was recorded extracellularly. The stimulus intensity was set at a level of 50% of the population spike of the maximum amplitude and a brief suprathreshold (30 pulses, 60 Hz), or subthreshold (20 pulses, 60 Hz) tetanic stimulation was applied at the same stimulus intensity to evoke LTP after the response became stable. The evoked-field potential was recorded for 60 min after tetanus. LTP was considered to occur when tetanus-potentiated spike amplitude maintained at a level of 20% higher than the baseline and lasted for more than 30 min. [20,21]

AGE was dissolved in saline and given to rats in a single dose of 500 mg/kg (p.o.), 30 min before tetanus. Ethanol (30% in saline) was given 20 min before tetanus by p.o. administration (2 ml/kg) through a cannula inserted into the stomach.

Statistics

The data of neuronal survival were analyzed with ANOVA and Dunnett's test (n=4); the parameters of neuronal morphology with Student's *t*-test (n=20); and the data of LTP experiment with ANOVA and Duncan's multiple range test (n=4).

RESULTS

Effects of AGE and Its Fractions on Primary Cultured Brain Neurons

Addition of AGE to the culture medium significantly supported the survival of cultured neurons derived from cerebral cortex, septum and hippocampus, but not the cerebellar neurons (Fig. 16.1). The survival-promoting effects of AGE were concentration-dependent and the minimum effective concentrations were 100-1000 μg/ml. A high-molecular-weight protein fraction

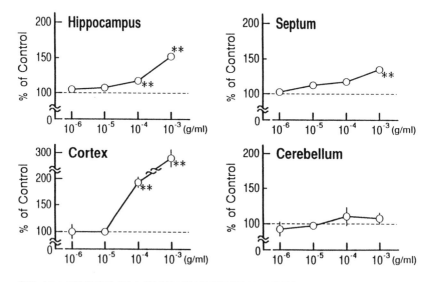

FIG. 16.1. EFFECTS OF AGE ON THE NEURONAL SURVIVAL IN CELL CULTURE
Neurons from 4 different brain regions, indicated in the figure, were cultured for 3 days in the presence of AGE and the number of surviving neurons were counted under microscope after staining with cresyl violet. Results were calibrated as a percentage of control and expressed as the mean$\pm$SEM of 4 observations. ** $P<0.01$ vs control (Dunnett's test).

of AGE (fraction 4) also enhanced the neuronal viability in the primary culture of septum, hippocampus and cerebellum, but not of cerebral cortex (Fig. 16.2). The survival-promoting effects of this fraction 4 were concentration-dependent and the minimum effective concentrations were 100-1000 μg/ml. Quite contrary to the protein fraction, the fructan fraction of AGE (10-1000 μg/ml) did not support the neuronal survival in any brain regions tested.

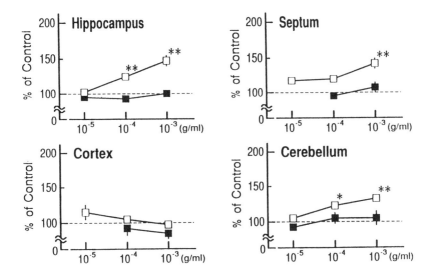

FIG. 16.2. EFFECTS OF FRACTION 3 AND FRACTION 4 OF AGE ON THE NEURONAL SURVIVAL IN CELL CULTURE

Neurons from 4 different brain regions, indicated in the figure, were cultured for 3 days in the presence of fraction 3 (■;fructan fraction) or fraction 4 (□;high molecular weight protein fraction) of AGE, and the number of surviving neurons were counted under microscope after staining with cresyl violet. Results were calibrated as a percentage of control and expressed as the mean±SEM of 4 observations. * P<0.05, ** P<0.01 vs control (Dunnett's test).

The effects of AGE on the morphology of cultured hippocampal neurons also were investigated. Morphological parameters were measured in 20 neurons. The number of branching points per axon gradually increased in the control culture, while AGE at a concentration of 1 mg/ml significantly potentiated branching, which in 48 h was more than twice that of the control culture value. On the other hand, neither the axon length nor the number of processes sprouted from the neuronal soma was affected by the presence of AGE (Fig. 16.3).

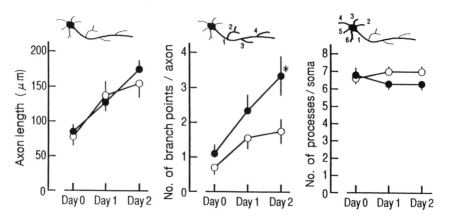

FIG. 16.3. EFFECTS OF AGE ON THE MORPHOLOGY OF HIPPOCAMPAL NEURONS IN
CELL CULTURE

More than 20 neurons bearing processes were randomly assigned to control ($\bigcirc$) or AGE-treatment ($\bullet$), and photographed 0, 1 and 2 days after addition of AGE. The axon length (left panel),
the number of branching points per axon (middle) and the number of processes per soma (right)
were measured. All data are expressed as mean $\pm$ SEM. * $p < 0.05$ (Student's t-test).

Effects of AGE on Long-Term Potentiation in Hippocampal Dentate Gyrus of Rats

First, the influences of AGE on the field potential evoked by low-fre-
quency test stimulation was examined. Oral administration of AGE (500 mg/Kg)
did not significantly affect the response evoked by test stimulation, suggesting
that AGE did not influence the basal synaptic responses (data not shown).

Next we investigated the effects of AGE on the LTP induced by
application of tetanic stimulation. When tetanic stimulation of 30 pulses at 60 Hz
was applied to a control rat, the amplitude of the population spike was greatly
potentiated and LTP was generated in all rats. AGE (500 mg/Kg, p.o.) did not
affect the magnitude of potentiation induced by strong tetanus (data not shown).

When ethanol (30%, 2 ml) was given orally 20 min prior to application
of the strong tetanus stimulation (30 pulses, 60 Hz), the basal synaptic responses
were not significantly changed, but the induction of LTP after tetanus was
completely blocked (data not shown). AGE (500 mg/Kg) given orally 10 min
before ethanol administration did not affect the population spike amplitude,
which was markedly attenuated by ethanol (data not shown).

Furthermore, the effect of AGE on the potentiation of evoked potential
induced by weak tetanic stimulation was investigated. When tetanic stimulation
of 20 pulses at 60 Hz was applied to the control rat, the population spike
amplitude was increased up to 120% of the basal level immediately after the

tetanus, but this potentiation declined to basal level within 10 min (data not shown). Since tetanic stimulation of 30 pulses at 60 Hz produced LTP in the same experimental system, the tetanus of 20 pulses, 60 Hz is regarded as the subthreshold stimulation in inducing LTP. Oral administration of AGE (500 mg/Kg) did not affect the potentiation induced by the subthreshold tetanus (data not shown).

DISCUSSION

AGE showed survival-promoting effects on a wide range of brain neurons in primary culture. Although a considerable number of investigators have reported that AGE activated or modulated immune cell functions *in vitro*, our results gave the first clear evidence that AGE acted directly on brain neurons. However, AGE did not support neuronal viability in cerebellar neurons, the reason for which is unknown at the moment. Neurotrophic effects of AGE on other neurons than cerebellum were concentration-dependent, suggesting that the action of AGE on primary-cultured neurons was achieved by a certain specific mechanism. Numerous factors have been known to enhance neuronal survival, such as fibroblast growth factors, epidermal growth factor, spermine, and interleukin 2.[11] Further experiments should reveal whether AGE interacts with neurons directly or indirectly via activating the above-mentioned factors.

In order to clarify the active components of neurotrophic AGE, we used two of its fractions, namely a high molecular weight protein fraction (fraction 4) and a fructan fraction (fraction 3). Fraction 4 promoted the neuron survival of all brain regions except cerebral cortex in a concentration-dependent manner. This fraction was neurotrophic on cerebellar neurons, which was insensitive to AGE. Conversely, cerebral cortex did not respond to fraction 4, although AGE enhanced the viability of cortical neurons. These regional differences indicate that distinct mechanisms were activated by AGE and fraction 4. Minimum effective concentration of fraction 4 on neuronal survival fell into the same range of AGE, 100-1000 μg/ml. It is impossible to postulate that the main active component of AGE is fraction 4, because the extraction yield of fraction 4 was less than 4%. Although fraction 3, whose extraction yield is 15-16%, did not show any neurotrophic activity, it would be interesting to study the interaction of fraction 3 and fraction 4 in this experimental system. Further studies on the effect of low molecular weight fraction would give more information about the active components of AGE.

Addition of AGE promoted the axonal branching of hippocampal neurons in low cell density culture. Since neuronal-glial interactions were strongly restricted in this low-density culture, AGE might affect neuronal morphology by directly acting on the neuronal cells. Although we used only one concentration of AGE (1000 μg/ml), it fits in the effective concentration range of AGE on the neuronal survival experiment. It is plausible that the same mechanism is operating in both process branching formation and neuronal survival. We observed the effects of AGE on neuronal morphology using hippocampal culture only, because hippocampal neurons survived and made processes in the low-density culture condition with more stability than neurons obtained from other brain neurons. However, its effect on the morphology of other brain neurons is an interesting subject and worth further investigation.

Miyagawa *et al.* reported that astrocyte-conditioned medium promoted the process elongation without affecting the process branching of cultured hippocampal neurons, while basic fibroblast growth factor (bFGF) increased the branching number rather than neurite extension.[12] This finding suggests that neurite elongation and branching are independently regulated by different mechanisms. AGE promoted only the axonal branching and did not alter the axon length or the number of processes. These results indicate that the pharmacological profile of AGE on hippocampal neuronal morphology resembled that of bFGF. Further studies using neutralizing antibody against bFGF or fractionated AGE would provide a clearer view of the mechanisms of AGE on brain neurons and synaptogenesis.

To get a deeper understanding of the functional role of AGE on synapse formation, we next observed the effect of AGE on hippocampal LTP in the dentate gyrus of anesthetized rat. The LTP in the hippocampus is one of the experimental models of activity-dependent synaptic plasticity. AGE (500 mg/Kg, p.o.) did not affect the potentiation of evoked potential induced by suprathreshold- or subthreshold-tetanic stimulation. AGE did not have an ameliorative effect on the ethanol-induced impairment of LTP generation. We gave AGE orally to rats in this study. It is possible that some absorption or distribution problems might interfere with the access of the active AGE component to the central nervous system. Further investigations utilizing intracerebroventricular application of AGE would clarify the latter hypothesis.

In conclusion, we have shown for the first time that AGE and/or high molecular weight protein fraction promote neuronal survival and axon branching of rat brain neurons in culture. Further fractionation of AGE may lead to the discovery of novel substances regulating synaptogenesis in the brain.

REFERENCES

1   Nakagawa, S., Masamoto, K., Sumiyoshi, H., Kunihiro, K. and Fuwa, T. Effect of raw and extracted-aged garlic juice on growth of young rats and their organs after peroral administration. J. Toxicol. Sci. 1980; 5:91–112.

2   Sumiyoshi, H., Kanezawa, A., Masamoto, K. *et al.* Chronic toxicity test of garlic extract in rats. J. Toxicol. Sci. 1984; 9:61–75.

3   Takasugi, N., Kotoo, K., Fuwa, T. and Saito, H. Effect of garlic on mice exposed to various stresses. Ovo Yakuri-Pharmacometrics 1984; 28:991–1002.

4   Nakagawa, S., Yoshida, S., Hirao, Y., Kasuga, S. and Fuwa, T. Cytoprotective activity of components of garlic, ginseng and ciuwjia on hepatocyte injury induced by carbon tetrachloride in vitro. Hiroshima J. Med. Sci. 1985; 34:303–309.

5   Nishino, H., Iwashima, A., Itakura, Y., Matsuura, H. and Fuwa, T. Antitumor-promoting activity of garlic extracts. Oncology 1989; 46:277–280.

6   Sumiyoshi, H. and Wargovich, M.J. Chemoprevention of 1,2-dimethylhydrazine-induced colon cancer in mice by naturally occurring organosulfur compounds. Cancer Res. 1990; 50:5084–5087.

7   Hirao, Y., Sumioka, I., Nakagami, S. et al. Activation of immunoresponder cells by the protein fraction from aged garlic extract. Phytother. Res. 1987; 1:161–164.

8   Hisajima, H., Saito, H. and Nishiyama, N. Human acidic fibroblast growth factor has trophic effects on cultured neurons from multiple regions from brain and retina. Jap. J. Pharmacol. 1991; 56:495–503.

9   Chida, N., Saito, H. and Abe, K. Spermine facilitates the generation of long-term potentiation of evoked potential in the dentate gyrus of anesthetized rats. Brain Res. 1992; 593:57–62.

10   Mizutani, A., Saito, H. and Abe, K. Involvement of nitric oxide in long-term potentiation in the dentate gyrus in vivo. Brain Res. 1993; 605:309–311.

11   Sarder, M., Saito, H. and Abe, K. Interleukin-2 promotes survival and neurite extension of cultured neurons from fetal rat brain. Brain Res 1993; 625:347–350.

12   Miyagawa, T., Saito, H. and Nishiyama, N. Branching enhancement by basic fibroblast growth factor in cut neurite of hippocampal neurons. Neurosci. Lett. 1993; 153:29–31.

CHAPTER 17

DIETARY AGED GARLIC EXTRACT INHIBITS SUPPRESSION OF CONTACT HYPERSENSITIVITY BY ULTRAVIOLET B (UVB, 280-320 NM) RADIATION OR *CIS* UROCANIC ACID

VIVIENNE E. REEVE, MEIRA BOSNIC, EMILIA ROZINOVA and
CHRISTA BOEHM-WILCOX

Department of Veterinary Pathology
University of Sydney
NSW 2006, Australia

ABSTRACT

Semi-purified powdered diets containing concentrations of 0.1%, 1% and 4% by weight of lyophilized aged garlic extract were fed to hairless mice. At the highest concentration, significant protection was observed from the oedema component of erythema induced by a moderate exposure to UVB radiation. Under UVB irradiation conditions resulting in 58% suppression of the systemic contact hypersensitivity response in control-fed mice, a dose-responsive protection was observed in the garlic-fed mice, such that suppression was reduced to 19% in the UVB-exposed mice fed 4% garlic extract. If suppression of contact hypersensitivity was achieved by topical application of the putative epidermal immunosuppressive mediator, cis urocanic acid, then suppression responsive to the dose of this natural photoproduct was demonstrated in control-fed mice, and this was prevented increasingly by increasing concentrations of aged garlic extract in the diet. Thus aged garlic extract contains ingredient(s) that protect both from the UVB-induced erythema reaction, and from photoimmunosuppression, for which the mechanism of protection is by antagonism of the cis urocanic acid mediation of this immune impairment.

INTRODUCTION

The induction of non-melanoma skin tumors by ultraviolet (UV; 280-400nm) radiation is the outcome of both the initiating event, which is the direct formation of cyclobutyl pyrimidine dimers in the DNA of epidermal keratinocytes and is independent of oxygen, and the promotion phase, which is subject to modulation by a number of regulatory mechanisms. UV radiation-initiated skin tumor promotion is responsive to the formation and persistence of active oxygen species[1], prostaglandins and related compounds[2], and the suppression of T cell-mediated immune recognition of the tumor cell.[3,4] UV radiation is a complete carcinogen; however additionally, UV irradiation of mice and man induces a specific defect in T cell-mediated activity, impairing the recognition and rejection of the antigenic UV radiation-initiated tumor cell in the host.

This immunological defect, photoimmunosuppression, has been revealed in a variety of other T cell-dependent immunologic functions in mice. UV-irradiated mice have been shown to be susceptible to parasitic (Leishmania[5]), fungal (Candida[6]), and viral (Herpes simplex[7]) infections, and to be unable to respond normally to alloantigens[8] and contact sensitizing agents.[9] The contact hypersensitivity (CHS) response has been well-exploited in attempting to clarify the mechanism by which photoimmunosuppression is achieved, and its significance to UV radiation-induced skin carcinogenesis. The search for the primary epidermal photoreceptor for the induction of this immunological impairment has identified *trans* urocanic acid, a normal constituent of the stratum corneum of mammalian skin, as one possibility.[10] *Trans* urocanic acid (deaminated histidine) undergoes photoisomerization when exposed to UVB (280-320nm) radiation to form *cis* urocanic acid, and *cis* urocanic acid but not the *trans* isomer, has been found to mimic many of the immunosuppressive properties of UV radiation in mice.[11] Structurally-related molecules such as histamine receptor antagonists[12-14] or carnosine[15] have been shown to inhibit the immunosuppressive properties of both UVB radiation and *cis* urocanic acid; inhibition of prostaglandin synthesis by indomethacin has likewise been shown to protect mice from the suppression of CHS by UV radiation.[16]

It is of interest that prostaglandins and reactive oxygen species are also involved in the damage and erythema reaction induced by UV exposure.[17,18] A reduction of endogenous enzymatic and non-enzymatic scavengers of oxidative free radicals occurs in UV-irradiated skin and cultured skin cells[19-21], whereas the cytotoxicity of UV radiation has been reversed by antioxidants such as glutathione and superoxide dismutase.[22-24] Thus an abundance of reactive oxygen species, as are released following UV irradiation, is associated with cytotoxicity, erythema, and with enhanced tumor promotion in the skin, but how pro-oxidant states might affect UV radiation-modified cutaneous immunity has not been described.

Garlic and onion oils, as well as aged garlic extract, have protected from a potent tumor promoter[25], from lipid peroxidation[26], and from tumor outgrowth in DMBA-initiated mice.[27,28] The anti-promoter activity was associated with the inhibition both of lipoxygenase activity[29,30] and of prostaglandin E_2 synthesis.[31] Aged garlic extract has also stimulated cell-mediated immune responsiveness[32-34], and has been successful in experimental tumor immunotherapy in mice, with histological evidence of stimulated macrophage and lymphocyte activity at the tumor margin.[35,36] It was therefore of interest to examine the role of garlic in photoimmunosuppression, particularly as it relates to the induction of skin carcinogenesis by UV radiation. Lyophilized aged garlic extract was selected for its consistency and stability of composition, and its acceptability as a dietary ingredient for mice. This study describes the effect of dietary aged garlic extract on UVB radiation-induced erythema, and on the suppression of CHS by either UVB radiation or by *cis* urocanic acid in the hairless mouse.

MATERIALS AND METHODS

Mice

Female inbred Skh:HR-1 mice from the Veterinary Pathology breeding colony were used when 10-14 weeks old. The mice were housed in wire-topped plastic boxes on vermiculite bedding (Boral Ltd., Camellia, NSW), under gold lighting supplied by GEC F40GO tubes on 12 h on/off cycle, at an ambient temperature of 25°C. Normal mouse stock pellets were supplied (LabFeed Northbridge, NSW) until the experimental diets were fed, and tap water ad libitum.

Semi-Purified Diets

The diets were prepared from the base ingredients at two-weekly intervals and stored at 2°C, as previously described.[37] The carbohydrate component was provided by sucrose (65% by weight), the protein by isolated soy protein (20% by weight; Supro 500E Protein Technologies International, St. Louis, MO, USA) and the fat by sunflower oil without added antioxidant (5% by weight; kindly donated by Mr. R. Berry, Vegetable Oils Ltd., Marrickville, NSW), which is stored at 2°C in the dark under nitrogen gas. Ground hay provided the fiber, and vitamins and minerals were added according to the published nutritional requirements of the mouse.[38] The diets were fed isocalorically in daily-weighed amounts providing 57.74 Kj per mouse, to maintain constant body weight.

Lyophilized aged garlic extract was kindly donated by Dr. H. Sumiyoshi, Wakunaga Pharmaceutical Co., Mission Viejo, CA USA as a single batch, which was stored in the dark at 2°C, and was incorporated into the powdered diets at the expense of the equivalent weight of sucrose.

UVB Irradiation

This was provided by a single unfiltered Oliphant FL40SE fluorescent tube, providing 2.6×10^{-4} W/cm^2 UVB radiation at the mouse dorsum, measured between 250-315 nm using an International Light IL 1700 radiometer, which had been calibrated to the emission spectrum of the UVB light source. Mice were exposed unrestrained with the wire cage tops removed, to a minimal erythemal dose previously determined to be 0.118 J/cm^2, delivered in 7.5 min exposure of the dorsum at a distance of 19 cm from the UVB tube, on each of three consecutive days. The treatment resulted in a moderate erythema without desquamation. The degree of erythema was quantitated by measuring the oedema component of this reaction as the double skinfold thickness at the mid-dorsum, using a spring micrometer (Mercer, St. Albans, UK) at 48 h after the first UVB exposure, at which time the erythema reaction is maximal in these mice. Statistical significance of the differences in skinfold thickness was assessed by Student's t test.

Cis Urocanic Acid-Containing Lotions

A solution (4% w/v) of trans urocanic acid (Sigma Chemical Co., St. Louis, MO, USA) in dimethyl sulfoxide was exposed to UVB radiation to cause photoisomerization to the photostationary state, as previously described.[15,39] The irradiated solution (UV-UCA) contained approximately 52% cis, 48% trans isomers, determined by thin-layer chromatographic resolution[40], and was diluted with phosphate-buffered normal saline (0.15M sodium chloride buffered at pH 7.3 with 10 mM potassium phosphate) and an innocuous cosmetic base lotion (an oil-in-water emulsion stabilized with isostearic acid kindly donated by Dr. M. Nearn, Colgate-Palmolive, Pty. Ltd., Villawood NSW) to finally contain 5% (v/v) dimethyl sulfoxide and photoisomerized urocanic acid at 0, 0.025%, 0.05%, 0.1% or 0.2% (w/v).

Induction of Contact Hypersensitivity

Groups of 10 mice were pre-fed the experimental diets for two weeks. Five of the mice in each group were then exposed to UVB radiation on three consecutive days (Days 1, 2, 3), the remaining mice constituting the non-UVB controls. On Days 8 and 9, the mice were sensitized by the application of either 0.05 ml 0.3% (v/v) 2,4-dinitrofluorobenzene (DNFB; Sigma Chemical Co., St. Louis, MO, USA) freshly prepared in acetone, or 0.1 ml 3% (w/v) oxazolone (Sigma Chemical Co., St. Louis, MO, USA) freshly prepared in ethanol, to the abdominal skin. On Day 15, the pre-challenge ear thickness was measured using a spring micrometer (Mercer, St. Albans, UK), and the mice were challenged by the application of 5 μl 0.2% (v/v) DNFB freshly prepared in acetone, or in mice sensitized with oxazolone, 5 μl 3% fresh oxazolone in ethanol, to each surface of both pinnae. Ear thickness was measured again at 16-20 h, and the maximum ear thickness recorded. Net ear swelling was calculated as the difference between the average pre-challenge and post-challenge ear thicknesses.

The experimental diets were fed continuously until challenge. When the effect of *cis* urocanic acid was to be assessed, groups of five mice were pre-fed the experimental diets for two weeks. On Days 0, 1 and 2, 0.1 ml of the appropriate base or UV-UCA lotion was applied to the dorsal skin. On Days 1 and 2 three h following the lotion, the mice were sensitized by applying 0.05 ml 0.3% DNFB (fresh) to the abdominal skin, and on Day 8, after measuring the pre-challenge ear thickness, mice were challenged with 5 μl 0.2% DNFB applied to each surface of the pinnae as before. Ear thickness was again measured at 18 h, and net ear swelling calculated. Diets were fed continuously until challenge.

Statistical significance of the net ear swelling between treatment groups was assessed using Student's *t* test.

RESULTS

The mice found all of the prepared diets palatable, and maintained a healthy demeanor and normal body weight comparable with stock pellet-fed mice.

Effect of UVB Irradiation

The moderate erythema resulting from these UVB exposures is demonstrated in Table 17.1 as the average dorsal skinfold thickness before and at 48 h after the first UVB exposure. Unexposed skinfold thickness was 77.0 to 82.0 mm $\times$ 0.01, and a marked thickening of approximately 40-70% was observed at 48 h. Mice fed 4% garlic extract responded with significantly less

TABLE 17.1

MEASUREMENT OF OEDEMA/ERYTHEMA BY THE AVERAGE DOUBLE SKINFOLD THICKNESS IN GROUPS OF 5 MICE FED CONTROL DIET OR GARLIC EXTRACT AT 0.1%, 1% OR 4%, BEFORE AND AT 48 H AFTER THE FIRST UVB EXPOSURE

Diet	Average Skinfold Thickness mm $\times$ 0.01 $\pm$ SD	
	Pre-UVB	Post-UVB
Control	77.0 $\pm$ 3.7	129.4 $\pm$ 7.4
Garlic Extract		
0.1%	77.3 $\pm$ 4.4	126.6 $\pm$ 13.7
1%	81.0 $\pm$ 4.0	126.4 $\pm$ 7.1
4%	82.0 $\pm$ 4.5	116.7 $\pm$ 3.2

(P < 0.05) skinfold thickening (116.7 mm × 0.01) compared to control-fed mice and mice fed 0.1% or 1% garlic extract (126.4 - 129.4 mm × 0.01). Thus at the highest concentration of dietary garlic extract, a mild but significant anti-erythema effect was evident.

When control-fed mice were sensitized to oxazolone, challenge resulted in a strong response seen as an approximate doubling of the ear thickness, resulting in an average net ear swelling of 25.3 ± 1.9 mm × 0.01 (Fig. 17.1). Exposure to UVB radiation suppressed the net response in control-fed mice to

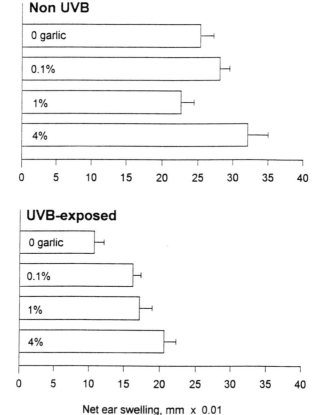

Net ear swelling, mm x 0.01

FIG. 17.1. THE CONTACT HYPERSENSITIVITY RESPONSE TO OXAZOLONE IN GROUPS OF 5 UNEXPOSED (NON UVB) AND 5 UVB-EXPOSED MICE FED DIETS CONTAINING 0, 0.1%, 1% OR 4% AGED GARLIC EXTRACT EXPRESSED AS THE AVERAGE NET EAR SWELLING ± SEM (CHALLENGED EAR THICKNESS MINUS PRE-CHALLENGE EAR THICKNESS)

10.7 $\pm$ 1.4 mm $\times$ 0.01, or 42% of the unexposed control. In UVB-exposed mice fed garlic diets, there was increasing abrogation of the suppression as the garlic extract concentration increased. At 0.1% garlic extract, the response was 64% of control, significantly less severe than in control-fed mice (P < 0.01). With 4% garlic extract, the response, 81% of control, was highly significantly ameliorated (P < 0.001), and was not significantly different from the unirradiated control. Mice fed 1% garlic extract also responded less severely (68% of control), although not significantly differently from mice fed 0.1% garlic extract.

Effect of *cis* Urocanic Acid-Containing Lotions

In control-fed mice the UV-UCA lotions caused a dose-responsive suppression of the contact hypersensitivity response to DNFB. Even at the lowest dose of 25 μg UV-UCA per mouse, the response was significantly (P < 0.01) reduced from 19.3 $\pm$ 0.7 to 15.2 $\pm$ 1.0 mm $\times$ 0.01, or to 78% of control, and at the highest dose of 200 μg, the response was reduced to 10.4 $\pm$ 0.9 mm $\times$ 0.01, or to 54% of control (Fig. 17.2).

Mice fed diets containing garlic extract were resistant to this immuno-suppressive action. At 1% garlic extract, 25 μg UV-UCA was not significantly suppressive (92% of control); 50 and 100 μg UV-UCA appeared to cause some suppression (77% of control), less than in control-fed mice, however this was not statistically significant. Only at 200 μg UV-UCA was significant suppression evident, to 59% of control. In mice fed 4% garlic extract, UV-UCA failed to suppress contact hypersensitivity, at all concentrations tested. The apparent potentiation to 121% and 127% of the control response in these mice treated with 25 μg or 50 μg UV-UCA, respectively was not statistically significant. It was evident, however, that dietary garlic extract had counteracted the immuno-suppressive action of UV-UCA.

DISCUSSION

These experiments have shown that aged garlic extract administered to mice as a dietary additive, offers protection from some of the damaging effects of UVB radiation, in a dose-responsive way. Feeding aged garlic extract at the highest concentration, 4% of the diet, caused a reduced erythema/oedema reaction following exposure to UVB radiation, consistent with an anti-inflammatory action. Prooxidant states contribute to UV radiation-induced erythema, and it is of interest that glutathione and other sulfhydryl compounds have been shown to protect from acute UVR damage.[22,23] A very recent study describes the UV dose-dependent destruction of glutathione, glutathione peroxidase and glutathione reductase[41] in the skin. The sulfur-rich chemistry of garlic may provide analogous protection from the erythema reaction. It has been shown that garlic constituents preserve the reduced state of glutathione and activate those enzymes involved in maintaining the antioxidant protectivity.[25,42]

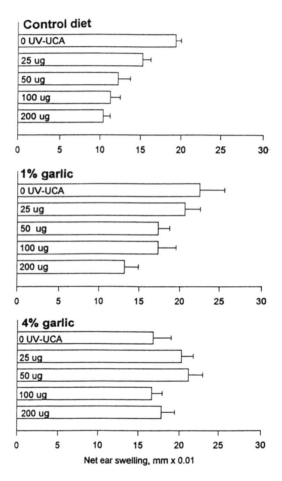

FIG. 17.2. SUPPRESSION OF THE CONTACT HYPERSENSITIVITY RESPONSE TO DNFB BY TOPICALLY APPLIED *CIS* UROCANIC ACID (UV-UCA)-CONTAINING LOTIONS OF INCREASING CONCENTRATION IN GROUPS OF 5 MICE, AND THE EFFECT OF FEEDING DIETS CONTAINING 0 (CONTROL DIET), 1% OR 4% AGED GARLIC EXTRACT ON THIS RESPONSE, EXPRESSED AS THE AVERAGE NET EAR SWELLING ± SEM

Prooxidant mechanisms are not known to be involved in the suppression of immunity by *cis* urocanic acid, nor does the topical application of UV-UCA lotions providing up to 200 μg urocanic acid appear to cause cutaneous inflammation. It was therefore unexpected to find marked and dose-responsive protection from the suppressive effect of *cis* urocanic acid by aged garlic extract. Not only was garlic extract at 1% of the diet able to reduce the greater suppression evident with greater UCA doses, but the increase to 4% in the dietary garlic concentration totally prevented suppression by *cis* urocanic acid. This suggests that increasing concentrations of garlic extract antagonize *cis* urocanic acid by increasingly blocking some receptor or reaction involving *cis* urocanic acid, as yet unknown.

In view of the current hypothesis that one mode of action of *cis* urocanic acid is as a histamine antagonist[12,13], it would be interesting to examine the sensitivity to garlic extract of the histamine receptors on some subpopulations of T lymphocytes.[43] Furthermore, there is evidence that prostaglandin E_2 is involved in the suppression of contact hypersensitivity in mice by UV radiation[16], but this has not been demonstrated for *cis* urocanic acid. Since garlic extracts have been shown to modify prostanoid synthesis[29-31], the photoimmunoprotection by aged garlic extract may result from a reduction of prostaglandin E_2 synthesis. As protection by garlic has also been observed here from *cis* urocanic acid-induced immunosuppression, it would be consistent if the action of *cis* urocanic acid were also prostanoid-dependent.

In a study of urocanic acid analogues, Norval *et al.*[44] have shown that the essential molecular structure of *cis* urocanic acid required for the capacity to suppress delayed type hypersensitivity in haired mice is the 5-membered heterocyclic ring. It may be significant that recent analytical studies of garlic and garlic oil[45-47] have revealed, in addition to the array of aliphatic sulfide and polysulfide compounds, the presence of thiophenes, which bear a 5-membered heterocyclic ring containing one or two sulfur atoms instead of the nitrogen atoms found in the imidazole ring of histamine, histidine and urocanic acid. Norval *et al.*[44] have demonstrated that immunosuppressive activity is retained by one such analogue, 2-thiopheneacrylic acid, but not by 3-thiopheneacrylic acid, compounds in which a sulfur atom occurs in different positions in the 5-membered ring. This suggests that there might be competition on a structural basis from the garlic thiophenes with the reactivity of *cis* urocanic acid. It would be revealing to identify the garlic constituents possessing the protective functions. In view of the potential link between contact hypersensitivity and immune surveillance, and the relationship of immune surveillance to carcinogenesis, it would not be surprising that this garlic extract might ameliorate photocarcinogenesis also.

ACKNOWLEDGMENTS

This study was supported by International Health Promotions Pty. Ltd., St. Leonards NSW, Australia, and by the University of Sydney Cancer Research Fund. The authors are grateful to Dr. H. Sumiyoshi, Wakunaga Pharmaceutical Co., for helpful discussions. We thank Ms. L. Blyth for breeding and maintaining the experimental mice.

REFERENCES

[1]   Black, H.S., Lenger, W.A. Gerguis, J. and Thornby, J.I. Relation of antioxidants and level of dietary lipid to epidermal lipid peroxidation and ultraviolet carcinogenesis. Cancer Res. 1985; 45:6254–6259.

[2]   Orengo, I.F., Black, H.S., Kettler, A.H. and Wolf, J.E. Influence of dietary menhaden oil upon carcinogenesis and various cutaneous responses to UV radiation. Photochem. Photobiol. 1989; 49:71–77.

[3]   Fisher, M.S. and Kripke, M.L. Systemic alteration induced in mice by UV light irradiation and its relationship to UV carcinogenesis. Proc. Natl. Acad. Sci. USA 1977; 74:1699.

[4]   Fisher, M.S. and Kripke, M.L. Suppressor T lymphocytes control the development of primary skin cancers in UV-irradiated mice. Science 1982; 216:1133–1134.

[5]   Giannini, M.S.H. Suppression of pathogenesis in cutaneous Leishmaniasis by UV irradiation. Infect. Immun. 1986; 51:838–843.

[6]   Denkins, Y., Fidler, I.J. and Kripke, M.L. Exposure of mice to UVB radiation suppresses delayed type hypersensitivity to *Candida albicans*. Photochem. Photobiol. 1989; 49:615–620.

[7]   Howie, S.E.M., Norval, M. and Maingay, J. Exposure to low dose ultraviolet radiation suppresses delayed type hypersensitivity to *Herpes simplex* in mice. J. Invest. Dermatol. 1986; 86:125–128.

[8]   Ullrich, S.E. Suppression of the immune response to allogeneic histocompatibility antigens by a single exposure to ultraviolet radiation. Transplantation 1986; 42:287–291.

[9]   Noonan, F.P., De Fabo, E.C. and Kripke, M.L. Suppression of contact hypersensitivity by UV radiation and its relationship to UV-induced suppression of tumour immunity. Photochem. Photobiol. 1981; 34:683–690.

[10] Noonan, F.P., De Fabo, E.C. and Morrison, H. *cis*-urocanic acid, a product formed by ultraviolet B irradiation of the skin initiates an antigen-presenting cell defect *in vivo*. J. Invest. Dermatol. 1988; 90:92–99.

11 Norval, M., Simpson, T.J. and Ross, J.A. Urocanic acid and immunosuppression (review article). Photochem. Photobiol. 1989; 50:267-275.

12 Norval, M., Gilmour, J.W. and Simpson, T.J. The effect of histamine
receptor antagonists on immunosuppression induced by the *cis* isomer of
urocanic acid. Photodermatol. Photoimmunol. Photomed. 1990; 7:243-248.

13 Matheson, M.J. and Reeve, V.E. Effect of the antihistamine cimetidine on
UV radiation induced tumorigenesis. Immunol. Cell. Biol. 1990;68
(Suppl.):22 (Abstract).

14 Gilmour, J.W., Norval, M., Simpson, T.J., Neuvonen, K. and Pasanen,
P. The role of histamine-like receptors in immunosuppression of delayed
hypersensitivity induced by *cis*-urocanic acid. Photodermatol. Photoimmunol. Photomed. 1992/3; 9:250-254.

15 Reeve, V.E., Bosnic, M. and Rozinova, E. Carnosine (β-alanylhistidine)
protects from the suppression of contact hypersensitivity by ultraviolet B
(280-320 nm) radiation or by *cis* urocanic acid. Immunology 1993; 78:99-
104.

16 Chung, H.-T., Bumham, D.K., Robertson, B., Roberts, L.K. and Daynes,
R.A. Involvement of prostaglandins in the immune alterations caused by the
exposure of mice to ultraviolet radiation. J. Immunol. 1986; 137:2478-
2484.

17 Black, A.K., Greaves, M.W., Hensby, C.N., Plummer, N.A. and Warin,
A.P. The effects of indomethacin on arachidonic acid and prostaglandins E_2
and $F_{2\alpha}$ in human skin 24 h after UVB and UVC irradiation. Br. J. Clin.
Pharmacol. 1978; 6:216-266.

18 Farr, P.M. and Diffey, B.L. A quantitative study of the effect of topical
indomethacin on cutaneous erythema induced by UVB and UVC radiation.
Br. J. Dermatol. 1986; 115:453-454.

19 Fuchs, J., Huflejt, M.E., Rothfuss, L.M., Wilson, D.S., Carcamo, G. and
Packer, L. Impairment of enzymic and nonenzymic antioxidants in skin by
UVB irradiation. J. Invest. Dermatol. 1989; 93:769-773.

20 Tyrrell, R.M. and Pidooux, M. Correlation between endogenous glutathione content and sensitivity of cultured human skin cells to radiation at
defined wavelengths in the solar UV range. Photochem. Photobiol. 1988;
47:405-412.

21 Punnonen, K., Puntala, A. and Ahotupa, M. Effects of ultraviolet A and
B irradiation on lipid peroxidation and activity of the antioxidant enzymes
in keratinocytes in culture. Photodermatol. Photoimmunol. Photomed.
1991; 8:3-6.

22 Chan, J.T. and Black, H.S. Antioxidant-mediated reversal of ultraviolet
light toxicity. J. Invest. Dermatol. 1977; 68:366-368.

23 Tyrrell, R.M. and Pidoux, M. Endogenous glutathione protects human skin
fibroblasts against the cytotoxic action of UVB, UVA and near-visible

radiation. Photochem. Photobiol. 1986; 44:561–564.

[24] Bissett, D.L., Chatterjee, R. and Hannon, D.P. Photoprotective effect of superoxidescavenging antioxidants against ultraviolet radiation-induced chronic skin damage in the hairless mouse. Photodermatol. Photoimmunol. Photomed. 1990; 7:56–62.

[25] Perchellet, J.P., Perchellet, E.M., Abney, N.L., Zirnstein, J.A. and Belman, S. Effects of garlic and onion oils on glutathione peroxidase activity, the ratio of reduced oxidised glutathione and ornithine decarboxylase induction in isolated mouse epidermal cells treated with tumor promoters. Cancer Biochem. Biophys. 1986; 8:299–312.

[26] Unnikrishnan, M.C., Soudamini, K.K. and Kuttan, R. Chemoprotection of garlic extract towards cyclophosphamide toxicity in mice. Nutr. Cancer 1990; 13:201–207.

[27] Nishino, H., Iwashima, A., Itakura, Y., Matsuura, H. and Fuwa, T. Antitumor-promoting activity of garlic extracts. Oncology 1989; 46:277–280.

[28] Athar, M., Rarn, H., Bickers, D.R. and Mukhtar, H. Inhibition of benzoyl peroxide mediated tumor promotion in 7,12-dimethylbenz(a)anthracene-initiated skin of Sencar mice by antioxidants nordihydroguaiaretic acid and diallyl sulfide. J. Invest. Dermatol. 1990; 94:162–165.

[29] Belman, S., Solomon, J., Segal, A., Block, E. and Barany, G. Inhibition of soybean lipoxygenase and mouse skin tumor promotion by onion and garlic components. J. Biochem. Toxicol. 1989; 4:151–160.

[30] Block, E. Lipoxygenase inhibitors from the essential oil of garlic. Markovnikov addition of the allyldithio radical to olefins. J. Am. Chem. Soc. 1988; 110:7813–7828.

[31] Shalinsky, D.R., McNamara, D.B. and Agrawal, K.C. Inhibition of GSH-dependent PGH-2 isomerase in mammary adenocarcinoma cells by allicin. Prostaglandins 1089; 37:135–148.

[32] Lau, B.H.S., Yamasaki, T. and Gridley, D.S. Garlic compounds modulate macrophage and T-lymphocyte functions. Mol. Biother. 1991; 3:103–107.

[33] Hirao, Y., Sumioka, I., Nakagami, S., Yamamoto, M., Hatono, S., Yoshida, S., Fuwa, T. and Nakagawa S. Activation of immunoresponder cells by the protein fraction from aged garlic extract. Phytotherapy Res. 1987; 1:161–164.

[34] Kandil, O.M., Abdullah, T.H. and Elkadi, A. Garlic and the immune system in humans: its effect on natural killer cells. Fed. Proc. 1987; 46:441.

[35] Lau, B.H.S., Woolley, J.L., Marsh, C.L., Barker, G.R., Koobs, D.H. and Torrey, R.R. Superiority of intralesional immunotherapy with *Corynebacterium parvum* and *Allium sativum* in control of murine transitional cell carcinoma. J. Urol. 1986; 136:701–705.

[36] Marsh, C.L., Torrey, R.R., Woolley, J.L., Barker, G.R. and Lau, B.H.S. Superiority of intravesical immunotherapy with *Corynebacterium parvum* and *Allium sativum* in control of murine bladder cancer. J. Urol. 1987; 137:359-362.

[37] Reeve, V.E., Matheson, M., Greenoak, G.E., Canfield, P.J., Boehm-Wilcox, C. and Gallagher, C.H. Effect of dietary lipid on UV light carcinogenesis in the hairless mouse. Photochem. Photobiol. 1988; 48:689-696.

[38] American Institute of Nutrition. Report of the American Institute of Nutrition ad hoc committee on standards for nutritional studies. J. Nutr. 1977; 107:1340-1348.

[39] Reeve, V.E., Greenoak, G.E., Canfield, P.J., Boehm-Wilcox, C. and Gallagher, C.H. Topical urocanic acid enhances UV-induced tumour yield and malignancy in the hairless mouse. Photochem. Photobiol. 1989; 49:459-464.

[40] Hais, L.M. and Strych, A. Increase in urocanic acid concentration in human epidermis following insolation. Collection Czechoslov. Chem. Commun. 1969; 34:649-655.

[41] Shindo, Y., Witt, E., Han, D. and Packer, L. Dose-response effects of acute ultraviolet irradiation on antioxidants and molecular markers of oxidation in murine epidermis and dermis. J. Invest. Dermatol. 1994; 102: 470-475.

[42] Sparnins, V.L., Barany, G. and Wattenberg, L.W. Effects of organosulfur compounds from garlic and onions on benzo(a)pyrene-induced neoplasia and glutathione S-transferase activity in the mouse. Carcinogenesis 1988; 9:131-134.

[43] Badger, A.M., Griswold, D.E., Poste, G. and Hanna, N. *In* Histamine and H2 Antagonists in Inflammation and Immunodeficiency, (Rocklin, R.E., ed.) pp. 3-22. Marcel Dekker, New York, 1990:3-22.

[44] Norval, M., Simpson, T.J., Bardishiri, E. and Howie, S.E.M. Urocanic acid analogues and the suppression of the delayed type hypersensitivity response to *Herpes simplex* virus. Photochem. Photobiol. 1988;49:633-639.

[45] Block, E. The organic chemistry of garlic sulfur compounds. First World Congress on the Health Significance of Garlic and Garlic Constituents, Washington, DC, 1990:12.

[46] Barany, G. and Baloga, D.W. The bio-organic chemistry of garlic sulfur compounds. Ibid:13.

[47] Block, E. Chemistry of garlic and onions. Sci. Am. 1985;252:114-119.

CHAPTER 18

GARLIC AND PREVENTION OF PROSTATE CANCER

JOHN T. PINTO and RICHARD S. RIVLIN

Clinical Nutrition Research Unit & Nutrition Research Laboratory
Department of Medicine
Memorial Sloan-Kettering Cancer Center &
New York Hospital — Cornell Medical Center
New York, New York 10021

INTRODUCTION

Prostate cancer ranks first among the most frequently diagnosed cancers and second among the most common causes of cancer-related deaths in American and European males.[1] To some extent the prominence of prostate cancer may reflect advances in early diagnosis.[2] Despite poor understanding of its etiology, epidemiological studies have characterized a number of dietary factors that correlate positively with the promotion and progression of the disease.[3,4] Chief among these is total calorie intake, particularly that derived from fat of animal origin. In addition, over-consumption of red meat correlates with increased prevalence of prostate cancer.[3,4] In a recently developed animal model using explants of human prostate carcinoma cells, dietary fat intake was shown clearly to affect the rate of prostate tumor growth.[5] In addition to influencing the overall prevalence of prostate tumors, total caloric intake from animal fat and red meat appear to be primary factors that contribute to altering their biological characteristics, namely the degree of invasiveness and spread. Some authors have suggested that linoleic acid is the critical fatty acid in prostate cancer pathogenesis.[6]

A number of studies suggest that dietary factors may be particularly useful in controlling prostate malignancy. Prostate cancer has a feature in which it differs from other malignant transformations, namely that the earliest lesions, known as prostatic intraepithelial neoplasia (PIN), exhibit generally similar prevalence in men throughout the world.[7,8] By contrast, very wide geographic variations in deaths from prostate cancer occur. Among countries, such as Thailand, that consume very low amounts of fat, there is very little prostate cancer, whereas nations like Denmark and The Netherlands that consume large amounts of fat, display an alarmingly high prevalence of prostate cancer.[3,4]

This relationship suggests that dietary interventions earlier in life should be goals in order to attempt to suppress the evolution or progression of the initial lesions. The challenge to investigators should be to identify whether there are other specific dietary factors that may influence prostate cancer spread and aggressiveness and whether these agents can be manipulated to reduce tumor progression. We need to determine with certainty the relative roles of specific dietary constituents in human cancer causation and prevention. Some components may have to be reduced as in the case of specific fat calories, while intake of diet-derived chemoprotective agents present in vegetables (e.g., polyphenols, fibers, carbinols, and thioallyls) may have to be increased.

Relevance of Garlic Derivatives to Cell Proliferation, Carcinogenesis and Testosterone Metabolism

A number of epidemiological investigations to date have shown that risk of several kinds of cancers is inversely related to intake of garlic. In Linqu County, Shangdong Province, an area of northeastern China, risk of stomach cancer declines as consumption of garlic and garlic products increases.[9,10] In a multicenter investigation in Italy, risk of stomach cancer also declines significantly with increase in garlic intake.[11] These findings underscore the need for further evaluation of the potential anticancer effects of garlic components in susceptible populations. Investigators should examine whether these dietary constituents can be used to control growth of malignant cells, particularly prostate cancer cells.

In addition to the results of these population analyses, garlic extracts have been shown to display antitumor potential in several model systems utilizing carcinogens.[12] The biochemical mechanisms that underlie the antimutagenic, antitumorigenic, and antiproliferative effects of garlic are not precisely known. Studies have demonstrated that allylsulfides derived from garlic can modify specifically the activities of mixed function oxidases and thus inhibit chemically-induced tumors in a wide variety of tissues.[13-15] Diallylsulfide (DAS), a constituent of garlic, suppresses oxidative demethylation by competitively inhibiting cytochrome P450 2E1, an enzyme that activates carcinogens, such as nitrosamines, hydrazines, and carbon tetrachloride.[16] That the effect of DAS on mixed function oxidases may be relatively specific rather than general is supported by the observation that the activity of another demethylating cytochrome, namely P450 2B1, is actually elevated by garlic components.[17]

Relevant to alterations of select methylation processes, activated protooncogenes of the "ras" gene family have been detected in numerous spontaneous and chemically-induced rodent tumors and in human neoplasms.[18] A point mutation in the H-ras oncogene appears to involve formation of the O^6-methyl adduct to guanosine. The promutagenic O^6methylguanosine has been detected in DNA of certain cancers and in oncogenes of ras-transfected cells. Activation of this oncogene has been observed to coincide with early progression of neoplastic transformation. Dietary supplementation with garlic derivatives reduces by about 40% the occurrence of O^6-methylguanosine, as well as N^7-

methyldeoxyguanosine in liver DNA of rats treated with N-nitrosodimethyl-amine.[19]

Furthermore, specific garlic-derived compounds depress covalent binding of xenobiotic compounds to DNA and thus decrease binding of carcinogens, such as, 7,12-dimethylbenz[a]anthracene (DMBA) and aflatoxin B1 metabolites, to either deoxyguanosine or deoxyadenosine.[20] The level of covalent adducts in cells correlates with the extent of mutagenesis and tumorigenesis and binding of compounds, such as aflatoxin B1 to DNA, is considered to be a measure of initiation of hepatocellular carcinoma.[20] Flow cytometric analysis of tumor cells indicates that garlic derivatives prevent progression of cells from G1 to S phase and thus cause cells to accumulate in G1 phase.[21] These findings strongly suggest that garlic derivatives may have promise in delaying the progression of cancer cells through their cell cycle. Investigations on the potential action of garlic in prevention of prostate cancer at its later stages have been largely neglected even though all indications point to a high degree of promise.

The efficacy of garlic compounds as anti-initiators and antipromoters of chemical carcinogenesis have been previously examined.[22-24] Diallylsulfide (DAS), diallyltrisulfide (DAT), allylmethyldisulfide (AMD), and allylmethyltri-sulfide (AMT) each inhibit gastric malignancy induced by benzo[a]pyrene.[25,26] DAS[22,27] and S-allylcysteine (SAC)[20] blunt adduct formation of 1,2-dimethylhy-drazine in liver and colon. Furthermore, DAS, diallyldisulfide (DADS), AMT and SAC each reduce forestomach tumors[28], and inhibit the development of papilloma and squamous cell carcinoma in rodent esophageal tissue[15] induced by N-nitroso compounds.

In view of the many organ-specific carcinogen models, no adequate chemical model has been developed in which the efficacy of garlic compounds on prostate cancer initiation, promotion or progression has been systematically addressed. Thus, garlic-derived compounds depress covalent binding of xenobiotic compounds to DNA and specifically decrease binding of carcinogens to either deoxyguanosine or deoxyadenosine.[29] This finding is highly relevant inasmuch as an increase in covalent adducts correlates with increased mutagene-sis and tumorigenesis.[30]

In a fashion similar to that of environmental carcinogens, the active testosterone metabolite, dihydrotestosterone (DHT), can enhance proliferation of prostate cancer.[31] Since testosterone metabolism occurs through the mixed function oxidase cytochrome P450 system[32,33], it is essential to examine how garlic derivatives may affect induction of selective P450 enzymes as measured by changes in the rates of metabolism of testosterone and dihydrotestosterone. The metabolism of testosterone is catalyzed by inducible P450 1A1, 1A2, 3A3, 3A4, and 3A5 enzymes.[33] These mixed function oxidases can be modulated by a wide variety of environmental factors and dietary constituents.[32-35] The balance of testosterone metabolites generated through the action of these P450 enzymes may be a critical mechanism, whereby dietary constituents, such as those derived from garlic, may exert specific actions in modulating the development of prostate cancer.

Although abundant experimental evidence exists linking increased serum testosterone levels with enhanced development of prostate cancer in humans[31,36], definitive information on relating such changes to prostate cancer risk in a given individual still needs to be obtained. In fact, under certain conditions, complete testosterone ablation can cause cancer cells to adapt over time and grow independently of hormone stimulation.[37] The progressive changes that occur following ablation therapy involve a transient up-regulation of testosterone receptors that causes these cells to become overly sensitive to testosterone upon re-treatment and to induce apoptosis. Future progress in treatment of advanced prostate cancer may involve repeated cycling of testosterone therapies rather than complete removal of the hormone as is performed in therapies today.[37] Given the potential for alteration of testosterone metabolism through modification of P450 enzymes, little is known about the potential for modifying testosterone risk or its treatment modality by administration of garlic constituents.

Recent reports demonstrate that allylsulfide derivatives inhibit growth of transplantable tumors and have anti-promotional activity on a number of different tumor cell lines, such as canine breast[38], human melanoma[39] and human neuroblastoma[40]. Treatment of human melanoma cells with SAC reduces expression of cell-surface gangliosides, tumor-associated markers of differentiation and transformation.[39] S180 tumor cells exposed to a garlic extract exhibit delayed progression to S phase and accumulation in G1 phase.[21] In skin exposed to DMBA and phorbol ester, pretreatment with an ethanolic garlic extract suppresses phosphorylation of phospholipids, an early event induced by tumor promoters.[41] In similar fashion, reduction in phosphorylation of GTP-activating protein and tyrosine phosphatase-1β correlates with the antiproliferative activity of garlic constituents on normal smooth muscle and endothelial cells.[21]

The "ras" family of G-protein oncogenes is associated with malignant transformation of cells subsequent to its translocation to the plasma membrane. A post-translational farnesylation of the mutated p21^{ras} oncoprotein has been shown to be a critical step in this process. Athymic mice implanted with H-ras oncogene-transformed 3T3 cells and treated orally with DADS exhibited marked inhibition of the tumor growth.[42] These studies demonstrate that DADS diminished membrane association of p21^{ras} and produced a concomitant increase in the non-transforming cytosolic fraction of this oncoprotein. Measurements of the activity of HMG CoA reductase, the regulatory enzyme necessary for elevating the farnesyl pyrophosphate pool, was reduced approximately 80% in both liver and tumor compared to that in non DADS-treated tumor-bearing mice. These and other studies suggest that garlic derivatives not only act as chemopreventive agents during the initiation phase of cancer, but may impart retardation of the promotional phase of prostate cancer.

Examples of recent findings from our laboratory on effects of garlic on inhibiting hormone-sensitive and -insensitive human breast cancer cells in culture are shown in Fig. 18.1 and 18.2 and Table 18.1. These studies were performed in human estrogen-receptor positive MCF-7 cells grown in culture that respond to estrogen stimulation. With increasing concentrations of Aged Garlic Extract (AGE), a proprietary preparation of garlic constituents containing SAC and SAMC, as well as concentrations of SAC and SAMC alone, there was progressive decrease in the rate of growth of human breast cancer cells following a seven day exposure particularly with SAMC (Fig. 18.1). Similar findings were also observed in MCF-7 ras-transfected cells that have lost their sensitivity to estrogen stimulation (Fig. 18.2). By contrast to estrogen-sensitive cells (Fig. 18.1), estrogen receptor negative cells, MCF-7^{ras}, appear to exhibit

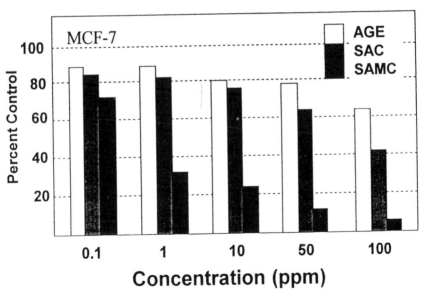

FIG. 18.1. EFFECT OF AGED GARLIC EXTRACT, S-ALLYLCYSTEINE, AND S-ALLYLMERCAPTOCYSTEINE ON GROWTH OF ESTROGEN RESPONSIVE HUMAN MAMMARY CARCINOMA CELLS

Cell number of MCF-7 after seven day exposure in culture. Values shown are mean of three independent experiments. Variation of the experimental results did not exceed 10% (from reference 44).

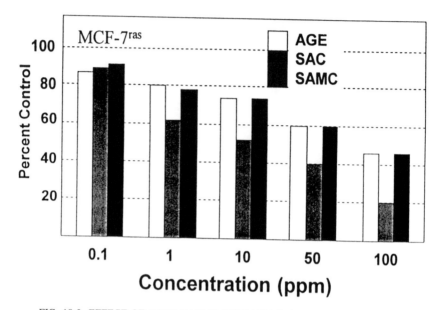

FIG. 18.2. EFFECT OF AGED GARLIC EXTRACT, S-ALLYLCYSTEINE, AND
S-ALLYLMERCAPTOCYSTEINE ON GROWTH OF ESTROGEN UNRESPONSIVE
HUMAN MAMMARY CARCINOMA CELLS

Cell number of MCF-7[ras] after seven day exposure in culture. Values shown are means of three
independent experiments. Variation of results did not exceed 10% (from reference 44).

TABLE 18.1
EFFECT OF GARLIC CONSTITUENTS ON ANCHORAGE-INDEPENDENT GROWTH
OF HUMAN BREAST CARCINOMA CELLS

	Number of Colonies per 1000 Cells	
Treatment	MCF-7	MCF-7[ras]
Control	162 ± 18	725 ± 32
AGE (10 ppm)	125 ± 20	628 ± 20
SAC (10 ppm)	92 ± 15	250 ± 32
SAMC(10 ppm)	102 ± 15	185 ± 20

(from reference 44)

a more pronounced effect and greater sensitivity to SAC. Both mammary cell lines are capable of growing in soft agar but their colony forming efficiencies differ, ~ 16% vs ~ 73% for MCF-7 and MCF-7 ras-transfected, respectively. Table 18.1 illustrates that colony forming ability or anchorage independent growth of the more aggressive ras-transfected cell is inhibited to the greatest degree by both SAC and SAMC.

These findings emphasize the usefulness of cell culture techniques, as well as of human cells implanted into nude mouse models to explore effects of dietary constituents on tumor growth. This technology is readily suited for performing similar studies of human prostate cancer cells and their response to garlic derivatives. Thus, garlic components are known to exert anti-proliferative effects on a number of normal and malignant cells and may promote differentiation to less malignant phenotypes, but little is known of these possible effects upon human prostate tissue.

As mentioned earlier, recent investigations at our institution have shown that growth of implanted LNCaP cells correlates positively with an increase in intake of dietary fat.[5] Whether co-administration of garlic derivatives can reduce or arrest the rate of LNCaP cell proliferation in this explant model is currently under investigation by our laboratory.

As inhibitors of the carcinogenic process, compounds derived from garlic can be grouped into one or more of three possible functional categories[43]:

(A)   Compounds that impede the generation of a carcinogen from its precursor;

(B)   Compounds that prevent a carcinogen from reacting with vulnerable cellular targets; and,

(C)   Compounds that delay or reverse expression of malignancy or that demonstrate antiproliferative activity on tumor cells.

SUMMARY

The results of investigations in human populations, animal models, cells in culture, and cell-free systems, all strongly suggest that garlic derivatives may hold considerable promise for primary prevention and inhibition of promotion of human prostate cancer. Because of the basic biology of prostate cancer, early intervention before tumor development is evident may be particularly effective in subsequently delaying or reducing the rate of spread of disease. Further research efforts should be intensified to investigate these hypotheses and to advance new dietary manipulations that alter metabolic responses to hormones or prevent oxidative stress.

ACKNOWLEDGMENTS

This work was supported in part by CA 39203 and the Clinical Nutrition Research Unit Grant CA 29502 from National Institutes of Health. Partial funding was also provided by grants from Nutrition International, Wakunaga of America Company, Ltd., The Frank J. Scallon Medical Science Foundation, The Stella and Charles Guttman Foundation, the Sunny and Abe Rosenberg Foundation, and the Rosenfeld Heart Foundation. The research was performed in coordination with the Nutrition Research Laboratory and George M. O'Brien Urology Research Center, Sloan-Kettering Institute for Cancer Research. A part of this study was presented at the Designer Foods Conference III held at Georgetown University Conference Center, May 24, 1994.

REFERENCES

[1] Wingo, P.A., Tong, T. and Bolden, S. Cancer Statistics. CA Cancer J. Clin. 1995; 45:8–30.

[2] Jacobsen, S.J., Katusic, S.K., Bergstralh, E.J., Oesterling, J.E., Ohrt, D., Klee, G.G., Chute, C.G. and Lieber, M.M. Incidence of prostate cancer diagnosis in the eras before and after serum prostate-specific antigen testing. JAMA 1995; 274:1445–1449.

[3] Berg, J.W. Can nutrition explain the pattern of international epidemiology of hormone-dependent cancers? Cancer Res. 1995; 35:3345–3350.

[4] Armstrong, B. and Doll, R. Environmental factors and cancer incidence and mortality in different countries, with special reference to dietary practices. Intern. J. Cancer 1975; 15:617–631.

[5] Wang, Y., Corr, J.G., Thaler, H.T., Tao, Y., Fair, W.R. and Heston, W.D.W. Decreased growth of established human prostate LNCaP tumors in nude mice fed a low-fat diet. J. Natl. Cancer Inst. 1995; 87:1456–1462.

[6] Rose, D.P. and Connolly, J.M. Effects of fatty acids and eicosanoid synthesis inhibitors on the growth of two human prostate cancer cell lines. Prostate 1991; 18:243-245.

[7] Boone, C.W., Kelloff, G.J. and Steele, V.E. Natural history of intraepithelial neoplasia in humans with implications for cancer chemoprevention strategy. Cancer Res. 1992; 52:1651–1659.

[8] Yatani, R., Kusano, I., Shiraishi, T., Hayashi, T. and Stemmermann, G.N. Latent prostatic carcinoma: Pathological and epidemiological aspects. Jpn. J. Clin. Oncol. 1989; 19:319–326.

[9] You, W.-C., Blot, W.J., Chang, Y.-S., Ershow, A., Yang, Z.T., An, Q., Henderson, B.E., Fraumeni, J.F., Jr. and Wang T.-G. Allium vegetables and reduced risk of stomach cancer. J. Natl. Cancer Inst. 1989; 81:162–164.

10 You, W.-C., Blot, W.J., Chang, Y.-S., Erchow, A., Yang, Z.T., An, Q., Henderson, B.E., Xu, G.W., Fraumeni, J.F., Jr. and Wang, T.-G. Diet and high risk of stomach cancer in Shangdong Province. Cancer Res. 1988; 48:3518-3523.

11 Buiatti, E., Palli, D., Decarli, A., Amadori, D., Avellini, C., Bianchi, S., Biserni, R., Cipriani, F., Cocco, P., Giacosa, A., Marubini, E., Puntoni, R., Vindigni, C., Fraumeni, J.F., Jr. and Blot, W. A case-control study of gastric cancer and diet in Italy. Intern. J. Cancer 1989; 44:611-616.

12 Sumiyoshi, H. and Wargovich, M.J. Garlic (Allium sativum): A review of its relationship to cancer. Asia Pacific J. Pharmacol. 1989; 4:133-140.

13 Brady, J.F., Li, D., Ishizaki, H. and Yang, C.S. Effect of diallyl sulfide on rat liver microsomal nitrosamine metabolism and other monooxygenase activities. Cancer Res. 1988; 48:5937-5940.

14 Reicks, M.M. and Crankshaw, D.L. Modulation of rat hepatic cytochrome P450 activity by garlic organosulfur compounds. Nutr. Cancer 1996; 25:241-248.

15 Wargovich, M.J., Woods, C., Eng, V.W.S., Stephens, L.C. and Gray, K. Chemoprevention of N-nitrosomethylbenzylamine-induced esophageal cancer in rats by naturally occurring thioether, diallyl sulfide. Cancer Res. 1988; 48:6872-6875.

16 Brady, J.F., Ishizaki, H., Fukuto, J.M., Lin, M.C., Fadel, A., Gapac, J.M. and Yang, C.S. Inhibition of cytochrome P-450 IIE1 by diallyl sulfide and its metabolites. Chem. Res. Toxicol. 1991; 4:642-647.

17 Pan, J., Hong, J.-Y., Ma, B.-L., Ning, S.M., Paranawithana, S.R. and Yang, C.S. Transcriptional activation of P-450 2B1/2 genes in rat liver by diallyl sulfone, a compound derived from garlic. Arch. Biochem. Biophys. 1993; 302:337-342.

18 Barch, D.H., Jacoby, R.F., Brasitus, T.A., Radosevich, J.A., Carney, W.P. and Iannaccone, P.M. Incidence of Harvey ras oncogene point mutations and their expression in methylbenzylnitrosamine-induced esophageal tumorigenesis. Carcinogenesis 1991; 12:2373-2377.

19 Lin, X.Y., Li, J.-Z. and Milner, J.A. Dietary garlic suppresses DNA adducts caused by N-nitroso compounds. Carcinogenesis 1994; 14:349-352.

20 Tadi, P.P., Teal, R.W. and Lau, B.H. Organosulfur compounds of garlic modulate mutagenesis, metabolism, and DNA binding of aflatoxin B1. Nutr. Cancer 1991; 15:87-95.

21 Lee, E.S., Steiner, M. and Lin, R. Thioallyl compounds: Potent inhibitors of cell proliferation. Biochim. Biophys. Acta 1994; 1221:73-77.

22 Sumiyoshi, H. and Wargovich, M. Chemoprevention of 1,2-dimethylhydrazine-induced colon cancer in mice by naturally occurring organosulfur compounds. Cancer Res. 1990; 50:5084-5087.

[23]   Wargovich, M.J. and Goldberg, M.T. Diallyl sulfide, a naturally occurring thioether that inhibits carcinogen induced nuclear damage to colon epithelial cells in vivo. Mutat. Res. 1985; 143:127–129.

[24]   Hong, J.-Y., Smith, T., Lee, M.-J., Li, W., Ma, B.-L., Ning, S.M., Brady, J.F., Thomas, P.E. and Yang, C.S. Metabolism of carcinogenic nitrosamines by rat nasal mucosa and the effect of diallyl sulfide. Cancer Res. 1991; 51:1509–1514.

[25]   Sparnins, V.L., Barany, G. and Wattenberg, L.W. Effects of organosulfur compounds from garlic and onions on benzo[a]pyrene-induced neoplasia and glutathione S-transferase activity in the mouse. Carcinogenesis (London) 1988; 9:131–134.

[26]   Sparnins, V.L., Mott, A.W., Barany, G. and Wattenberg, L.E. Effects of allylmethyltrisulfide on glutathione S-transferase activity and BP-induced neoplasia in the mouse. Nutr. Cancer 1986; 8:211–215.

[27]   Hayes, M.A., Rushmore, T.H. and Goldberg, M.T. Inhibition of hepatocarcinogenic responses to 1,2-dimethylhydrazine to diallyl sulfide, a component of garlic oil. Carcinogenesis 1987; 8:1155–1157.

[28]   Wattenberg, L.W., Sparnins, V.L. and Barany, G. Inhibition of N-nitrosodiethylamine carcinogenesis in mice by naturally occurring organosulfur compounds and monoterpenes. Cancer Res. 1989; 49:2689–2692.

[29]   Amagase, H. and Milner, J.A. Impact of various sources of garlic and their constituents on 7,12-dimethylbenz[a]anthracene binding to mammary cell DNA. Carcinogenesis 1993; 14:1627–1631.

[30]   Amagase, H., Schaffer, E.M. and Milner, J.A. Dietary components modify the ability of garlic to suppress 7,12 dimethylbenz[a]anthracene-induced mammary DNA adducts. J. Nutr. 1996; 126:817–824.

[31]   Wilding, G. The importance of steroid hormones in prostate cancer. Cancer Surv. 1992; 14:113–130.

[32]   Guengerich, F.P. Human cytochrome P450 enzymes. Life Sci. 1992; 50: 1471–1478.

[33]   Nelson, D.R., Koymans, L., Kamataki, T., Stegeman, J.J., Feyereisen, R., Waxman, D.J., Waterman, M.R., Gotoh, O., Coon, M.J., Estabrook, R.W., Gunsalus, I.C. and Nebert, D.W. P450 superfamily: update on new sequences, gene mapping, accession numbers and nomenclature. Pharmacogenetics 1996; 6:1–42.

[34]   Musey, P.I., Collins, D.C., Bradlow, H.L., Gould, K.G. and Preedy, J.R. Effect of diet on oxidation of 17 β-estradiol in vivo. J. Clin. Endocrinol. Metab. 1987; 65:792–795.

[35]   Relling, M.V., Lin, J.S., Ayers, G.D. and Evans, W.E. Racial and gender differences in N-acetyltransferase, xanthine oxidase, and CYP1A2 activities. Clin. Pharmacol. Ther. 1992; 52:643–658.

[36] Wilson, J.D. The pathogenesis of benign prostatic hyperplasia. Am. J. Med. 1980; 68:745–747.

[37] Umekita, Y., Hiipakka, R.A., Kokontis, J.M. and Liao, S. Human prostate tumor growth in athymic mice: Inhibition by androgens and stimulation by finasteride. Proc. Natl. Acad. Sci., USA 1996; 93:11802–11807.

[38] Sundaram, S.G. and Milner, J.A. Impact of organosulfur compounds in garlic on canine mammary tumor cells in culture. Cancer Lett. 1993; 74:85–90.

[39] Takeyama, H., Hoon, D.S.B., Saxton, R.E., Morton, D.L. and Irie, R.F. Growth inhibition and modulation of cell markers of melanoma by S-allyl cysteine. Oncology 1993; 50:63–69.

[40] Welch, C., Wuarin, L. and Sidell, N. Antiproliferative effect of the garlic compound S-allyl cysteine on human neuroblastoma cells in vitro. Cancer Lett. 1992; 63:211–219.

[41] Nishino, H., Iwashima, A., Itakura, Y., Matsuura, H. and Fuwa, T. Antitumor-promoting activity of garlic extracts. Oncology 1989; 46:277–280.

[42] Singh, S.V., Mohan, R.R., Agarwal, R., Benson, P.J., Hu, X., Rudy, M.A., Xia, H., Katoh, A., Srivastava, S.K., Mukhtar, H., Gupta, V. and Zaren, H.A. Novel anti-carcinogenic activity of an organosulfide from garlic: Inhibition of H-RAS oncogene transformed tumor growth in vivo by diallyl disulfide is associated with inhibition of p21^{H-ras} processing. Biochem. Biophys. Res. Commun. 1996; 225:660–665.

[43] Wattenberg, L.W. Chemoprevention of cancer. Cancer Res. 1985; 45:1–8.

[44] Li, G., Qiao, C.H., Pinto, J., Osborne, M.P. and Tiwari, R.K. Antiproliferative effects of garlic constituents in cultured human breast cancer cells. Oncology Repts. 1995; 2:787–791.

CHAPTER 19

POTENTIAL INTERACTION OF AGED GARLIC EXTRACT WITH THE CENTRAL SEROTONERGIC FUNCTION: BIOCHEMICAL STUDIES

GILLES M. FILLION, MARIE-PAULE FILLION,
FRANCISCO BOLAÑOS-JIMENEZ, HALA SARHAN
and BRIGITTE GRIMALDI

Unite de Pharmacologie Neuro-Immuno-Endocrinienne
Institut Pasteur
28 rue du Dr. Roux
F75015 Paris, France

ABSTRACT

The potential interaction of garlic extract (Kyolic) (Aged Garlic Extract-AGE) with several neurotransmitters system functions was studied using biochemical assays. More precisely, we determined the effect of AGE on the specific binding of various neurotransmitters to their receptors (serotonergic, α and β adrenergic, D_1 and D_2 dopaminergic, muscarinic and benzodiazepine receptors) in rat brain synaptosomal membrane preparations. AGE did not modify significantly the binding of most of the systems studied (and slightly affected that of $5\text{-}HT_{1A}$ and β_1 adrenergic receptor). Moreover, the binding of [^3H] 5-HT to $5\text{-}HT_1$ receptors, likely $5\text{-}HT_{1B/1D}$ receptors, was specifically enhanced in a non-competitive manner.

These results suggest that AGE might act at the CNS by modulating the activity of the serotonergic system via its interaction with $5\text{-}HT_{1B/D}$ receptors.

INTRODUCTION

It is known that aged garlic extract (AGE) possesses various interesting properties, e.g., interaction with vascular functions and immune activity. Potentially, it may also affect brain function (cf. Saito *et al.*, in this book). Therefore, in a series of preliminary experiments, the *in vitro* potential effects of garlic extract on various neurotransmitters receptors were examined.

MATERIALS AND METHODS

We measured the effect of increasing concentrations of AGE on the specific binding of various radiolabelled ligands to their receptors: adrenergic α_1, (^{3}H Prazozin), β_1 adrenergic (^{3}H dihydroalprénolol), Dopamine D_1 (^{3}H SCH 13390), Dopamine D_2 (^{3}H Spiperone), Muscarinic (^{3}H Quinuclidyl benzilate) and benzodiazepine receptors ([^{3}H] flunitrazepam) 5-HT$_{1A}$ (^{3}H-8-OH-DPAT), serotonergic ([^{3}H]5-HT). The binding of these ligands was measured on synaptosomal membranes preparations obtained from rat brain and incubated in the appropriate medium.

RESULTS

The results show that aged garlic extract, at the concentrations tested, did not interact with most of the studied receptors, namely, adrenergic, dopamine D_1 and D_2 muscarinic and benzodiazepine receptors. However, it slightly affected β_1 adrenergic and 5-HT$_{1A}$ receptors, inhibiting part of the specific binding of the corresponding ligand (maximal decrease of 30-50%) at the highest concentration of liquid aged garlic extract. It was interesting to note that the binding of [^{3}H]5-HT in the presence of 8-OH-DPAT and mesulergine, that mainly corresponds to 5-HT$_{1B}$ receptors was significantly enhanced in the presence of the lowest concentrations of AGE tested in these assays. Thus, at concentrations which correspond to circa 10-60 μg/ml of dry extract, the binding of [^{3}H]5-HT to 5-HT$_{1B}$ receptors was increased following a bell-shape curve with a maximal effect of > 250% (obtained at 10-20 μg/ml).

The increase in binding was markedly altered in the presence of 5-carboxyamidotryptamine, strongly suggesting that this increase actually affected the 5-HT$_1$ receptor subtype. The Scatchard plots of the saturation curves for the binding of [^{3}H]5-HT in the presence and in the absence of AGE indicate that AGE interacts in a non-competitive manner with 5-HT$_{1B}$ binding sites. Although these results have to be completed in order to characterize the type of interaction involved in the phenomenon, they indicate that AGE specifically modifies the binding of 5-HT to 5-HT$_{1B/1D}$ receptors.

DISCUSSION

The interaction of AGE with 5-HT$_{1B/1D}$ binding sites observed in the present study might have interesting consequences in pharmacology and therapeutics. Indeed, the serotonergic system is considered as a regulatory system of other neurotransmissions to maintain the homeostasis of the brain.[1,2] Thus, the 5-HT system is involved in a various number of physiological

TABLE 19.1
EFFECT OF AGED GARLIC EXTRACT ON THE BINDING OF [^{3}H]5-HT TO 5-HT$_{1B}$
RECEPTORS IN RAT BRAIN MEMBRANE PREPARATION

	CONTROL	GARLIC
Bmax (pmol/mg.prot.)	0.097	0.935
Kd (nM)	0.98	30.6

Rat brain synaptosomal membranes were prepared from striatum and incubated with [^{3}H]5-HT at increasing concentrations (1 to 20 nM) in the presence (GARLIC) or in the absence (CONTROL) of aged garlic extract (30 μg/ml). The maximal number of binding sites (Bmax) and the affinity constant of the sites for 5-HT (Kd) were calculated from the Scatchard plot of the saturation curves.

functions: sleep, thermoregulation, learning and memory, pain, behavior (feeding, sexual), neuroendocrine secretion, cell growth and also immune activity. 5-HT is also implicated in a number of pathological dysfunctions: stress, anxiety, depression, obsessive compulsive disorders, panic, migraine, alcoholism, etc.[3] The regulating activity of the 5-HT system exerted on other neurotransmissions is mediated by a number of different receptors.[4] The multiplicity of these receptors allows the 5-HT system to independently regulate different cells in various areas of the brain and confers to the serotonergic system the capacity to control finely the activity of various transmitters in the CNS. The function of the 5-HT system is controlled by afferent fibers innervating the raphe area, where most of the serotonergic cellular bodies are concentrated, and through axo-axonic interactions of afferent fibers regulating the serotonergic activity at the numerous 5-HT terminals present in almost all brain areas. The activity of the serotonergic system is also regulated by autoreceptors which are present either on the cellular bodies of 5-HT neurones (5-HT$_{1A}$ receptors) or on serotonergic terminals (5-HT$_{1B/1D}$ receptors).

The 5-HT system may be considered as a biological oscillator characterized by its frequency and its amplitude. The frequency corresponds to the number of discharges per unit of time (i.e., during active wake-up it corresponds to 5-7 spikes per second and during paradoxical sleep it is zero.[5] The amplitude of the oscillator corresponds to the amount of 5-HT released at the nerve terminals. The stimulation of 5-HT$_{1A}$ receptors decreases the frequency of discharge, whereas the release of 5-HT at the synaptic level is under the inhibitory control of 5-HT$_{1B/1D}$ receptors. Both, 5-HT$_{1A}$ and 5-HT$_{1B}$ receptors thus play a role in the negative feedback control of 5-HT activity. These receptors, which are members of the large multigenic family of the G-protein coupled receptors, are allosteric proteins able to exist under different transitional conformations corresponding to various states of activity. It was recently shown in our laboratory that an endogenous peptide present in the brain likely acts as a modulator of 5-HT$_{1B}$ receptors.[6] The functional consequences of such interaction are important since via this mechanism, the release of 5-HT is efficiently controlled. It is proposed that AGE may contain compound(s) which interact with 5-HT$_{1B/1D}$ receptors in a similar way. Although this has to be further investigated, the fact that AGE appears to interact non-competitively with 5-HT$_{1B/1D}$ receptors suggests that these compounds may act as allosteric modulators of 5-HT$_{1B/1D}$ receptors.

Changes in the amplitude of the 5-HT release result in important consequences in CNS function. Indeed, a number of antidepressants activities are based on the ability of these drugs to affect the availability of 5-HT in the synaptic cleft (IMAO, 5-HT uptake inhibitors, etc.). Therefore, it is plausible that AGE via its activity at 5-HT$_{1B/D}$ receptors may affect serotonergic release and thus, the control exerted by the 5-HT system on the activity of the CNS. Neuro-pharmacological properties are expected for AGE which could be the subject of future research.

REFERENCES

[1]   Coccaro, E.F. Central serotonin and impulsive aggression. Br. J. Psy. 1989; 155(Suppl. 8):52–62.

[2]   Van Praag, H.M., Asnis, G.M., Kahn, R.S., et al. Monoamines and abnormal behavior, a multi-aminergic perspective. Br. J. Psy. 1990;157: 723–734.

[3]   Zifa, E. and Fillion, G. 5-hydroxytryptamine receptors. Pharmacol. Rev. 1992;44:3:401–458.

[4]   Peroutka, S.J. Molecular Biology of Serotonin (5-HT) Receptors. Synapse 1994; 18:241–260.

[5]   Jacobs, B.L. and Fornal, C.A. 5-HT and motor control–A hypothesis. TINS 1993;16:346–352.

[6]   Massot, O., Rousella, J.C., Fillion, M.P., Grimaldi, B., Cloez-Tayarani, I., Fugelli, A., Prudhomme, N., Seguin, L., Rousseau, B., Plantesol, M., Hen, R. and Fillion, G. 5-HT-moduline, a new endogenous cerebral peptide controls the serotonergic activity via its specific interaction with 5-HT$_{1B/1D}$ receptors. Mol. Pharmacol., 1996; 50:752–762.

CHAPTER 20

GARLIC AND SERUM CHOLESTEROL

DALE D. SCHMEISSER

Department of Nutrition and Medical Dietetics
University of Illinois, Chicago
Chicago, Illinois

ABSTRACT

Several blinded, placebo-controlled studies have demonstrated cholesterol-lowering by garlic supplements in humans, and two recent meta-analyses suggest that cholesterol is reduced by 9-12% by processed garlic in hypercholesterolemics. A preparation containing allicin and another containing S-allyl-cysteine were effective in cholesterol-lowering; other preparations (steam-distilled, oil macerate) have not been adequately studied to date. In order to develop practical uses of garlic in treatment of high cholesterol, future studies should examine the relative efficacy of different garlic preparations and the bioactive compounds which they contain, garlic's effects under various dietary conditions, as well as garlic's simultaneous effects on other cardiovascular risk factors.

INTRODUCTION

Garlic has a long history of use as a "blood thinning" agent in China and India, but it is only in the last several years that a systematic study of the cardioprotective effects of garlic and the other *Allium* vegetables has been undertaken by Western medicine. Interest in the use of garlic for the prevention and treatment of cardiovascular and other circulatory diseases is based on the favorable effects which garlic has on several cardiovascular risk factors, namely, reductions in serum cholesterol and triglycerides[1-3], platelet aggregation[4,5], blood pressure[4], and oxidized LDL.[6]

Of garlic's multiple cardiovascular effects, the lipid-lowering properties have been the most extensively studied in humans. Much of the public's interest in garlic is attributable to its purported effect on cholesterol. This paper briefly summarizes the human research to date on garlic and cholesterol and poses some of the questions which have been raised by this research, in order to extend these findings to new areas, including practical uses for garlic.

Garlic Food Forms Used in Clinical Studies

Studies on garlic and cholesterol prior to the early 1980s used fresh garlic (whole or extracts), whereas the studies conducted since 1985 have tested processed garlic, either whole-garlic powders[2,7] or aged garlic extract.[1] Garlic's potency as a cholesterol-lowering agent is attributable to many of its organosulfur compounds, which can vary several-fold in different garlic forms.[8] The composition of fresh garlic varies somewhat based on where it is grown and growing conditions.[9] More importantly, the several different methods of processing garlic result in products with vastly different arrays of organosulfur compounds and consequently different biological activities among the preparations.[8,10,11]

The most widely-tested processed garlic in human trials is a powder produced from air-dried garlic, standardized to alliin content to yield allicin on consumption (Kwai® Lichtwer).[7] Aged garlic extract (Kyolic®, Wakunaga) is produced by extraction of garlic in 20% ethanol over several months, and is standardized to its S-allyl cysteine content.[1] Both preparations have been shown to decrease serum cholesterol in clinical trials.[1,7] The cholesterol-lowering effects of steam-distilled oils and oil-macerates have not been extensively evaluated in human trials, although many contain significant quantities of diallyl disulfide and ajoenes[10], which, *in vitro*, are inhibitors of cholesterol synthesis and thus are presumed to affect serum cholesterol.[12-14]

Results of Human Studies

Kleijnen, *et al.*[15] and Lawson[9] have recently reviewed the human studies on garlic and serum lipids conducted over the past 30 years. The 32 studies reported up to 1992 varied widely in terms of type of garlic used, study purpose and design, and subject characteristics, yet a cholesterol-lowering effect (5-20% reduction) was demonstrated in most of these studies. Studies on fresh garlic used very high doses (5-28 cloves/day or equivalent extract).[15] Studies with garlic preparations have generally produced cholesterol-lowering with much lower levels of garlic (as powder or extract); the processing techniques are thought to either preserve or concentrate the most bioactive compounds present in raw garlic (e.g., allicin, S-allyl-cysteine) or alternatively to produce particularly bioactive compounds which were not present in the unprocessed garlic (e.g., diallyl sulfides, ajoenes). The studies are somewhat difficult to compare because of differences in the garlic processing methods and thus concentration of bioactive components. The few studies which did not show a cholesterol-lowering effect used preparations which were not well-described; in these cases, the garlic supplements may have been given in too low a dose, or prepared in such a way that active agents were lost.[9]

Warshafsky *et al.*[16] have quantified garlic's cholesterol-lowering effect using meta-analysis. Of 28 published studies which they initially evaluated, only five met their criteria for inclusion: randomized and placebo-controlled, included subjects with initial cholesterol levels greater than 5.17 mmol/L (200 mg/dL) and provided enough information so that an effect size could be computed. Of the five, four studies showed cholesterol-lowering with garlic and the fifth did not.

Meta-analysis of the five studies showed a significant cholesterol reduction of approximately 9% with garlic supplementation of one-half to one clove-equivalent of garlic/day. All of these studies used processed garlic (the criterion of placebo-control effectively excluded studies with fresh garlic); three used Kwai®, one used Kyolic®, and the fifth study (the negative study) used a low dose of a spray-dried extract. Generally, cholesterol-lowering was most pronounced in subjects with the highest cholesterol levels (e.g., above 6.46 mmol/L; 250 mg/dl), with smaller reductions occurring in subjects with initial cholesterol levels in the 5.17 to 6.46 mmol/L range.

Another meta-analysis using somewhat less stringent criteria showed a mean cholesterol decrease of 0.77 mmol/L (12% reduction), across 16 trials including 952 subjects.[17] In the studies where other lipids were measured, a significant reduction in triglycerides was also observed, without a reduction in HDL. The studies which have evaluated lipoproteins have shown that the overall cholesterol reduction is mainly due to reductions in LDL cholesterol.[1,7,18]

Directions for Future Studies

In order to achieve practical applications of the finding that garlic forms lower cholesterol, future studies must address a number of unanswered questions about the conditions under which garlic forms may be most effective.

Notably absent from the studies conducted to date is any sort of dietary control. Hypercholesterolemic subjects were studied while on their usual diets; in a few studies, dietary intake was assessed, but no studies have evaluated garlic's effects in subjects consuming cholesterol-lowering diets (e.g., low saturated fat, low cholesterol, high soluble fiber, etc.). An important consideration for the hypercholesterolemic patient following a cholesterol-lowering diet is whether garlic intake will provide an additive effect beyond diet or other cholesterol-lowering modalities.

Because garlic has multiple cardiovascular effects as well as antioxidant and immune effects, it would be especially helpful to examine these other endpoints in the same study (e.g., hemostasis, blood pressure, measures of immune function, as well as more sophisticated lipid analysis), to provide a more comprehensive picture of garlic's effects. This is particularly important given that different garlic preparations may be producing effects by different mechanisms, and while they may have similar effects on cholesterol, they may have different endpoints of interest. To date, there have been no studies which directly compare different garlic preparations which contain different organosulfur or other compounds. Lawson[9] has suggested a study comparing an allicin-yielding preparation with, for example, an ajoene-containing preparation; a number of such studies using different preparations would be helpful, especially studies with multiple endpoints as described above. Only in this way can the most efficacious garlic forms be identified.

Placebo-controlled and double-blinded trials continue as the design of choice for studies on processed-garlic supplements. In future studies, subjects in all trials should (at the minimum) be counseled and monitored on cholesterol-lowering diets (e.g., AHA Step 1). Controlled-feeding studies using such diets may be a better approach, given the difficulties in assessing dietary intakes in free-living subjects. Feeding studies may even permit evaluations of fresh, as well as processed garlic in partially-blinded fashion, using dietary ingredients to hide garlic compounds and flavor. Future studies should be a minimum of four to six months in length, rely on multiple baseline and ending measurements, and incorporate multiple endpoints.

Garlic, in either natural or processed form, may well be a cardioprotective food. By addressing the above issues, garlic's cholesterol-lowering properties can be placed in context with its other biological properties, and practical guidelines for its use can be developed.

REFERENCES

[1]   Lau, B.H.S., Lam, F. and Wang-Cheng, R. Effect of an odor-modified garlic preparation on blood lipids. Nutr. Res. 1987; 7:139–149.

[2]   Mader, F.H. Treatment of hyperlipidaemia with garlic-powder tablets. Evidence from the German Association of General Practitioners' multicentric placebo-controlled double-blind study. Arzneim-Forsch/Drug Res. 1990; 40:1111–1116.

[3]   Bordia, A. Effect of garlic on blood lipids in patients with heart disease. Am. J. Clin. Nutr. 1981; 34:2100-2103.

[4]   Harenberg, J., Glese, C. and Zimmerman, R. Effect of dried garlic on blood coagulation, fibrinolysis, platelet aggregation and serum cholesterol levels in patients with hyperlipoproteinemia. Atherosclerosis 1988; 74:247–249.

[5]   Srivastava, K.C. and Tyagi, O.D. Effects of a garlic-derived principle (ajoene) on aggregation and arachidonic acid metabolism in human blood platelets. Prostaglandins Leukotrienes and Essential Fatty Acids 1993; 49: 587-595.

6   Phelps, J. and Harris, W.S. Garlic supplementation and lipoprotein oxidation susceptibility. Lipids 1993; 28:475–477.

7   Jain, A.K., Vargas, R., Gotzkowsky, S. and McMahon, F.G. Can garlic reduce levels of serum lipids? A controlled clinical study. Am. J. Med. 1993; 94:632–635.

8   Block, E. The organosulfur chemistry of the genus *Allium* — Implications for the organic chemistry of sulfur. Angew Chem. 1992; 31:1135–1178.

9   Lawson, L.D. Bioactive organosulfur compounds of garlic and garlic products: role in reducing blood lipids. *In* Human medicinal agents from plants, (Kinghorn, A.D. and Bolandrin, M.F. eds.) Am. Chemical Soc., Washington, DC, 1993: 306–330.

10   Lawson, L.D., Wang, J. and Hughes, B.G. Identification and HPLC quantification of the sulfides and dialk(en)yl thlosulfinates in commercial garlic products. Planta Med. 1991; 57:363–370.

11   Block, E., Naganathan, S., Putnam, D. and Zhao, S.H. Allium chemistry: HPLC analysis of thiosulfinates from onion, garlic, wild garlic (ramsoms), leek, scallion, shallot, elephant (great-headed) garlic, chive and Chinese chive. Uniquely high allyl to methyl rations in some garlic samples. J. Agric. Food Chem. 1992; 40:2418–2430.

12   Gebhardt, R. Inhibition of cholesterol biosynthesis by a water-soluble garlic extract in primary cultures of rat hepatocytes. Arzneim-Forsch/Drug Res. 1991; 41:800–804.

13   Om Kumar, R.V., Banerji, A., Ramakrishna Kurup, C.K. and Ramasarma, T. The nature of inhibition of 3-hydroxy-3-methylglutaryl CoA reductase by garlic-derived diallyl disulfide. Biochem. Biophys. Acta 1991; 1078: 219–225.

14   Sendl, A., Schliack, M., Loser, R., Stanislaus, F. and Wagner, H. Inhibition of cholesterol synthesis in vitro by extracts and isolated compounds prepared from garlic and wild garlic. Atherosclerosis 1992; 94:79–95.

15   Kleijnen, J., Knipschild, P. and Ter Reit, G. Garlic, onions and cardiovascular risk factors. A review of the evidence from human experiments with emphasis on commercially available preparations. Br. J. Clin. Pharmac. 1989; 28:535–544.

16   Warschavsky, S., Kamer, R.S. and Livak, S.L. Effect of garlic on total serum cholesterol: a meta-analysis. Ann. Intern. Med. 1993; 119:599–605.

17   Silagy, C. and Neil, A. Garlic as a lipid-lowering agent — a meta-analysis. J. Royal Coll. Physic. 1994; 28:39–45.

18   De Santos, O.S. and Grunwald, J. Effect of garlic powder tablets on blood lipid and blood pressure. A six-month placebo controlled double blind study. Br. J. Clin. Res. 1993; 4:37–44.

CHAPTER 21

MODULATION OF ARACHIDONIC ACID METABOLISM BY GARLIC EXTRACTS

NIKOLAY V. DIMITROV and MAURICE R. BENNINK

Departments of Medicine, and Food Science
and Human Nutrition
Michigan State University
East Lansing, Michigan

ABSTRACT

We have conducted a pilot study designed to evaluate the effect of an aged water-ethanol garlic extract (Kyolic) on serum prostaglandins (PGE_2 and PGF_2^a). Normal subjects (females) ingested garlic extract (10 ml daily) for 3 months. Serum PGE_2 and PGF_2^a levels were determined before and after consumption of the garlic extract. The results of this pilot study indicated that the amount of water-ethanol garlic extract used in this trial was tolerated well by all participants. The serum levels of PGE_2 and PGF_2^a at the end of the study were decreased in most of the subjects. A more extensive design including kinetic studies will be needed to evaluate the effect of garlic extracts on biosynthesis of PGE_2 and PGF_2^a.

INTRODUCTION

Garlic (Allium sativum) has been used as a folk remedy for a variety of ailments since ancient times. Recently many nutritionists' attention has been attracted toward garlic due to a variety of communications by lay and scientific sources.[1,2] The range of reporting quality is from careful observation to extreme and frequently speculative statements.[1-4] One of the most remarkable observations is the effect of garlic extracts on arachidonic acid metabolism and platelet aggregation.[3,4] Organosulfur compounds in garlic were identified as potential cancer prevention agents.[5,6]

Since garlic extract appears to inhibit arachidonic acid metabolism, which may play an important role in cancer prevention, we decided to design a pilot study to investigate the effect of ethanol-water garlic extract (Kyolic) on arachidonic acid metabolism in humans.

199

SUBJECTS AND METHODS

Normal healthy volunteers (only females), age from 28-64, participated in this pilot study. Subjects were considered eligible for this study if they had no acute or chronic illness and were not taking any medications, vitamins or mineral supplements. The participants were entered into the study after signing an appropriate informed consent form approved by the University Committee on Research Involving Human Subjects at Michigan State University. All studies were done in accord with the Helsinki Declaration.

The ethanol-water garlic extract (Kyolic) was supplied by NCI repository. The content of the concentrate was studied by our laboratory and outside analysis for comparison. The results of two batches (90001 and 910013) are presented in Table 21.1. Each participant took 10 ml of kyolic mixed with orange or vegetable (V8) juice daily in the morning. All subjects ingested a daily dose of 10 ml garlic extract for 3 months. Blood samples were collected before the treatment and at the end of the three-month consumption. Prostaglandin E_2 and F_2^a levels were determined using an American Magnetics Inc. $^3[H]$ radioimmuno assay kits. Separation of antibody-bound analyze from unbound analyze was done by centrifugation (AMI, Cambridge, MA).

TABLE 21.1
ANALYSIS OF ORGANOSULFUR COMPOUNDS IN KYOLIC CONCENTRATE
(Micrograms/ml Concentrate)

	Allylcysteine Sulfoxide (Alliin)	Glutamyl-5 Allyclysteine	S-Allylcysteine	Allicin	S-Allyl Mercaptocysteine
Concentrate					
900001	820	160	1280	<2	160
Concentrate					
910013	400	240	1050	<2	210

Allyl sulfide, diallyl sulfide, diallyl disulfide, diallyl trisulfide, methyl allyl sulfide, methyl allyl disulfide and methyl allyl trisulfide were present in minimal amounts (less than 10 micrograms/ml concentrate).

RESULTS

PGE$_2$ and PGF$_2^a$ were measured before consumption of garlic extract and after 3 months daily ingestion of 10 ml of Kyolic. The results obtained from eight healthy females are presented in Table 21.2. Although the number of subjects studied was small, the serum PGE$_2$ levels in the majority of the subjects after consumption of garlic extract were substantially decreased. Less convincing are the changes in the PGF$_2^a$ observed after 3 months consumption of garlic extract.

TABLE 21.2
SERUM PROSTAGLANDIN LEVELS

	PGE$_2$ pg/ml SERUM		PGF$_2^a$ pg/ml SERUM	
SUBJECT*	BASELINE	AFTER 3 MONTHS	BASELINE	AFTER 3 MONTHS
1	230	190	80	-
2	250	250	160	80
3	450	800	70	-
4	850	275	60	100
5	650	750	100	170
6	850	650	140	170
7	650	250	250	40
8	700	270	110	450

*FEMALES

DISCUSSION

Arachidonic acid, the precursor of PGE$_2$ and PGF$_2^a$, is considered as the most common precursor in mammalian tissue. PGE$_2$ differs from PGF$_2^a$ only in that there is a keto in position 9 in PGE$_2$ and a hydroxyl in position 9 in PGF$_2^a$. In order to produce any biological changes during the consumption of garlic extract, the product used should contain certain effective components. It has been shown that the most effective organosulfur compounds are those which contain an allyl group.[6] Some of the allyl group containing effective compounds are shown in Table 21.3. In our study we used an ethanol-water soluble garlic extract which contains substantial amounts of s-allyl-cystein and s-allyl-mercaptocysteine (Table 21.1).

The results presented in Table 21.2 indicate that ethanol-water soluble extract (Kyolic) is capable of modulating PGE$_2$ and PGF$_2^a$. This modulation appears to be more inhibitory, particularly for PGE$_2$. The levels of PGF$_2^a$ are substantially lower as compared to those of PGE$_2$.

TABLE 21.3
EFFECTIVE COMPONENTS IN GARLIC EXTRACTS

OIL SOLUBLE	WATER SOLUBLE	OIL AND WATER SOLUBLE
DIALLYL SULFIDE	S-ALLYL-CYSTEINE	S-OXO-DIALLYL DISULFIDE (ALLICIN)
DIALLYL DISULFIDE	S-ALLYL-MERCAPTOCYSTEINE	
DIALLYL TRISULFIDE		

The small number of subjects does not allow any interpretation at this time. A larger sample size is needed to obtain more information and to extend the scope of this pilot study to a more detailed clinical project. This should include studies on neutrophils, lymphocytes, platelets and buccal mucosa cells.

ACKNOWLEDGMENT

This study was supported by PHS-NIH-NCI Grant-Contract N01-CN-05236.

REFERENCES

[1] Fenwich, G.R. and Hanley, A.B. The genus allium. CRC Crit. Rev. Food Sci. Nutr. 1985; 22(3)199–271.

[2] Reuter, H.D. Knoblauch (Allium Sativum): Neue Pharmacologische Ergebnisse Einer "Uralten" Arzneipflanze. Zeitschrift Für Phytotherapie 1986; 7:99–106.

[3] Ariga, T., Oshiba, S. and Tamada, T. Platelet aggregaton inhibitor in garlic. Lancet 1981; 1:150–151.

[4] Apitz-Castro, R., Ladezma, E., Escalante, J. and Jain, M.K. The molecular basis for the anti-platelet action of ajoene: Direct interaction with the fibrinogen receptor. Biochem. Biophys. Res. Commun. 1986; 141:145–150.

[5] Sumiyoshi, H. and Wargovich, M.J. Garlic (Allium sativum): A review of its relationship to cancer. Asia Pacif. J. Pharmacol. 1989; 4:133–140.

[6] Sumiyoshi, H. and Wargovich, M.J. Chemoprevention 01,2-dimethylhydrazine-induced colon cancer in mice by naturally occurring organosulfur compounds. Cancer Res. 1990; 50:5084-5087.

PHYTOPHARMACOLOGY OF SOY FOOD FORMS

CHAPTER 22

ISOLATED SOY PROTEIN TECHNOLOGY — POTENTIAL FOR NEW DEVELOPMENTS

BELINDA H. JENKS, DOYLE H. WAGGLE and
E.C. HENLEY

Protein Technologies International Inc.
Checkerboard Square
St. Louis, MO 63164

ABSTRACT

Research indicates that isolated soy protein provides health benefits relative to the reduction of coronary heart disease. Cholesterol-lowering occurs when individuals with elevated plasma cholesterol consume 20-40 g of SUPRO® Brand isolated Soy Protein (SUPRO® is a registered trademark of Protein Technologies International, Inc.) in the daily diet. Epidemiologic research indicates an inverse relationship between soy intake and rates of certain cancers, including cancers of the breast, colon, lung, and stomach. Animal and human cancer cell tissue studies have demonstrated that specific phytochemicals in soy are responsible for the anti-cancer effect observed. The challenge is to respond to consumer requests to develop food products delivering the health benefits of isolated soy protein. New and emerging isolated soy protein technologies provide options for food companies interested in developing good-tasting foods offering these health benefits and complementing the existing eating patterns, practices, and lifestyles of Western cultures.

INTRODUCTION

Over the past two decades, researchers have documented significant health benefits in populations that consume soy protein in their daily diets. These benefits relate to the reduction of key chronic diseases, including certain types of cancer and cardiovascular disease.

In a recent article entitled "Soy Intake and Cancer Risk: A Review of the In Vitro and In Vivo Data"[1], a number of epidemiologic research publications were reviewed, indicating that consumption of soy is related to significantly lowered rates of breast, colon, and prostate cancer in countries such as China and Japan. The epidemiologic evidence indicates that, while the diet of Asian populations are generally low in fat, they are also higher in soy intake than that of Western cultures. Researchers speculate that soy in the diet may be responsible for the lower rates of certain types of cancers observed in these populations. Animal and human cell cancer tissue studies have further supported these observations, providing evidence that any naturally-occurring components found in soy products may be anticarcinogenic in nature.

While the apparent chemopreventive effects of soy require further study, the cholesterol-lowering effects of soy protein are well-documented.[2] Research has demonstrated that the protein itself promotes the effects observed, independent from fat intake. This research has far-reaching implications for dietary intervention as a key strategy to reduce cardiovascular disease risk.

Epidemiological Observations

From a historical perspective, soybeans have served as one of the most important sources of protein. As early as 450 A.D., the Chinese attached a value of health to soy.[3] More recently, key epidemiologic studies by Adlercreutz et al.[4] and Lee et al.[5] indicated that intake of soy foods can reduce cancer risk. Adlercreutz et al. have proposed that a low mortality rate from prostate cancer is found in Japanese men consuming a low-fat diet that is also high in soy intake. Lee et al.[5] found that, in premenopausal women in Singapore, high intakes of animal protein and red meat were associated with increased risk of breast cancer. Decreased risk of breast cancer was associated with high intakes of polyunsaturated fatty acids (PUFA), beta-carotene, soy protein, total soy products, a high PUFA to saturated fatty acid ratio, and a high proportion of soy protein to total protein. Soy protein was found to be one of the protective factors. Lee et al. indicated that soy products may protect against breast cancer in younger women and attributed this finding to their rich phyto-oestrogen content. Researchers speculate that soy as a component of the diet has a direct effect on cancer risk, independent of other dietary variables, such as fat and fiber intake.

The epidemiologic evidence has been convincing enough to capture the attention of leaders in the field of cancer research. In 1989, the National Cancer Institute (NCI) announced their intent to fund a $20 million, five-year program aimed at examining the potential role of common foods, including soybeans, in preventing cancer. Also, in June 1990, NCI held a workshop specifically focused on the relationship between soy and cancer. In the past four years, funding continues to be allocated by NCI and NIH to study the pharmokinetics and metabolism of compounds in soy that appear to have cancer prevention properties.[6]

Soy Consumption

Around the world, soy consumption varies. The daily intake of soy in several countries of Southeast Asia compared to the United States, is shown in Fig 22.1. The amount of soy consumed in the typical Asian diet is 4 to 9 times greater than that of the average individual in the United States.[7]

In Taiwan, Japan, Korea, and Indonesia, traditional soyfoods serve as one of the major sources of protein and come in a wide variety of forms. In communities in Hawaii, where a large number of southeast Asians have migrated, current nutritional epidemiologic studies are evaluating soyfood consumption in relationship to changes in cancer and heart disease rates. At the University of Hawaii Cancer Research Center, Dr. Jean Hankin is attempting to track changes in relationship to soyfood consumption patterns.[8]

In Western cultures, traditional soyfoods, such as soy milk, tofu, tempeh, and miso are available, primarily through health food stores, but are not widely consumed.[9] Even though their health values might be evident, it is presumed that the foods are too unfamiliar to be widely accepted in Western

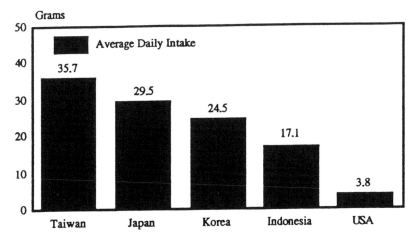

FIG. 22.1. PER CAPITA SOYBEAN CONSUMPTION AS REPORTED FROM DATA ON "SOYA USE FOR DIRECT HUMAN FOOD, WORLDWIDE AND IN MAJOR SOYBEAN CONSUMING NATIONS, AS REPORTED BY FAO FOOD BALANCE SHEETS, UNLESS OTHERWISE STATED" AS COMPILED BY SOYATECH INC.
From reference[7]

diets, at least in the short-term. Consumer studies indicate that a growing percentage of Americans choose foods based on taste, first and nutrition, second. Thus new food product development focused on the health benefits of soy needs to take into account taste preferences and expectations. Incorporating soy protein into common food products without affecting traditional taste and texture offers a significant opportunity for food manufacturers. Current and new technologies, particularly in isolated soy proteins, offer food manufacturers opportunities to meet this need.

Isolated Soy Protein Technology

In many food applications, isolated soy proteins are valued for their nutritional, functional, and economic benefits. Isolated soy protein, by definition, is over 90% protein, on a dry basis. It is simply the protein portion of the soybean, which has been separated or isolated from the rest of the soybean (Fig 22.2). The resulting product is bland in flavor and extremely versatile in food products. Isolated soy proteins can be specifically designed to meet the functional requirements of a variety of food systems. This flexibility allows their use in many food products, such as milk alternatives, powdered and liquid beverages, soups, and processed meats, as well as many other processed food products.

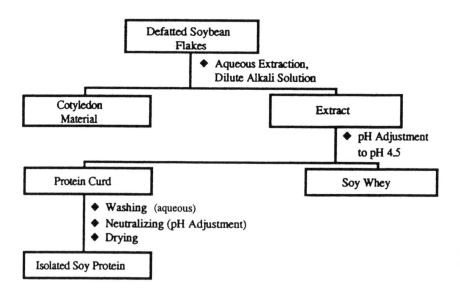

FIG. 22.2. ISOLATED SOY PROTEIN MANUFACTURING PROCESS
From reference[10]

Isolated soy proteins differ from other common soy ingredients, such as soy flour and soy concentrate, in their composition and use in food products (Fig 22.3). Isolated soy proteins, because of their high protein to carbohydrate ratio, offer food processors more flexibility in formulating to meet specific fat, carbohydrate, protein, and vitamin and mineral levels. They also offer flexibility for processors in the selection of carbohydrate sources to meet specific nutritional profiles. Finally, isolated soy proteins are far more versatile in food systems, because of their lower flavor profiles, relative to soy flour and concentrates. Of current soy ingredient technologies, isolated soy proteins offer the greatest promise for development of healthful food products that fit with current taste preferences, eating patterns and practices.

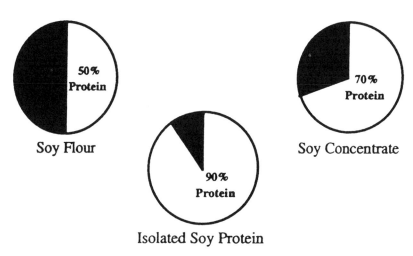

FIG. 22.3. SOY PROTEIN PRODUCTS
From reference[11]

Cardiovascular Disease Incidence

The American Heart Association, in their 1994 Statistical Summary, reports that cardiovascular disease continues to be the number one killer of American adults.[12] More than 2 out of every 5 Americans will die of cardiovascular disease each year. Of the current 252 million Americans, more than 56 million, or more than one in 5, have some form of cardiovascular disease.[12]

High blood cholesterol levels clearly play a causal role in coronary heart disease (CHD). High CHD rates occur among people with cholesterol levels of 240 mg/dL or above. However, an even larger number of cases occurs in Americans with blood cholesterol levels below 240 mg/dL.[13]

In the United States, an estimated 37 percent of children under the age of 19 have serum cholesterol levels of 170 mg/dL or above. Approximately 95 million American adults (52 percent of the population) have blood cholesterol levels above 200 mg/dL. Of those, 37 million (22 percent of the population) have levels above 240 mg/dL[12] (Table 22.1).

TABLE 22.1
INCIDENCE OF ELEVATED BLOOD CHOLESTEROL IN AMERICAN ADULTS

94.9 Million American Adults (52%) Have Blood Cholesterol Levels of 200 mg/dl and Higher

37.0 Million American Adults (20%) Have Levels of 240 mg/dl or Above

Heart and Stroke Facts: 1994 Statistical Supplement, American Heart Association, p. 16. From reference[12]

As high blood cholesterol is clearly a risk factor for CHD, intervention strategies based on diet modification are necessary and would have a tremendous impact on health care costs and CHD risk reduction. For children at risk, early intervention is necessary. For American adults and children, the Population Panel of the National Cholesterol Education Program (NCEP) recommends limiting fat intake to 30 percent of total calories, with less than 10 percent of the total calories as saturated fat, maintaining desirable body weight, and consuming less than 300 mg of cholesterol per day.[13]

Plasma Cholesterol-Lowering Effect with Isolated Soy Protein

Adding or partially substituting isolated soy protein for animal protein in the diet can have significant effects on lowering plasma cholesterol.[2] Studies by Potter et al.[14], Bakit et al.[15], and Widhalm et al.[16] using SUPRO® Brand Isolated Soy Proteins (Protein Technologies International, St. Louis, MO) have confirmed this effect most recently. Potter observed the effect when feeding mildly hypercholesterolemic men (~218 mg/Dl) 50 g of SUPRO Protein daily. Bakit reported similar observations at 25 g of SUPRO Protein fed daily. Widhalm observed at 20 g of SUPRO Protein per day a hypocholesterolemic effect with children with elevated cholesterol levels.

Carrol et al.[2] reviewed clinical studies on cholesterol-lowering response to soy protein in 28 studies of hypercholesterolemic subjects. Data from these studies indicated that plasma total cholesterol and low density lipoprotein (LDL) cholesterol levels in individuals with hypercholesterolemia declined between 2 and 34 percent (Fig. 22.4). These results have been obtained in many studies where the effect of soy protein is measured independent of the fat-modified-diet effects. Research has indicated that the soy protein is responsible for the effects observed, independent of changes in fat intake.

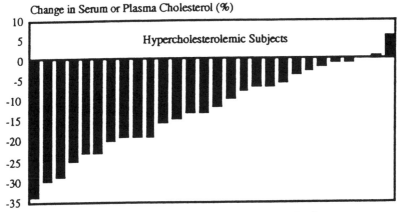

FIG. 22.4. CHANGE IN SERUM PLASMA CHOLESTEROL IN HUMAN HYPERCHOLESTEROLEMIC SUBJECTS FOLLOWING SUBSTITUTION OF SOYBEAN PROTEIN FOR ANIMAL PROTEIN IN THE DIET
From reference[2]

Proposed Cholesterol-Lowering Mechanisms

Several mechanisms have been proposed for the cholesterol-lowering effect of isolated soy protein. A well-known hypothesis by Kritchevsky et al.[17] is related to the amino acid profile of isolated soy protein versus animal proteins. Arginine, a conditionally essential amino acid is found in soy in higher amounts than in casein or other animal proteins. Kritchevsky points out that the ratio of lysine to arginine in a protein could be an important determinant of its atherogenic capacity as observed in studies on both rabbits and rats. Studies in which mixtures of amino acids were fed to rabbits showed little correlation between the lysine to arginine ratio and plasma cholesterol.

Other properties of proteins, such as their digestibility may also influence the results. Although a recent study on effects of casein, formalde-hyde-treated casein, and soybean protein on serum cholesterol in relation to their digestibility in rabbits suggested that this may not be an important determinant.[18]

Another possible mechanism has been proposed by Sanchez et al.[19] They observed that, when a plant-protein diet was fed to human volunteers, the results showed that there were significant reductions in serum cholesterol and in the ratio of lysine to arginine in plasma. Plasma concentrations of a number of other amino acids have also been correlated with the serum cholesterol level. Sanchez suggested that plasma amino acids may alter the secretion of hormones such as glucagon and insulin, which in turn might affect serum cholesterol levels by altering the activity of HMGCoA reductase, the rate-limiting enzyme in the synthesis of cholesterol.[19]

Thyroid hormones have been implicated in studies on gerbils and pigs. Forsythe et al.[20] found that gerbils fed a soy protein diet had lower plasma cholesterol levels than those fed a casein diet. Plasma levels of insulin and thyroid-stimulating hormone (TSH) were also higher in soy-fed gerbils. Pigs fed a soy-protein diet supplemented with one percent cholesterol also had lower plasma cholesterol levels and higher plasma thyroxine levels than pigs fed a casein diet. In this study there were no significant differences in plasma levels of insulin, glucagon, growth hormone, triiodothyronine or cortisol.[21]

Implication of Dietary Modification on Heart Disease Risk

It has been noted in the Report of the Expert Panel on Population Strategies for Blood Cholesterol Reduction, USDHHS, November 1990, that a reduction of 10 percent or more in the average blood cholesterol level of the U.S. population would lead to an approximate reduction of 20 percent or more in coronary heart disease.[13] On a smaller scale, it has been reported that for every one percent reduction in plasma cholesterol, there is a two percent reduction in coronary heart disease risk.[22]

Based on results observed in studies by Potter et al., Bakit et al., and Widhalm et al., all of whom observed reductions of 10-32% total cholesterol, it can be assumed that a daily intake of 20 and 50 grams of isolated soy protein could result in a 20-30% reduction in heart disease risk. This level represents an intake level that is practical and achievable and would have significant impact in reducing coronary heart disease risk.

Cancer Incidence in Western Populations

It is estimated that one in 9 women in the U.S. will develop breast cancer in their lifetime, while one in 11 men in the U.S. will have prostate cancer.[23] In Southeast Asian countries, the death rates for these cancers are 2-6 times lower. Diet and other environmental factors are believed to be responsible for this difference.[24]

Both epidemiological and experimental studies suggest that the effect of dietary influences is limited to the promotion-progression stage of the natural history of breast or prostate cancer. The low mortality from cancer of the prostate in Japanese men and cancer of the breast in Southeast Asian women may be attributed to a high intake of soy foods in the diet. The nutrition intervention studies using SUPRO Protein as part of the diet currently being conducted by Herman and Adlercruetz et al.[25], Petrakis[26], and Barns[27] will help to identify the components of soy that play a role in chemoprevention.

The Mechanism of the Anti-Cancer Effect

Animal and human cancer cell tissue studies have demonstrated that naturally-occurring compounds, found in soy and isolated soy protein, have anti-cancer effects. The isoflavones, specific amino acid profile, protease inhibitors and phytate are among the compounds being studied for their potential positive effects in delaying or preventing cancer tumors.

Soy is a rich source of genistein, an isoflavone and a tyrosine-specific protein kinase inhibitor[28] that also has been shown to inhibit angiogenesis, the development of blood vessels around the cancer cell[29], and is an essential event in tumor growth and metastasis. Genistein is an isoflavone that is found in soy, clover and only a few other green plants[30]; however, soy-based foods represent the only practical way consumers can incorporate genistein into their diets.

Scientists are still attempting to define the mechanism by which genistein may inhibit cancer. In addition to inhibiting angiogenesis and protein tyrosine kinase, a significant number of research studies lead to the theory that genistein acts to reduce the binding of estrogen to estrogen receptor sites, thus inhibiting cancer promotion and particularly those related to promotion-propagation of hormone-sensitive cancers of the breast and prostate.[31]

Two of the other well-known compounds that have been identified as having anti-carcinogenic activity include the protease inhibitors and phytic acid. Research by Kennedy et al.[32] suggests that it may be the protease inhibitor, Bowman-Birk Inhibitor (BBI), which may suppress the development of cancer cells.[33] Graf et al.[34] suggest that phytic acid may block absorption of some minerals and prevent the oxidation of dietary iron, thereby preventing the formation of free radicals that may be carcinogenic in nature.

Another theory by Hawrylewicz et al.[35] suggests that it may be the more ideal methionine content of soy that may be responsible for the anti-cancer effect he observed in his study of rats fed either soy or a casein-based diet.

Challenge to the Food Industry

The NCEP/NIH, in their recommendations on CHD risk reduction through dietary intervention, encouraged food manufacturers to develop new food options for consumers to help in the achievement of more acceptable cholesterol levels. Today, with increasing consumer concern and awareness of the effect of diet on health and disease, combined with emerging evidence around the benefits of soy-based foods relative to CHD and cancer, the industry is faced with a great opportunity and challenge. As consumers become aware of more of these benefits, good-tasting foods delivering the health benefits of soy will be in greater demand and should become a high priority for the food industry in the future. Newer, innovative technologies, such as isolated soy protein technology, will provide the industry with the capabilities to respond to these challenges.

Technology for the Future

SUPRO Brand Isolated Soy Protein is one of the technologies that will be key in the development of good-tasting mainstream, soy-based foods. Research has demonstrated that SUPRO Protein is a high-quality, complete protein for food products, equal in protein quality to meat, milk, and egg proteins.[36,37] Its essential amino acid profile meets or exceeds the FAO/WHO/UNU requirement pattern for children and adults[38,39] (Fig. 22.5).

As discussed previously, clinical studies have demonstrated the positive impact of SUPRO Proteins on blood lipids. Additionally, SUPRO Proteins contain many of the naturally-occurring compounds demonstrating anti-cancer effects in research studies. Research is currently underway to understand its precise role in cancer prevention.

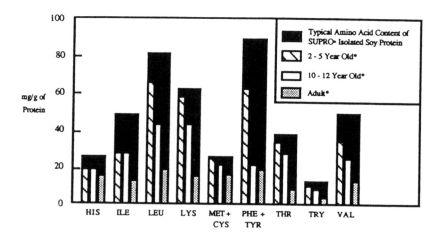

FIG. 22.5. AMINO ACID CONTENT OF SUPRO® ISOLATED SOY PROTEIN
AND RECOMMENDED PATTERNS FOR HIGH-QUALITY PROTEIN
Compiled from references[38,39]

SUPRO Proteins are now used extensively in the food industry for their nutritional, functional, and economic benefits. In many applications, they provide a high-quality, low-fat, cholesterol-free, lactose-free protein source (Table 22.2). In other applications, they are used for their functional properties, i.e., emulsification, gelling, water or fat absorption, aeration, or structure texture building.[10] Across many food systems, they provide an economic alternative to meat, milk, and egg protein, while delivering comparable performance and nutrition in the food system.

TABLE 22.2
NUTRITIONAL ASPECTS OF SUPRO® BRAND ISOLATED SOY PROTEIN

High-Quality Protein (PDCAAS - 1.0)[39]	Cholesterol Free
Plant Based	Lactose Free
Low Fat	

From reference[40]

Future developments are certain to focus on SUPRO Proteins' fit in healthy foods designed to provide specific health benefits. Application research, already underway and available to the food industry, is looking at various food forms where SUPRO Proteins can be used at levels that will help consumers add a significant amount of soy protein to their diet. The food forms being explored today are complementary to today's consumers' taste preferences, eating practices, and lifestyles. Continuing advancements in technologies, such as these, will focus on how to build the health benefits of soy into familiar, good-tasting food products which will be key to the future success and acceptance of soy-based foods in the American diet.

CONCLUSION AND SUMMARY

Epidemiologic studies suggest a strong link between soy intake and reduction in breast, colon and prostate cancers. Current research is now documenting that key components in soy appear to have anti-cancer effects. Strong evidence already exists indicating that isolated soy protein lowers elevated blood cholesterol levels.

Evidence from human clinical research studies, as well as epidemiologic studies, indicate that between 20 to 50 g/day of isolated soy protein provides protective effects against heart disease and cancer.[2,40,41] Using today's innovative soy protein technology, this level is considered practical and achievable through the enhancement of familiar food products, while maintaining traditional taste and texture. SUPRO Isolated Soy Protein is one of the technologies that offers the greatest promise for building the benefits of soy into food products in a manner that is accepted by mainstream consumers and meet taste expectations.

With new public health policies encouraging mainstream consumers to incorporate more healthful foods into their diets and with growing public awareness of the health value of soy-based foods, the food industry is now challenged to provide a variety of good-tasting, soy-based foods. Familiar, good-tasting, traditional food forms that have been nutritionally enhanced through the addition of soy will be the most readily accepted by mainstream, health-conscious consumers.

REFERENCES

[1] Messina, M.J., Persky, V., Setchell, K.D.R. and Barnes, S. Soy intake and cancer risk: a review of the in vitro and in vivo data. Nutr Cancer. 1994; 21:113–134.

[2] Carroll, K.K. Review of clinical studies on cholesterol-lowering response to soy protein. J. Am. Diet. Assoc. 1991; 91:820–827.

[3] Smith, A.K. and Circle, S.J. Historical background. In Soybeans: Chemistry and Technology, (Smith, A.K. and Circle, S.J., eds.) Chapman Hall (AVI), New York, 1972:1–26.

[4] Adlercreutz, H., Markkanen, H. and Watanabe, S. Plasma concentrations of phytoestrogens in Japanese men. Lancet 1993; 342:1209–1210.

[5] Lee, H.P., Gourley, L., Duffy, S.W., Esteve, J., Lee, J. and Day, N.E. Dietary effects on breast-cancer risk in Singapore. Lancet 1991; 337:1197–1200.

[6] Messina, M. and Barnes, S. The role of soy products in reducing risk of cancer. J. Natl. Cancer Inst. 1991; 83:541–546.

[7] As reported from data on "Soya Use For Direct Human Food, Worldwide and in Major Soybean Consuming Nations, As Reported By FAO Food Balance Sheets, Unless Otherwise Stated" as compiled by Soyatech, Inc., Bar Harbor, ME. 1990.

[8] Kolonel, L., Hankin, J.H. and Nomura, A.M.Y. Multiethnic studies of diet, nutrition and cancer in Hawaii. In Diet, Nutrition and Cancer: Proceedings of the 16TH International Symposium of the Princess Takamatsu Research Fund, (Hayashi, Y., Nagao, M., Sugimura, T., Takayama, S., Tomatis, L., Wattenberg, L.W. and Wogan, G.N., eds.) Scientific Societies Press, Tokyo, 1985; 16:29–40.

[9] Messina, M. and Messina, V. Increasing use of soy foods and their potential role in cancer prevention. J. Am. Dietet. Assoc. 1991; 91:836–840.

[10] Wilke, H. Isolated Soy Protein. *In* New Protein Foods, (Altschul, A.M., ed.) Academic Press, San Diego,, CA, 1985; 5:261.

[11] Hoogenkamp, H.W. Vegetable Protein: Technology Value in Meat, Poultry, and Vegetarian Foods. Protein Technologies International, St. Louis, MO. 1992:14.

[12] American Heart Association. Heart and Stroke Facts: 1994 Statistical Supplement. American Heart Association. 1993.

[13] National Cholesterol Education Program. Report of the Expert Panel on Population Strategies for Blood Cholesterol Reduction. National Institutes of Health. 1990.

[14] Potter, S.M., Bakhit, R.M., Essex-Sorlie, D.L., Weingartner, K.E., Chapman. K.M., Nelson. R.A. *et al.* Depression of plasma cholesterol in men by consumption of baked products containing soy protein. Am. J. Clin. Nutr. 1993; 58:501–506.

[15] Bakhit, R.M., Klein, B.P., Essex-Sorlie, D., Ham, J.O., Erdman, J.W. Jr. and Potter, S.M. Intake of 25 g of soybean protein with or without soybean fiber alters plasma lipids in men with elevated cholesterol concentrations. J. Nutr. 1994; 124:213–222.

[16] Widhalm, K., Brazda, G., Schneider, B. and Kohl, S. Effect of soy protein diet versus standard lowfat, low cholesterol diet on lipid and lipoprotein levels in children with familial or polygenic hypercholesterolemia. J. Pediatr. 1993; 123:30–34.

[17] Kritchevsky, D., Tepper, S.A. and Klurfeld, D.M. Dietary protein and atherosclerosis. J. Am. Oil Chem. Soc. 1987; 64:1167–1171.

[18] Kuyvenhoven, M.W., Roszkowski, W.F., West, C.E., Hoogenboom, R.L.A.P., Vos, R.M.E., Beynen, A.C. and Van der Meer, R. Digestibility of casein, formaldehyde-treated casein and soya-bean protein in relation to their effects on serum cholesterol in rabbits. Br. J. Nutr. 1989; 62:331–342.

[19] Sanchez, A., Horning, M.C., Shavlik, G.W., Wingeleth, D.C. and Hubbard, R.W. Changes in levels of cholesterol associated with plasma amino acids in humans fed plant proteins. Nutr. Rept. Intern. 1985; 32:1047–1056.

[20] Forsythe, W.A., Green, M.S. and Anderson, J.M. Dietary protein effects on cholesterol and lipoprotein concentrations. A review. J. Am. Coll. Nutr. 1986; 5:533–549.

[21] Forsythe, W.A. III. Comparison of dietary casein or soy protein effects on plasma lipids and hormone concentrations in the gerbil (*Meriones unguiculatus*). J. Nutr. 1986; 116:1165–1171.

[22] National Cholesterol Education Program. Second Report of the Expert Panel on Detection, Evaluation and Treatment of High Blood Cholesterol in Adults (Adult Treatment Panel II). Circulation 1994; 89:1329–1445.

[23] American Cancer Society. Cancer Facts & Figures — 1994. American Cancer Society, Atlanta, GA. 1994.

24 Wynder, E.L., Rose, D.P. and Cohen, L.A. Nutrition and prostate cancer: a proposal for dietary intervention. Nutr. Cancer. 1994;22:1 -10.

25 Herman, C., Adlercreutz, T., Golden, B.R., Gorback, S.L., Hockerstedt, K.A.V., Watanabe. *et al.* Soybean phytoestrogen intake and cancer risk. J. Nutr. 1995; 757S-770S.

26 Petrakis, N., Winkie, J., Coward, L., Kirk, M. and Barnes, S. A clinical trial of the chemopreventive effect of a soy beverage in women at high risk for breast cancer. First International Symposium on the Role of Soy in Preventing and Treating Chronic Disease. 1994. (Abstr.)

27 Barnes, S. Effect of genistein on in vitro and in vivo models or cancer. J. Nutr. 1995; 777S-783S.

28 Akiyama, T., Ishida, J., Nakagawa, S., *et al.* Genistein, a specific inhibitor of tyrosine-specific protein kinases. J. Biol. Chem. 1987; 262:5592-5595.

29 Fostis, T., Pepper, M., Adlercreutz, H., Fleischmann, G., Hase, T., *et al.* Genistein, a dietary-derived inhibitor of in vitro angiogenesis. Proc. Natl. Acad. Sci. USA. 1993; 90:2690-2694.

30 Bennetts, H.W., Underwood, E.J. and Sheir, F.L. A specific breeding problem of sheep on subterranean clover pastures in Western Australia. Aust. Vet. J. 1946; 22:2-12.

31 Rose, D.P. Diet, hormones, and cancer. Ann. Rev. Publ. Health 1993; 14:1-17.

32 Kennedy, A.R. Anticarcinogenic activity of protease inhibitors: overview. *In* Protease Inhibitors as Cancer Chemopreventive Agents (Troll, W. and Kennedy, A., eds.) Plenum Press, New York, 1993.

33 Workshop Report from the Division of Cancer Etiology, National Cancer Institute, National Institutes of Health. Protease inhibitors as cancer chemopreventive agents. Cancer Res. 1989; 49:499-502.

34 Graf, E. and Eaton, J.W. Antioxidant functions of phytic acid. Free Radical Biol. Med. 1990; 8:61-69.

35 Hawrylewicz, E.J., Huang, H.H. and Blair, W.H. Dietary soybean isolate and methionine supplementation affect mammary tumor progression in rats. 1991; 121:1693-1698.

36 Young, V.R., Wayler, A., Garza, C., Steinke, F.H., Murray, E., Rand, W.M. and Scrimshaw, N.S. A long-term metabolic balance study in young men to assess the nutritional quality of an isolated soy protein and beef proteins. Am. J. Clin. Nutr. 1984; 29:8-15.

37 Torun, B., Pineda, O., Viteri, F.E. and Arroyave, G. Use of amino acid composition data to predict nutritive value for children with specific reference to new estimates of their essential amino acid requirement. *In* Protein Quality in Humans: Assessment and In Vitro Estimation, (Bodwell, C.E., Adkins, J.S. and Hopkins, D.T, eds.) Chapman Hall (AVI), New York, 1981:347-389.

[38] Energy and Protein Requirements. Report of a Joint FAO/WHO/UNU consultation. World Health Organ. Tech. Rept. Ser. 1985;724.

[39] Henley, E.C. and Kuster, J.M. Progein quality evaluation by protein digestibility-corrected amino acid scoring. Food Technology. 1994; 48:74-77.

[40] Nutritional Aspects of SUPRO® Brand Isolated Soy Protein. Protein Technologies International, St. Louis, MO. 1993.

[41] Coward, L., Barnes, N.C., Setchell, K.D.R. and Barnes, S. Genistein, daidzein, and their β-glycoside conjugates: antitumor isoflavones in soybean foods from American and Asian diets. J. Agric. Food Chem. 1993; 41:1961–1967.

CHAPTER 23

ROLE OF SOY FOOD FORMS IN PREVENTION OF HUMAN VASCULAR DISEASE

TAKEMICHI KANAZAWA, TOMOHIRO OSANAI, TSUGUMICHI UEMURA,
TAKAATSU KAMADA and KOGO ONODERA

The Second Department of Internal Medicine
Hirosaki University School of Medicine
5 Zaifu-cho, Hirosaki, Aomori
Japan 036

HIROBUMI METOKI and YASABURO OIKE

Reimeikyo Rehabilitation Hospital
30 Ikarigaseki-mura, Aomori
Japan 038-01

ABSTRACT

In vitro *and* in vivo *human and rabbit experiments investigated whether the administration of soy protein was effective in the reduction of plasma cholesterol, the inhibition of LDL peroxidation, and the suppression of LDL molecular size enlargement.*
(1) Peroxidized LDL and large molecular size LDL are injurious to arterial vessels. (2) Soy protein suppressed the peroxidation of LDL both in vitro *as well as* in vivo. *(3) The increase in plasma LDL-cholesterol after cholesterol feeding was suppressed (and that of LDL-apo B) by the feeding of soy protein. Namely, LDL-cholesterol/LDL-apo B ratio became less with soy protein administration. The ratio after cholesterol feeding increased with an enlargement of LDL molecules. Thus, it appears that the enlargement of LDL molecules due to cholesterol feeding was inhibited by soy protein administration. (4) Soy as a food appears to provide protection from blood vessel injuries, and thus aids in the prevention of cardiovascular disease.*

INTRODUCTION

A mechanism of familial hypercholesterolemia was explained by the discovery of the LDL receptor[1,2], as well as the importance of modified low-

219

density lipoprotein (LDL) in the formation of foam cells.[3,4,5] Thus, the mechanism of lipid accumulation into the arterial wall was presumably clarified. However, the role of modified LDLs *in vivo* was not clear.

Recently, peroxidized LDL has been identified in plasma[6,7,8] and tissue[9,10]; and accordingly attention to the mechanism of atherosclerosis. We reported that peroxidized LDL accelerated[11] platelet aggregation, enhanced[12] an internalization into macrophages, and LDL infusion into the rabbit auricular vein injured[13] the arterial wall.

Further, the role of LDL molecular size in atherosclerosis is recognized.[14,15,16] Kanazawa *et al.* reported[14] that although normal LDL is antiatherogenic in the rabbit, large molecular size LDL is injurious to arterial endothelial cells. Blood corpuscles, such as lymphocytes, red cells, white cells and platelets significantly adhere on the arterial endothelium after the infusion of large molecular size LDL into venous vessels.

Cholesterol feeding for four weeks in the rabbit promotes the growth of LDL particles to a larger molecular size as evidenced by negative staining by transitional electron microscope. It was thus speculated that the inhibition of the production of peroxidized LDL and large molecular size LDL was very important for the prevention of atherosclerotic diseases.

MATERIALS AND METHODS

This chapter examines whether soy protein affects two types of LDL and atherogenesis.

Experiment I: suppression of peroxidized LDL production by soy protein *in vitro* and *in vivo* experiment.

(1)   Definition of peroxidized LDL

Kanazawa *et al.*[14] reported that there is a clear difference in the lipid constituents of peroxidized LDL in plasma of the patients with atherosclerotic diseases and peroxidized LDL in the plasma of the patients with lung cancer. Therefore the measurement of lipid peroxide (LPO) or thiobarbituric acid reactive substances in LDL are not sufficient to demonstrate the specific effects of peroxidized LDL in atherosclerotic disease. Thus, in this chapter, the designation of peroxidized LDL denotes hydroperoxidized cholesteryl linoleate (HPO-CL). According to Kanazawa *et al.*[16] normal LDL lipids on thin-layer chromatography (TLC), consists of cholesteryl ester (CE), triglycerides (TG), free fatty acids (FFA), free cholesterol (FC), and phospholipids (PL). Peroxidized LDL reveals a different lipid profile on TLC: namely, the appearance of spot X_1 between TG and FFA, and spot X_2 between FFA and FC.

Spot X_1 and spot X_2 were induced by the peroxidation of cholesteryl linoleate. Namely, spot X_1 consists of HPO-CL and spot X_2 has a structure in which cholesteryl ester in HPO-CL is fragmented. Spot X_1 differentiates clearly from other lipids spots.

Therefore, the production of HPO-CL is a marker for peroxidized LDL on TLC, and the HPO-CL/CE ratio can serve as the indicator of the peroxidizability of CE in LDL. As shown in Fig. 23.1, spot X_1 and spot X_2 appear when peroxidation of normal LDL occurs. But, these spots are not produced when ethylene diamine tetra acetic acid (EDTA) or butylated hydroxy toluene (BHT) was added in the course of peroxidation. Thus the existence of spot X_1 or spot X_2, especially spot X_1, indicates peroxidized LDL.

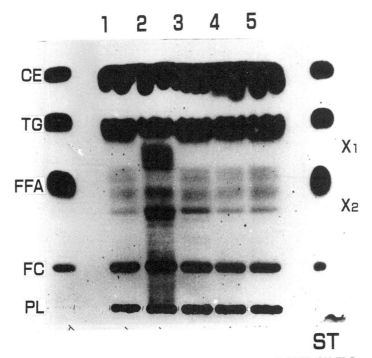

FIG. 23.1. LIPIDS EXTRACTED FROM LDL WERE DEVELOPED ON TLC
Lane 1: normal LDL, Lane 2: LDL dialyzed into Cu + + or tap-water (tap-UWS-LDL), Lane 3: LDL dialyzed into Cu + + or tap-water with EDTA-2Na, Lane 4: LDL dialyzed into Cu + + or tap-water with BHT. CE: cholesterol ester; TG: triglycerides; FFA: free fatty acid; FC: free cholesterol; PL: phospholipids; X_1 and X_2: peroxidized products of LDL; EDTA: ethylene diamine tetra acetic acid; BHT: butylated hydroxytoluene; Solvent: petroleum ether 75-ethyl ether - 25 acetic acid 1.

(2)   The effect of soy protein on the production of HPO-CL *in vitro*. The soy protein solution was prepared by the method shown in Fig. 23.2.

Normal LDL dialyzed into filtered tap-water for 24-48 h, spot X_1 and X_2 were produced. In this experiment, filtered tap-water was better than $CuCl_2$ solution, because apo B was not so fragmentated in filtered tap-water on SDS electrophoresis, compared to a $CuCl_2$ solution. So, normal LDL (300 μg as cholesterol) was dialyzed into tap-water or tap-water with soy protein (1 mg as protein/ml). As shown in Fig. 23.3, although spot X_1 was clearly recognized on TLC of lipid extracted from LDL dialyzed into tap-water, spot X_1 was barely stained.

The results suggest that soy protein suppresses the *in vitro* peroxidation of LDL.

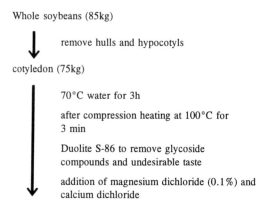

Whole soybeans (85kg)

↓   remove hulls and hypocotyls

cotyledon (75kg)

70°C water for 3h

after compression heating at 100°C for 3 min

Duolite S-86 to remove glycoside compounds and undesirable taste

addition of magnesium dichloride (0.1%) and calcium dichloride

Pressing fluid (100kg) was soaked with yeast (50kg), water (50kg), sodium chloride (4.2kg), ethanol (18kg), at 4°C for 1 mo.

↓

soycream

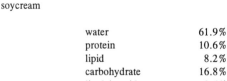

water	61.9%
protein	10.6%
lipid	8.2%
carbohydrate	16.8%
linoleic acid	3.6%
ash	2.5%

FIG. 23.2. PREPARATION OF SOY PROTEIN SOLUTION

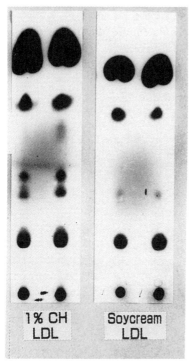

Lane 1: normal LDL
Lane 2: Cu^{++}-treated normal LDL
Lane 3: Cu^{++}-treated normal LDL with soycream
Lane 4

FIG. 23.3. COMPARISON OF TLC PROFILES BETWEEN LDL FED 1% CHOLESTEROL
DIETS FOR 7 WEEKS (1% CH LDL) AND LDL WITH 1% SOYCREAM (SOYCREAM
LDL) AFTER DIALYSIS INTO 5μm CuCl$_2$ FOR 24 H

(3)   The effect of soy protein on the production of HPO-CL *in vivo.*

To confirm the suppressive ability of soy protein on LDL peroxidation,
180 ml of soy drink (1 mg as protein) was daily provided orally for 6 months
to patients with cerebrovascular disease (CVD).

Aliquots of blood of 18 healthy persons, 26 CVD patients and 23 CVD
patients consuming soy protein daily for 6 months were drawn in the early
morning before breakfast. Immediately after drawing, plasma was separated and
LDL (1.006-1.063) was isolated by sequential ultracentrifugation. LDL was

dialyzed into saline (pH 7.0) for 48 h. Thereafter the LDLs of each subject were dialyzed into 5 μM $CuCl_2$ solution. The lipid was extracted by Bligh-Dyer's method[16] and TLC was carried out using the solvent petroleum ether 75: ethyl ether 25: acetic acid 1. The lipids were measured by a TLC charring method.[17]

As indicator of Cu+ + induced LDL peroxidizability, spot X_1/CE ratios were calculated. See Tables 23.1 and 23.2 and the hourly lipid percent changes illustrated in Fig. 23.4.

TABLE 23.1

COMPARISON OF THE INFLUENCE OF DIALYZING INTO CU+ + ON PERCENTAGES OF LIPID CONSTITUENTS IN LDL AMONG CVD PATIENTS (N=24), CVD PATIENTS+SOY (N=14) AND HEALTHY PERSONS (N=11)

	0 hour	6 hours	24 hours	48 hours	72 hours
CE					
Patients	36.0±5.0	29.1±5.5[##]	22.5±7.2[#]	16.3±5.3	13.4±6.2
Patients+Soy	35.8±5.1	34.5±7.0	31.8±6.4	20.8±4.5	16.8±3.8[#]
Healthy	37.8±4.1	36.4±5.6	30.1±5.3	19.8±4.1	13.4±3.9
TG					
Patients	20.6±7.2	15.1±5.2	9.5±4.7	6.9±4.0	5.8±4.1
Patients+Soy	19.8±6.8	18.7±6.8	16.8±4.8	10.9±4.2[#]	9.8±4.0[##]
Healthy	16.9±4.8	13.6±4.7	11.2±4.1	6.8±2.7	5.3±2.5
Spot X_1					
Patients	1.1±1.2[#]	5.7±2.3[##]	9.6±2.7[##]	10.2±2.5	11.7±1.7[#]
Patients+Soy	1.2±0.8[#]	2.8±1.2	5.2±1.9	6.2±2.1	8.0±2.2
Healthy	0.7±0.5	2.7±1.0	5.8±1.6	8.3±1.6	9.4±2.1
FFA					
Patients	0.7±0.3	2.4±1.2	4.3±1.7[#]	4.7±1.4	4.8±1.4
Patients+Soy	1.0±0.4	1.6±0.6	2.2±1.0	2.7±1.1[#]	3.0±1.8[#]
Healthy	1.0±0.9	0.9±0.6	2.5±1.6	3.7±1.4	4.2±1.4
Spot X_2					
Patients	1.1±0.8[#]	3.1±0.8[##]	5.4±3.4[##]	6.1±2.8[##]	7.1±2.3[##]
Patients+Soy	0.8±0.4[#]	1.6±0.6[#]	2.0±0.6	2.4±0.8	2.9±1.0[#]
Healthy	0.5±0.4	0.9±0.6	1.6±0.8	2.6±0.9	3.5±1.6
FC					
Patients	14.7±2.9	15.2±2.1	17.0±3.1	17.6±2.6	17.9±2.7[##]
Patients+Soy	14.0±2.2	16.2±2.4	17.0±4.2	19.2±2.1	21.2±1.5
Healthy	14.7±2.0	15.5±1.3	16.3±2.1	19.6±1.9	21.2±2.1
PL					
Patients	25.5±4.0	29.6±4.7	31.9±2.8	37.6±5.1	40.8±8.5
Patients±Soy	26.5±4.0	24.6±5.0[#]	25.0±5.2[##]	37.8±6.0	38.3±5.4[#]
Healthy	27.4±1.8	29.2±4.3	32.4±4.5	39.3±4.6	43.5±6.6

(%, Mean ± S.D.). Compared with healthy persons, [#]:P<0.05 [##]:P<0.01

Percent of Spot X1 Production from VLDL, LDL and HDL due to Dialysis into 5μM CuCl2

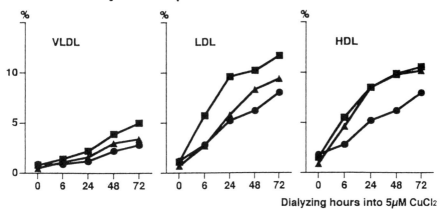

Dialyzing hours into 5μM CuCl2

Percent of Spot X2 Production from VLDL, LDL and HDL due to Dialysis into 5μM CuCl2

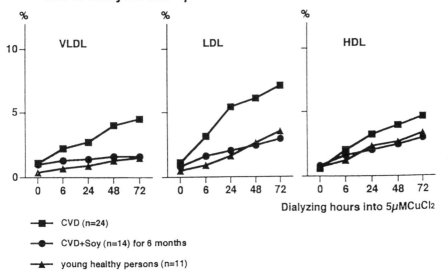

Dialyzing hours into 5μMCuCl2

—■— CVD (n=24)

—●— CVD+Soy (n=14) for 6 months

—▲— young healthy persons (n=11)

FIG. 23.4. CVD: CEREBROVASCULAR DISEASE

The percent of spot X_1 and X_2 increased markedly hour by hour in the LDL of CVD patients. However, the production of spot X_1 and X_2 were suppressed in LDL derived from the CVD patients administered the soy protein

TABLE 23.2
COMPARISON OF LDL

Comparison of the influence of dialyzing into Cu++ on percentages of lipid constituents in HDL among CVD Patients (n=24), CVD Patients+Soy (n=14) and Healthy Persons (n=11)

	0 hour	6 hours	24 hours	48 hours	72 hours
CE					
Patients	32.2±5.1	25.5±6.2#	13.0±4.6	6.7±2.6	4.8±1.9
Patients+Soy	35.8±5.1	34.5±7.0	31.8±6.4##	20.8±4.5##	16.8±3.8#
Healthy	37.0±6.0	31.1±4.8	15.5±3.6	7.4±3.2	5.4±2.2
TG					
Patients	12.4±7.0	7.5±6.1	3.6±3.3	2.8±2.5	2.7±2.2
Patients+Soy	15.8±6.8#	18.7±6.8##	16.8±4.8##	10.9±4.2##	9.8±4.0##
Healthy	8.9±3.3	8.0±4.0	3.4±1.1	2.7±0.8	2.5±1.0
Spot X1					
Patients	1.5±2.1#	5.5±1.7	8.5±1.8	9.9±2.9	10.6±3.4
Patients+Soy	1.8±0.8#	2.8±1.2#	5.2±1.9##	6.2±2.1##	8.0±2.2##
Healthy	0.9±0.4	4.6±1.9	8.5±1.9	9.8±2.2	10.2±1.7
FFA					
Patients	1.3±0.5	2.4±0.8	3.2±0.9	3.4±1.0	3.9±1.6
Patients+Soy	1.0±0.4	1.6±0.6	2.2±1.0#	2.7±1.1#	3.0±1.8
Healthy	0.8±0.5	1.8±0.9	3.5±0.9	3.5±1.0	3.5±0.9

Comparison of the influence of dialyzing into CU++ on percentages of lipid constituents in VLDL among CVD Patients (n=24), CVD Patients+Soy (n=11) and Healthy Persons (n=11)

	0 hours	6 hours	24 hours	48 hours	72 hours
CE					
Patients	18.6±6.2	14.3±3.8	12.0±3.2	7.9±2.7	6.3±2.3
Patients+Soy	18.8±4.8	17.4±3.8#	16.8±4.0#	14.8±3.9##	10.6±3.0##
Healthy	15.5±3.4	13.9±5.2	12.5±4.8	7.1±2.0	5.5±3.0
TG					
Patients	49.2±5.2#	39.8±6.4##	33.2±8.8	20.6±7.8#	17.8±8.0
Patients+Soy	50.2±5.2#	49.2±6.0	46.2±5.6	40.8±5.0#	25.8±4.8#
Healthy	57.0±7.6	51.8±9.7	44.6±16.1	33.8±13.8	24.8±12.5
Spot X1					
Patients	0.8±0.7	1.4±0.8	2.2±1.1	3.9±1.4	5.0±2.1#
Patients+Soy	0.9±0.4	0.9±0.5	1.2±0.4	2.2±0.6#	2.8±0.8#
Healthy	0.5±0.6	1.1±1.1	1.6±1.2	3.0±1.5	3.4±1.4
FFA					
Patients	1.2±1.1	2.4±0.9#	3.4±1.6##	3.9±2.2	4.4±2.6
Patients+Soy	1.3±1.0	1.7±1.2	2.0±1.0	2.4±1.1	2.6±1.2#
Healthy	0.9±0.6	1.3±0.8	1.8±0.8	2.8±1.4	3.7±1.5

TABLE 23.2 Continued

Spot X2					
Patients	0.6 ± 0.6	2.0 ± 1.5	3.2 ± 1.7	3.9 ± 2.3	4.6 ± 2.1
Patients+Soy	0.8 ± 0.4	1.6 ± 1.0	2.0 ± 0.6	2.4 ± 0.8	2.9 ± 1.0
Healthy	0.7 ± 0.4	1.2 ± 0.7	2.3 ± 1.4	2.6 ± 1.5	3.3 ± 1.6
FC					
Patients	11.1 ± 1.2	12.7 ± 1.9	15.3 ± 1.3	$16.3\pm1.8^{\#}$	17.7 ± 2.5
Patients+Soy	$14.0\pm2.2^{\#}$	$16.2\pm2.4^{\#}$	17.0 ± 4.2	19.2 ± 2.1	21.2 ± 1.5
Healthy	10.8 ± 2.8	11.9 ± 3.3	16.5 ± 3.3	18.1 ± 1.9	19.6 ± 1.7
PL					
Patients	40.6 ± 3.5	44.3 ± 6.7	52.9 ± 6.0	56.9 ± 7.4	57.0 ± 8.3
Patients+Soy	$26.5\pm4.8^{\#}$	$24.6\pm5.0^{\#}$	$25.0\pm5.2^{\#\#}$	$37.8\pm6.0^{\#}$	$38.3\pm5.4^{\#}$
Healthy	40.3 ± 3.4	41.4 ± 6.1	50.3 ± 7.4	55.9 ± 6.6	55.4 ± 5.4

($\%$, Mean ± S D.).
Compared with healthy persons, $\#$:P<0.05, $\#\#$:P<0.01.

Spot X2					
Patients	$1.1\pm0.6^{\#\#}$	$2.2\pm0.8^{\#}$	$2.7\pm1.2^{\#}$	$4.0\pm2.2^{\#}$	$4.5\pm2.3^{\#}$
Patients+Soy	$1.0\pm0.4^{\#\#}$	1.3 ± 0.5	1.4 ± 0.5	1.6 ± 0.6	1.6 ± 0.6
Healthy	0.4 ± 0.3	0.7 ± 0.6	0.9 ± 0.5	1.3 ± 0.7	1.5 ± 0.6
FC					
Patients	$10.4\pm2.2^{\#}$	12.3 ± 2.7	$14.2\pm2.3^{\#}$	14.8 ± 3.9	17.0 ± 4.7
Patients+Soy	$11.3\pm2.4^{\#}$	11.8 ± 2.8	11.5 ± 2.9	12.8 ± 3.0	13.8 ± 3.4
Healthy	8.2 ± 2.3	10.0 ± 2.7	10.7 ± 2.4	13.4 ± 2.9	15.5 ± 3.6
PL					
Patients	19.9 ± 2.3	26.6 ± 8.0	32.7 ± 10.4	45.0 ± 12.3	46.6 ± 12.4
Patients+Soy	16.7 ± 4.0	17.7 ± 4.1	$20.9\pm4.8^{\#}$	$25.4\pm5.0^{\#\#}$	$32.8\pm5.2^{\#}$
Healthy	17.4 ± 4.1	20.4 ± 6.8	27.4 ± 13.3	39.1 ± 11.0	45.4 ± 10.3

($\%$, Mean ± S.D.).
Compared with healthy persons, $\#$:P<0.05, $\#\#$:P<0.01.

solution. As shown in Tables 23.1 and 23.2, the peroxidizability of LDL of CVD patients differed from that of the LDL of healthy persons. However, the LDL lipid of CVD patients with soy protein did not show significant differences to those of healthy persons at any dialysis hour.

Thus, soy protein exerted suppressive effects *in vivo* on the peroxidizability of LDL. In addition, the data suggest that the ingestion of soy protein makes the LDL of the CVD patient similar to the LDL of healthy persons.

Experiment II: Suppression on production of large-molecular-sized LDL by soy protein in the rabbit.

(1)   Injurious activity of large molecular sized LDL on arterial endothelia

Four hundred milliliters of normal LDL diluted by physiological saline to the cholesterol level of normal plasma LDL saline was infused into rabbit auricular vein. Thereafter, 300 ml of saline was perfused with bleeding, as well as 200 ml of 2% phosphotungstic acid. The bilateral carotid arteries were removed and the intimas with media exfoliated.

The surface of the carotid arteries was observed by scanning electron microscopy. The experiment was repeated but with an infusion of large molecular sized LDL from rabbits fed 1% cholesterol. Kanazawa *et al.*[14], demonstrated that the molecular size of LDL enlarged week by week after 1% cholesterol feeding in the rabbit. The infusate of molecular-sized LDL was harvested from the rabbit in a 1% cholesterol feeding for 10 weeks. For infusion of the large-molecular-sized LDL, the LDL after 1% cholesterol feeding was diluted to normal plasma LDL cholesterol level with physiological saline. Thus, two kinds of LDL (normal LDL and large-molecular-sized LDL) with the same cholesterol concentration and volume were infused into the auricular vein of rabbit.

As shown in Fig. 23.5 and 23.6, the bilateral carotid arterial surface did not change after the infusion of normal LDL, but the infusion of large-molecular-sized LDL, resulted in red cells, white cells and platelets adhering to the arterial endothelial surface. In addition, the number of endothelial endocytotic vesicles after the infusion of large-molecular-sized LDL was greater than that as a result of the infusion of normal LDL (Fig. 23.5 and 23.6).

For the prevention of atherosclerotic vessel injuries, it is critical to know whether soy proteins suppress an enlargement of LDL molecules provoked by cholesterol feeding.

In experiments feeding 1% cholesterol to rabbits, LDL molecular size calculated from photograph by negative staining of LDL, and LDL-cholesterol/LDL-apo B ratio showed significant positive correlation in cholesterol feeding for seven weeks.

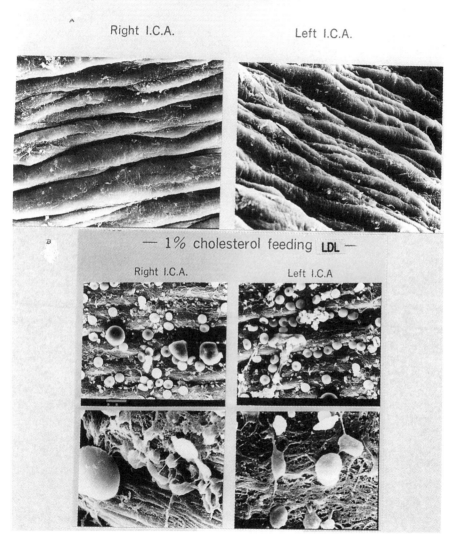

FIG. 23.5. SCANNING ELECTRON MICROSCOPIC FINDINGS OF INTERNAL CAROTID
ARTERY (I.C.A.) AFTER INFUSION WITH (A) NORMAL LDL AND (B) 1%
CHOLESTEROL FEEDING LDL
The infusion was carried out in the left carotid artery.

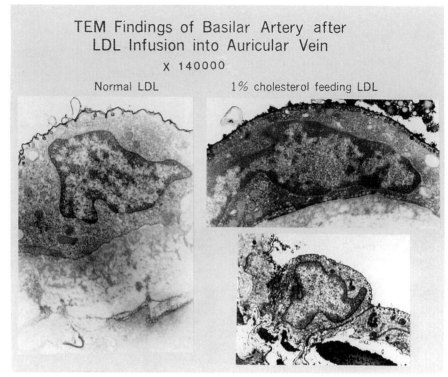

TEM Findings of Basilar Artery after
LDL Infusion into Auricular Vein
X 140000

Normal LDL   1% cholesterol feeding LDL

FIG. 23.6. TRANSMISSION ELECTRON MICROSCOPIC FINDINGS OF BASILAR
ARTERY AFTER INFUSION INTO THE AURICULAR VEIN OF NORMAL LDL AND
1% CHOLESTEROL FEEDING LDL

Therefore it was thought that LDL-cholesterol/LDL-apo B ratio would be a good indicator of LDL molecule size.

In addition, as shown in Figs. 23.7 and 23.8, LDL-cholesterol levels and LDL-apo B levels increased week by week. Correlation coefficients for LDL-apo B and LDL-cholesterol levels were calculated for the onset period as shown in Fig. 23.8. The equation was $y = 4.00 x + 23.2$, $\gamma = 0.90$ (first experiment), or $y = 3.50 x + 33.9$, $\gamma = 0.97$ (second experiment) at the starting period. Two weeks after the cholesterol feeding began, the correlation coefficient provided smaller values ($y = 0.27 x + 212.2$, $\gamma = 0.91$) compared to that of the onset values.

Subsequently, 8 rabbits were fed 1% cholesterol plus 10% soy protein. Two weeks after this feeding regimen, the values of the correlation coefficient got near to the values for the correlation coefficient in the onset period.

As shown in Fig. 23.9 the reduction in LDL-cholesterol due to feeding soy protein was greater than that for LDL-apo B. Thus, the LDL-cholesterol/LDL-apo B ratio was decreased by soy protein administration and it will be considered that soy protein administration diminished the molecular size of the LDL.

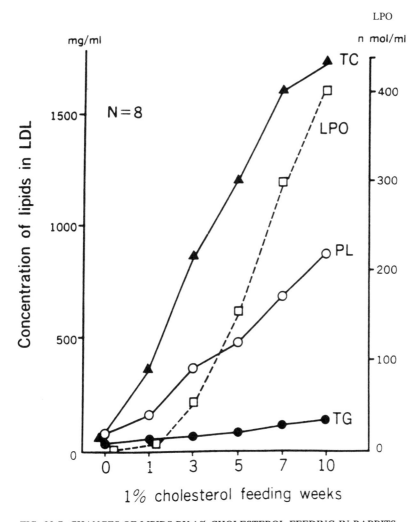

FIG. 23.7. CHANGES OF LIPIDS BY 1% CHOLESTEROL FEEDING IN RABBITS

DISCUSSION

To date, many investigators have reported the reduction of plasma cholesterol[18,19] and blood pressure[20] by the feeding of soy protein. However research on peroxidized LDL or the molecular size of LDL could not be found.

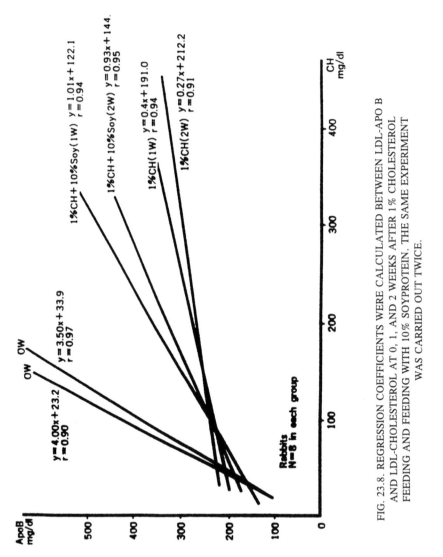

FIG. 23.8. REGRESSION COEFFICIENTS WERE CALCULATED BETWEEN LDL-APO B AND LDL-CHOLESTEROL AT 0, 1, AND 2 WEEKS AFTER 1% CHOLESTEROL FEEDING AND FEEDING WITH 10% SOYPROTEIN. THE SAME EXPERIMENT WAS CARRIED OUT TWICE.

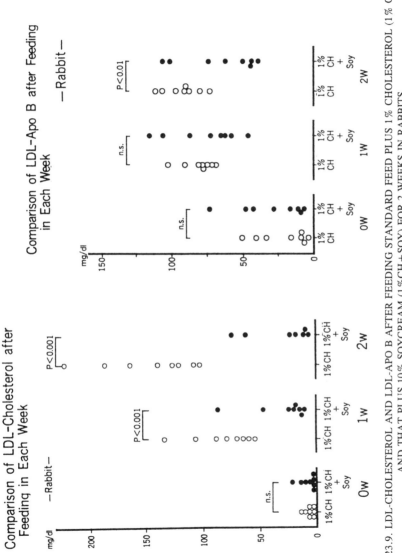

FIG. 23.9. LDL-CHOLESTEROL AND LDL-APO B AFTER FEEDING STANDARD FEED PLUS 1% CHOLESTEROL (1% CH) AND THAT PLUS 10% SOYCREAM (1%CH+SOY) FOR 2 WEEKS IN RABBITS

W: Week

It is now recognized that in the pathogenesis of atherosclerosis, peroxidized LDL or large-molecular-sized LDL may play an important role in foam cell formation and vessel injury.[14,21,22,23]

In the research reported here, the reduction of plasma LDL cholesterol levels, the decreased accumulation of cholesterol on arterial vessel walls, the suppression in the production of peroxidized LDL, and the suppression in the enlargement of LDL molecules were observed when soy protein was incorporated into the diet. Although the mechanism(s) are still unclear, the following hypotheses are offered.

Soy protein administration accelerates the excretion of fecal sterols.[24] Therefore, the administration of soy protein may decrease cholesterol levels by suppressing the absorption of bile acids produced in the liver. Sirtori et al.[24] suggest that one reason for plasma cholesterol reduction by soy protein may be that it promotes the synthesis of LDL receptors. Khosla et al.[25] have suggested that plasma cholesterol reduction when soy protein is ingested is related to cholesterol turnover or cholesterol transit time. In prior research[26], we found the amount of cholesterol excreted was high in rats administered soy and that the conjugation of bile acid and cholesterol was stronger in the soy-treated rats.

In any event all these results support the reduction of plasma cholesterol by soy protein. Soy protein suppresses an increase of LDL-apo B to a greater extent than LDL-cholesterol. Thus, the amount of cholesterol-apo B in LDL decreased after soy feeding. This means that the enlargement of LDL molecular size was suppressed.

Finally, peroxidizability of LDL induced by $Cu++$ was suppressed by soy protein both in vitro and in vivo. The mechanism is unknown. Soy protein contains vitamin E, lysine and phytic acid, which might inhibit the peroxidation of LDL. Although we did not provide details in this chapter, some fractions of soy protein suppressed platelet aggregation and some fractions showed a lowering of blood pressure.

Thus, soy protein is a food recommended for the prevention of cardiovascular disease.

REFERENCE

[1] Anderson, R.G.W., Goldstein, J.L. and Brown, M.S. Fluorescence visualization of receptor-bound low density lipoprotein in human fibroblasts. J. Receptor Res. 1980; 1:17–39.

[2] Beisiegel, U., Schneider, W.J., Goldstein, J.L., Anderson, R.G.V. and Brown, M.S. Monoclonal antibodies to the low density lipoprotein receptor as probes for study of receptor-mediated endocytosis and the genetics of familial hypercholesterolemia. J. Biol. Chem. 1981; 256:11923–11931.

[3] Goldstein, J.L., Ho, Y.K., Basu, S.K. and Brown, M.S. Binding site on macrophages that mediates uptake and degradation of acetylated low density lipoprotein, producing massive cholesterol deposition. Proc. Natl. Acad. Sci. (USA) 1979; 76:333–337.

[4] Fogelman, A.M., Schechter, I., Seager, J., Hokom, M., Child, S. and Edwards, P.A. Malondialdehyde alteration of low density lipoprotein leads to cholesteryl ester accumulation in human monocyte-macrophages. Proc. Natl. Acad. Sci. (USA) 1980; 77:2214–2218.

[5] Mahley, R.W., Innerarity, T.L., Weisgraber, K.H. and Sy, O. Altered metabolism (in vivo and in vitro) of plasma lipoproteins after selective modification of lysine residues of the apoproteins. J. Clin. Invest. 1979; 64: 743–750.

[6] Avogaro, P., Bon, G.B. and Cazzolato, G. Presence of a modified low density lipoprotein in humans. Arteriosclerosis 1988; 8:79–87.

[7] Tertov, W., Orekhov, A.N., Martseny, O.N. Perova, N.V. and Smirnov, V.N. Low-density lipoproteins isolated from the blood of patients with coronary heart disease induce the accumulations of lipids in human aortic cells. Exp. Mol. Pathology 1989; 50:337–347.

[8] Kanazawa, T., Uemura, T., Konta, Y., Tanaka, M., Fukushi, Y., Onodera, K., Metoki, H. and Oike, Y. A new approach to prevention and treatment of atherosclerosis by dyslipoproteinemia, Ann. NY Acad. Sci. 1990; 598:281–300.

[9] Yla-Herttuala, S., Palinski, W., Rosenfeld, M.E., Steinberg, D. and Witztum, J.L. Lipoproteins in normal and atherosclerotic aorta. European Heart J. 1990; 11:88–99.

[10] Yla-Herttuala, S., Palinski, W., Rosenfeld, M.E., Steinberg, D., and Witztum, J.L. Evidence for the presence of oxidatively modified low density lipoprotein in atherosclerotic lesions of rabbit and man. J. Clin. Invest. 1989; 84:1086–1095.

[11] Uemura, T. A study of the biological function and lipid constituents of oxidatively modified lipoproteins. J. Jap. Atheroscler. Soc. 1990; 18:81–90.

[12] Kaneko, H. The causes of the arteriosclerosis – Internalization of ultrawater-soluble LDLs into mouse peritoneal macrophages and their cell injury potential. J. Jap. Atheroscler. Soc. 1990; 18:51–61.

[13] Kanazawa, T., Chui, D.H., Tanaka, M., Fukushi, Y., Onodera, K., Metoki, H. and Oike, Y. Endothelial cell injuries by an infusion of various low density lipoproteins into rabbit auricular vein. Ann. NY Acad. Sci. 1990; 598:512–513.

[14] Kanazawa, T., Osanai, T., Uemura, T., Onodera, K. and Oike, Y. Evaluation of oxidized low-density lipoprotein and large molecular size low-density lipoproteins in atherosclerosis. Pathobiology 1993; 61:200–210.

[15] Tajima, S., Yokoyama, S. and Yamamoto, A. Effect of lipid particle size on association of apolipoproteins with lipid. J. Biol. Chem. 1983; 25:10073–10082.

[16] Bligh, E.G. and Dyer, W.J. A rapid method of total lipid extraction and purification. Can. J. Biochem. Physiol. 1959; 37:911–917.

[17] Kritchevsky, D., Davidson, L.M., Kim, H.K. and Malhotra, S. Quantitation of serum lipids by a simple TLC-charring method. Clin. Chem. Acta. 1973; 46:63–68.

[18] Witztum, J.L. Intensive drug therapy of hypercholesterolemia. Am. Heart J. 1987; 113:603–609.

[19] Kuo, P.T. Management of blood lipid abnormalities in coronary heart disease patients. Clin. Cardiol. 1989; 12:553–560.

[20] Imura, T., Kanazawa, T., Watanabe, T., Fukushi, Y., Kudou, S., Uchida, T., Osanai, T. and Onodera, K. Hypotensive effect of soy protein and its hydrolysate. Ann. NY Acad. Sci. 1993; 676:327–330.

[21] Steinbrecher, V.P., Zhang, H. and Lougheed, M. Role of oxidatively modified LDL in atherosclerosis. Free Radic. Biol Med. 1990; 09:155–168.

[22] Witztum, J.L. and Steinberg, D. Role of oxidized low density lipoprotein in atherogenesis. J. Clin. Invest. 1991; 88:1785–1792.

[23] Aviram, M. Modified forms of low density lipoprotein and atherosclerosis. Atherosclerosis 1993; 98:1–9.

[24] Sirtori, C.R., Even, R. and Lovati, M. Soybean protein diet and plasma cholesterol: From therapy to molecular mechanisms. Ann. NA Sci. 1993; 676:188–201.

[25] Khosla, P., Samman, S., Carroll, K.K. and Huff, M.W. Turnover of 1251-VLDL and 1311-LDL apolipoprotein B in rabbits fed diets containing casein or soy protein. Biochim Biophys Acta. 1989; 1002:157–163.

[26] Tanaka, M., Kanazawa, T., Imura, T., Watanabe, T., Fukishi, Y., Osanai, T., Uchida, T. and Onodera, K. Effect of soy protein on the excretion amount of fecal sterols in Jcl:Wistar rats. Ann. NY Acad. Sci. 1993; 676: 359–363.

CHAPTER 24

COMMERCIAL PHYTOCHEMICALS FROM SOY

JAMES P. CLARK

Henkel Corporation
5325 South 9th Avenue
LaGrange, IL 60525

ABSTRACT

The soybean is a rich source of phytochemicals, as well as protein and lipids. Approximately 75% of the world's steroids (both corticosteroids and sex hormones) are produced from the phytosterols (sitosterol, stigmasterol, and campesterol) isolated from soybeans. Mixed tocopherols produced from soybeans are used as food antioxidants and are also converted into natural-source vitamin E. Natural-source vitamin E is a single stereoisomer which has greater biological activity than the racemic mixture produced by total synthesis. There is growing evidence that vitamin E protects against cardiovascular disease and certain types of cancer, as well as improving the immune function in some individuals.

INTRODUCTION

The soybean is a remarkable plant. It is the most important human source of protein and edible oil in human nutrition, providing more protein and triglycerides than any other single plant. Soybeans were first cultivated over 5,000 years ago in China. Today, the United States grows more than 50% of the total soybean crop for the entire world, although we did not start planting them until 1829.

Soybeans are a very prolific source of phytochemicals. The beans themselves serve as a food for animals and humans. The protein isolated from soybeans is a major animal feed source. The oil serves as a source of glycerine and fatty acids. Lecithin, a very important phospholipid used as a food emulsifier, is isolated from soybeans. Soybeans are also the source of some very important, but less well-known phytochemicals. Two of these, phytoestrogens and protease inhibitors, may have the ability to protect humans against cancer. These benefits are still being investigated.

Two other phytochemicals, tocopherols and sterols, have been isolated from soybeans commercially for more than 40 years. The soybean serves as the key raw material for production of natural source vitamin E. It also provides the vast majority of raw materials used throughout the world for steroid production. These phytochemicals are isolated from a by-product of edible oil production called deodorizer distillate. Most soybeans that are grown in the world are crushed, and the oil is extracted. The meal, which is very high in protein, is used primarily for animal feed.

MATERIALS AND METHODS

The oil is processed through a series of steps to provide an edible oil (Fig. 24.1). The first step is degumming, which removes the lecithin and other phospholipids. The second step is refining, usually using aqueous alkali, which removes the free fatty acids. The free fatty acids are then used as a raw material for production of soap. The crude oil at this stage still contains chlorophyll and other colored components, which are removed by bleaching, normally with an activated clay. The final step is deodorization. Deodorization removes the last

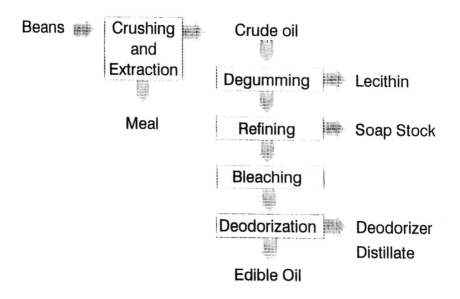

FIG. 24.1. SOYBEAN PROCESSING

traces of free fatty acids and other odoriferous components. It is conducted by heating the oil to a relatively high temperature under vacuum and sparging steam through the oil, so it is essentially a steam distillation conducted under vacuum. The distillate produced by this process is the raw material for production of natural source vitamin E and soy phytosterols.

Soy phytosterols consist of three different sterols in approximately the following concentrations: 60% sitosterol, 20% stigmasterol, and 20% campesterol (Fig. 24.2). These three sterols are found throughout the plant kingdom and are chemical analogues of cholesterol, which is ubiquitous in the animal kingdom. Phytosterols are very similar in chemical structure to cholesterol, but have an additional alkyl group in the side chain. These compounds have found some use in cosmetics, but are essentially inactive physiologically, although sitosterol has been used to treat hypercholesterolemia.

Stigmasterol has an unsaturated side chain, which allows its facile chemical cleavage. Consequently, stigmasterol was used as a raw material for the commercial production of cortisone as early as 1950. At that time, two other materials were commercially important raw materials for steroid production: diosgenin, a saponin isolated from the barbasco plant; and bile acids, which were isolated from the gall bladders of beef cattle collected at slaughtering plants. Today all three of these raw materials have been largely replaced by

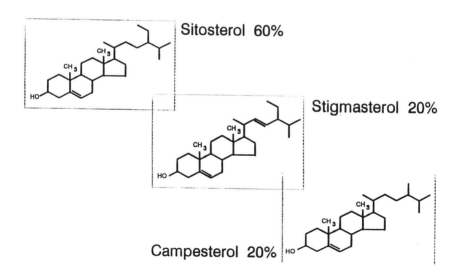

FIG. 24.2. PHYTOSTEROLS FROM SOY

phytosterols from soy. The phytosterol mixture is fermented with bacteria, which cleave the alkyl side chain to produce a steroid compound, which is then converted by subsequent chemical and microbiological processing into cortisone and other medically useful steroids. Today phytosterols from soy serve as raw materials for over 75% of the world's steroid production.

Soy deodorizer distillate is also the major source of alpha-, beta-, gamma- and delta-tocopherols (Fig. 24.3) which differ from each other only by the degree and position of the methyl groups on the aromatic ring. These mixed tocopherols are used as food grade antioxidants or they can be converted into vitamin E. The beta, gamma, and delta tocopherols are effective food-grade antioxidants, but alpha-tocopherol is the most biologically effective source of vitamin E. The non-alpha tocopherols are converted to alpha tocopherol using methylation chemistry which preserves the stereochemistry of the tocopherol molecule. Preserving this stereochemistry is critical because natural RRR-alpha-tocopherol, also called natural-source vitamin E, is more biologically active on a weight basis than is synthetic all-rac-alpha-tocopherol and the stereochemistry is responsible for this enhanced activity.

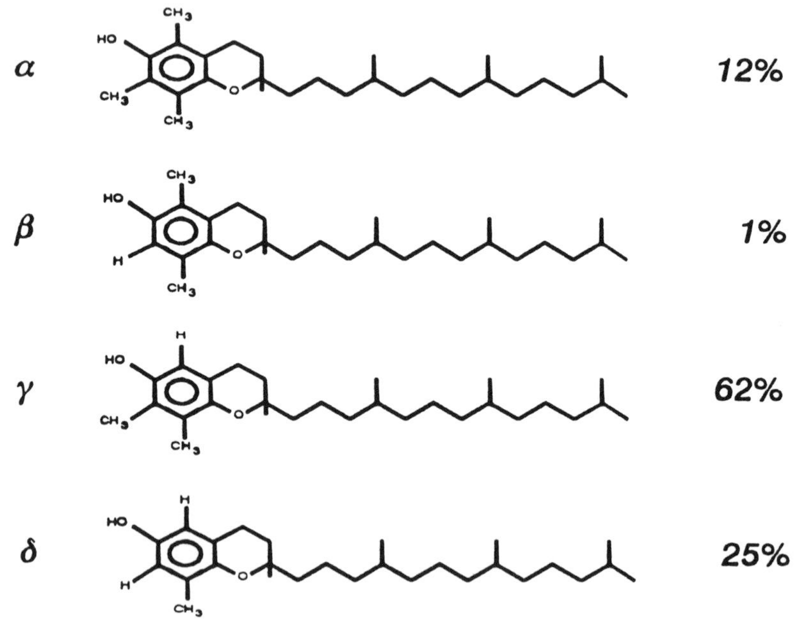

Crude soybean oil contains ~ 1000 ppm total tocopherols

FIG. 24.3. TOCOPHEROLS FROM SOY

It has been shown that the human body has a preference for the natural stereoisomer. Synthetic vitamin E is an equal molar mixture of eight different stereoisomers, only one of which has the same stereochemistry as natural-source vitamin E isolated from soybeans. All eight stereoisomers present in synthetic vitamin E are absorbed equally, but the natural stereoisomer is retained longer consequently and is more biologically effective.

Beneficial effects of vitamin E in human health have been demonstrated in recent medical research. Most of this evidence comes from epidemiological studies, which are supported by biochemical and animal research. Large human intervention trials are in progress which use randomized placebo controls and should give definitive answers on the protective effects of vitamin E against cardiovascular disease and certain types of cancer. However, these studies will not be completed for 4 to 6 years. In the meantime, the consistency of the biochemical, animal and human data shows a clear cardioprotective benefit for vitamin E in human nutrition.

Two of the most convincing papers were published in 1993 in the *New England Journal of Medicine* last year. One paper written by Stampfer[1] and co-workers showed that intake of 100 IU or more per day of vitamin E for at least two years reduced the risk of serious cardiovascular disease by about 40%. This study is based on data collected from nearly 90,000 nurses over an eight-year period. The companion paper written by Rimm[2] and co-workers reviewed data on 40,000 male health care professionals collected over a four-year period. They reached an almost identical conclusion, namely, that the risk of major coronary disease was reduced by approximately 40% for those men who ingested at least 100 IU per day of vitamin E for at least 2 years. It is important to note that this dose is substantially higher than anyone can achieve from a well-balanced diet alone. The typical diet only provides 10-12 IU of vitamin E per day. Obtaining protective levels of vitamin E from the diet alone is particularly difficult for those individuals attempting to reduce their fat intake, because most dietary vitamin E comes from vegetable oil and foods which are high in oil, such as nuts and sunflower seeds. Consequently, decreasing fat consumption also reduces vitamin E intake.

The second observation of this data is that vitamin E supplements had no cardioprotective effect unless they were consumed for at least two years. This is very consistent with biochemical and animal studies which have shown that vitamin E is a protective agent, not a therapeutic agent. Vitamin E appears to function primarily by preventing the formation of atherosclerosis, although there is some animal evidence for reversing atherosclerosis. Research is continuing on the role of vitamin E in prevention of certain degenerative diseases, including cardiovascular disease and certain cancers.

CONCLUSION

In conclusion, phytochemicals from soy (primarily tocopherols and phytosterols) have been shown to have a beneficial role in human health and disease. Research continues to identify more of these useful materials and to determine their role in preventing and curing a wide variety of human diseases. However, tocopherols and phytosterols are the only commercially significant materials isolated from soybeans, which are used in human health and nutrition today.

REFERENCES

[1] Stampfer M.J., Hennekens, C.H., Manson, J.E., Colditz, G.A., Rosner, B. and Willett, W.C. Vitamin E consumption and the risk of Coronary Disease in Women. N. Engl. J. Med. 1993; 328:1444–1449.
[2] Rimm, E.B., Stampfer, M.J., Ascherio, A., Giovannucci, E., Colditz, G.A. and Willett, W.C. Vitamin E consumption and the risk of coronary heart disease in men. N. Engl. J. Med. 1993; 328:1450-1456.

PHYTOPHARMACOLOGY OF LICORICE FOOD FORMS

CHAPTER 25

PHYTOCHEMISTRY OF LICORICE HORTICULTURAL AND PROCESSING PROCEDURES

PETER S. VORA and LUCIA C.A. TESTA

Macandrews and Forbes Co.
3rd Street and Jefferson Avenue
Camden, New Jersey 08104

ABSTRACT

Licorice root phytochemistry is greatly impacted by both the licorice species and the growth cultivation practices. Additionally, phytochemicals in licorice extracts are affected by raw material selection, processing conditions, and parameters. Even though there are many factors affecting consistent quality in licorice extracts, standardization can be achieved through analysis and controlled manufacturing. There are various applications for standardized licorice products in the food, pharmaceutical, and cosmetic industries. It is therefore necessary to understand the impact of cultivation and processing on the phytochemistry of licorice products.

INTRODUCTION

Background

The history of licorice is as old as the history of medicine and of confection. Licorice was recommended by the Indian prophet Brahma[1], was found in the tombs of Egyptian kings[2], and was extensively formulated in traditional Chinese medicine centuries ago.[2,3] Accordingly, these products are currently found in many modern pharmaceutical preparations, as well as food products and personal care products throughout Europe and Asia. In the United States, licorice products are traditionally used for flavoring in the confectionery and tobacco industries. However, in the last 4 or 5 years, there has been a significant amount of research conducted by NCI's Chemoprevention Branch, Diet and Cancer Branch, and various other institutions to evaluate the pharmacological activities of the crude form and the individual components of licorice.[4-7]

The Glycyrrhiza Plant and Its Phytochemicals

The commercially utilized species of licorice plants are *Glycyrrhiza glabra, uralensis, echinata* and *inflata*.[8] There are a number of other species which have less commercial value. The plant species can be identified by the leaf structure configuration.[2] The licorice plant is found between the 30th and 45th parallel[9], starting from the northwest part of China, followed by the former Soviet Republics, the eastern part of Turkey, central Syria, Iraq, Iran, the central part of Spain and the southern part of Italy. The licorice plant is also found in the southern hemisphere, specifically in Australia and South Africa. According to the analytical data and flavor profile, the southern hemisphere root was reported to be acceptable and comparable to licorice root from the northern hemisphere.

There are several factors affecting the phytochemistry of the licorice root.[10] Some of these factors include the soil condition, the environmental factors (temperature variation, humidity, rainfall, the source and availability of water), the variation in the different species, geographical location, and the time of harvesting.

The commercially valuable plants grow in areas with specific environmental conditions.[9,11] They prosper in very hot climates with long dry spells and brief periods of high water availability. For approximately 2 to 3 months, the plains are flooded from river floods or melting snow from the mountains. Then, for approximately nine months, there is no water due to the dry climate. During this time, licorice plants undergo a stress cycle caused by lack of moisture and temperature variations, forcing them to produce secondary metabolites. One of the commercially valuable metabolites produced is glycyrrhizic acid.

The above-ground portion of the plant has no commercial value and is normally discarded during harvesting. The licorice root extends both vertically and horizontally. It may extend as much as 25 feet in search of moisture during the dry, hot weather. The rhizomes extending horizontally propagate themselves and create additional plantlets, forming a very strong matrix underground. The licorice root reaches maturity in 3 to 4 years; at this stage, the root is approximately 1/4 in. to 3/4 in. thick. The horizontal portion of the licorice root is harvested. The taproot is normally untouched, allowing the farmer to harvest the root again the following year. If the entire field is destroyed and the plant is completely removed, a regeneration time of 3 to 4 years is necessary before again harvesting licorice root.

The licorice plant grows wild and is not cultivated commercially. Since it grows in a very hostile environment, there is no other type of vegetation in the area. *Glycyrrhiza inflata* grows approximately 2 to 3 feet in height and has a thick leaf structure with a waxy layer to maximize moisture retention.

Glycyrrhiza glabra grows approximately five feet tall and has smaller leaves.[8,12,13] In the spring, licorice plants have violet flowers which form seeds. A number of studies have been conducted in the past 15-20 years to cultivate the licorice plant from seeds in the above mentioned countries.[12,14,15] It has not been successful, because the yield is poor, and thus is not commercially feasible.

A number of studies have also been conducted to cultivate licorice from root cuttings.[9] MacAndrews & Forbes Co. has conducted experimental trials both in China and in Turkey. These techniques of cultivating from the seeds or cuttings will not have significant effects in terms of maximizing, altering, or modifying the phytochemistry of licorice products, since very little is known regarding the root germplasm found in these areas.

The only way to maximize the plants phytochemical yield is to employ one of the modern plant cultivation techniques, such as micropropagation[16,17] and biofermentation.[18] MacAndrews & Forbes Co. has studied both of these procedures and is currently working with a plant biotechnology firm in India in a joint cooperative project.

Harvesting

When the licorice root is removed from the ground, it has approximately 40-60% moisture. It is necessary to dry the root before it is shipped to the bailing plant. Since in most of the areas where licorice root is found, the climate is very dry, with a very low humidity, the licorice root is simply air-dried. It is then compressed into bales or bagged.

Table 25.1 lists the composition of licorice root. The water-soluble portion of the root is approximately 26-38%. Soluble fiber is almost nonexistent. The insoluble dietary fiber is approximately half of the mass. The other components are glycyrrhizic acid, chalcones, flavonoids, tannins, amino acids, carbohydrates (such as sucrose, glucose, and fructose), and inorganic salts.[19,20] Due to its pharmacological activity, flavoring, and chemical characteristics, glycyrrhizic acid is one of the most important constituents from the licorice plant. Traditionally, the content of glycyrrhizic acid is used to determine the cost and quality of the licorice product.

Glycyrrhizic Acid

Glycyrrhizic acid is a triterpene molecule with an aglycone associated with the two sugar moieties of glucuronic acid[1] as shown on Fig. 25.1. The licorice plant is the only known vegetation in the world with a significant amount of this particular compound.[1] It is present in the licorice root as the sodium/magnesium/potassium salt of glycyrrhizic acid. It is approximately 50 times sweeter than sucrose and has a synergistic effect with other sweeteners.[21]

TABLE 25.1
COMPOSITION OF LICORICE ROOT

	%
Water Soluble Solids	26-38
Soluble Fiber	0.5-1.0
Dietary Fiber, Insoluble	50-60
Glycyrrhizic Acid	1-5
Carbohydrates	18-22
Sucrose	4-8
Amino Acids	0.1-1.5
Chalcones	2-19
Flavonoids/Isoflavones	0-12
Saponins	2-10
Inorganic Salts	6-10

When hydrolyzed with mineral acid or treated with enzymes, it forms glycyrrhetinic acid and two glucuronic acids moieties.[22-24] Neither glycyrrhetinic acid nor glucuronic acid displays any sweetness or other flavoring characteristics. Glycyrrhetinic acid can be used for pharmaceutical applications, and there are various derivatives that can be produced from it which also exhibit pharmacological activities.

Phytochemical Variations in Licorice Root

As mentioned previously, various factors affect the phytochemistry of licorice root. Since glycyrrhizic acid is the most commonly monitored compound in licorice products, this standard phytochemical will be used as a marker to describe the effects of horticulture practices and manufacturing processes.

The geographical area has a significant impact on glycyrrhizic acid content[19,20,25] and yield (the water-soluble portion of the licorice root).[10] Even within the same country, there is a considerable variation in glycyrrhizic acid content and yield. For example, the glycyrrhizic acid content in Turkish licorice root can vary by as much as 2.5% and the yield can vary by as much as 16.0%, as seen in Table 25.2.

Similarly, two different species of *Glycyrrhiza* may vary considerably in phytochemical content.[25] For example, Table 25.3 compares the chemical constituents found in *Glycyrrhiza glabra*, which grows in one of former Soviet Republics to *Glycyrrhiza inflata*, which grows in the Xinjiang Province in the northwestern part of China.[26] On the average, glycyrrhizic acid levels are similar amongst the two species; however, the *Glycyrrhiza glabra* contains the phenolic compounds glabridin and glabrene. These two compounds have antioxidant and antimicrobial properties, and are not found in *Glycyrrhiza inflata*. On the other hand, *Glycyrrhiza inflata* contains licochalcone A and licochalcone B; there are undetectable levels of these two compounds in *Glycyrrhiza glabra*.

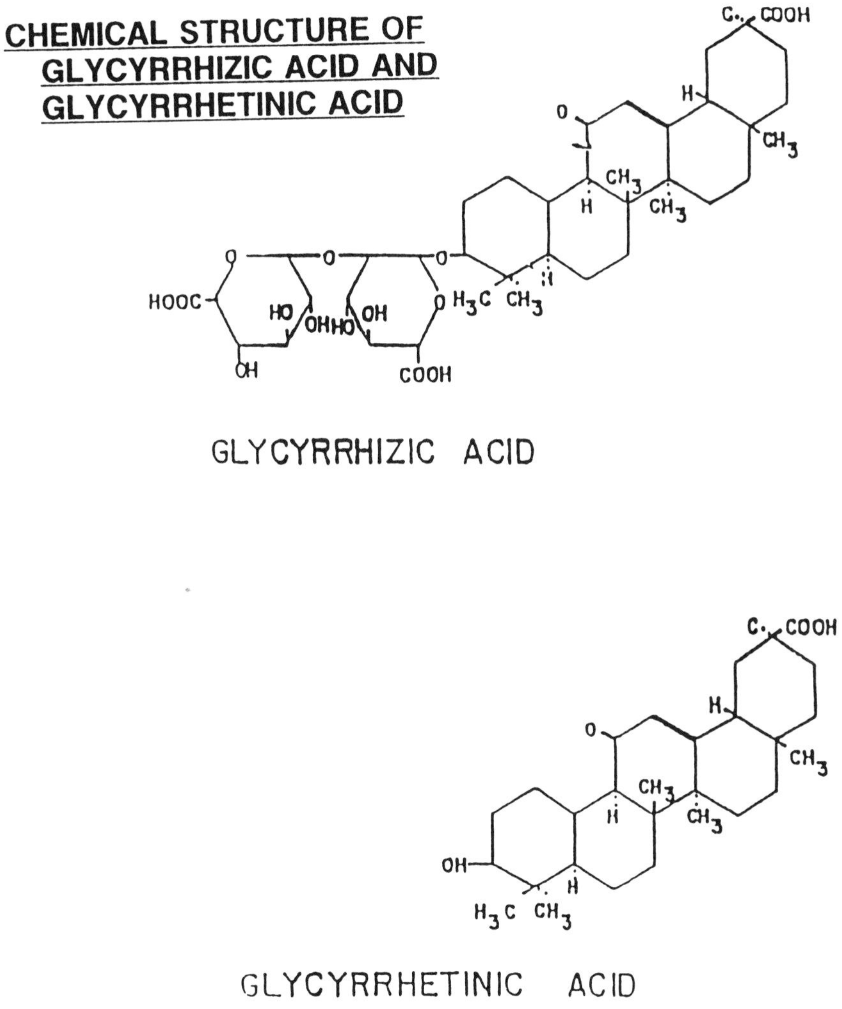

CHEMICAL STRUCTURE OF GLYCYRRHIZIC ACID AND GLYCYRRHETINIC ACID

GLYCYRRHIZIC ACID

GLYCYRRHETINIC ACID

FIG. 25.1. GLYCYRRHIZIC ACID, ONE OF THE ACTIVE MOLECULES FOUND IN LICORICE, IS A TRITERPENE

Glycyrrhizic acid is composed of the aglycone molecule, glycyrrhetinic acid and two glucuronic units

TABLE 25.2
EFFECT OF GEOGRAPHICAL AREA AND ENVIRONMENTAL FACTORS
ON PHYTOCHEMISTRY YIELD OF LICORICE ROOT

TURKISH ROOT	GLYCYRRHIZIC ACID	YIELD
Aegean District	7.6	39.0
Batman District	5.1	31.0
Mus District	6.0	23.0

TABLE 25.3
EFFECT OF SPECIES ON PHYTOCHEMISTRY OF LICORICE ROOT

Components	Glycyrrhiza glabra var. glandulifera	G. Inflata
Glycyrrhizic acid	4.0	4.0
licochalcone A*	---	9.1
licochalcone B*	---	2.8
glabriden*	18.2	---
glabrene*	2.7	---

*Okada *et al.* 1989 Chem. Phar. Bull. 37(9)2528–2530.

The leading countries which produce licorice root are China, Afghanistan, the Soviet Republics, and Iran. Of all the licorice root produced in the world, approximately 30-40% is used in traditional Chinese medicine in countries throughout Asia. The rest is used to produce licorice extract and derivatives for applications in the food, pharmaceutical, cosmetic, and tobacco industries.

Manufacturing Process of Licorice Extract and Its Effects on Phytochemical Content

This section discusses the manufacturing process[9] to produce licorice extract and some of its derivatives. The aqueous extraction is discussed in detail. Licorice root can also be extracted using hydro-alcoholic, alcoholic, or chemical extraction procedures.[19,20] Table 25.4 lists the steps involved in manufacturing licorice extract using the aqueous extraction method. Since there are significant variations in the phytochemical content and other chemical constituents of the licorice root, the selection and blending are important.[10] By blending licorice root from different areas, the end-product always has a consistent quality. The next step is the grinding, where the root is cut to expose the maximum surface area for the extraction step. Extraction is accomplished with water where its parameters play a very important role in determining the yield, chemical property, and flavoring characteristics of the licorice extract. The resultant product is then subjected to filtration, evaporation, and conditioning steps for further processing. Finally, it is finished into different forms of licorice extract, such as spray-dried powder, spray-dried powder with block juice flavor, block form, or semifluid. The total extract can be further processed to isolate or separate some of the individual compounds or fractions depending upon the desired application.

TABLE 25.4
LICORICE MANUFACTURING PROCESS

Selection and Blending

Grinding

Extraction

Filtration/Blending

Evaporation

Conditioning/Blending

Spray Drying, Block Juice Finishing, or
Semi-Fluid Finishing

Treatment or Total Extract To Make Various
Licorice Derivatives

Table 25.5 shows the effects on licorice extract analysis and flavor by varying one of the processing conditions, such as the extraction temperatures. When extraction is accomplished under low temperatures, the glycyrrhizic acid, sugar content, and inorganic salts are significantly higher, compared to extraction accomplished on the same root using significantly higher temperatures and pressure. The yield and flavor characteristics are also significantly impacted by the extraction parameters. Low temperature extraction yields a mild, sweet licorice extract, while higher temperature extraction yields a stronger-flavored extract.

Nutritionally, licorice extract is composed of approximately 70.0% carbohydrates, some vitamin A and a trace quantity of iron and zinc. Licorice products contain no fat or cholesterol. Other components found in licorice extract include glycyrrhizic acid (which depending on the extraction parameters and root selection, can vary from as low as 1.5% to as high as 12%), sugars, starch, gums, inorganic salts, and amino acids.

TABLE 25.5
EFFECT OF EXTRACTION PARAMETERS ON YIELD, FLAVOR, AND
PHYTOCHEMISTRY OF LICORICE ROOT

	Low Temperature Extraction (150-180°F)	High Temperature Extraction Under Pressure
Glycyrrhizic Acid, %	8-11	6-8
Total Sugars, %	20-30	14-22
Inorganic Salts, %	7-14	7-10
Yield, %	24-28	28-35
Flavor	Mild, Very Sweet	Strong Licorice Character with Balanced Sweetness

Table 25.6 lists the various amino acids found in licorice extract. There are approximately 18 different amino acids found in licorice extract, with a total concentration of approximately 5.3%. These amino acids play an important role in licorice extract production. If the extract is subjected to the proper processing conditions, the amino acids can react with the reducing sugars through the Maillard reaction and form furan derivatives[27] and Amadori compounds.[28] These compounds have a strong impact on the aroma and licorice flavor character.

Licorice Derivative Manufacturing

Several compounds can be produced from the licorice extract obtained from the aqueous extraction. When licorice extract is treated with mineral acids, it separates into a crude glycyrrhizic acid and a deglycyrrhizinated licorice extract (DGLE) portion. The DGLE has been used in Europe[29,30] and Asia[31] as an anti-ulcer preparation. The crude glycyrrhizic acid is then treated with alkali to form various salts of glycyrrhizic acid. One of the commonly utilized salts in the U.S. is ammonium glycyrrhizinate (AG). When treated with organic solvents and subjected to a series of separation and isolation steps, the resultant product is monoammonium glycyrrhizinate (MAG), the other commonly utilized salt in the U.S. Both AG and MAG are Generally Regarded As Safe (GRAS)[32] by the FDA. MAG is a white crystalline powder, which can be further converted to dipotassium glycyrrhizinate, disodium glycyrrhizinate, and glycyrrhizic acid. These compounds have various applications in Japan and Europe for their flavoring characteristics and pharmacological activity.[10,33-35] Additionally, MAG can be further hydrolyzed, using mineral acid or enzymes, into glycyrrhetinic acid and then reacted with sodium succinate to form carbenoxolone.[36] This compound is also prescribed in Europe for treatment of ulcers. Some of these derivatives have recently been studied in the U.S. in animal models to determine their cancer preventive properties.[4-7,37]

TABLE 25.6
AMINO ACID CONTENT OF LICORICE EXTRACT

Amino Acids	% (Dry Basis)
Tryptophan	0.31
Aspartic acid	1.17
Threonine	0.20
Serine	0.21
Glutamic acid	0.51
Proline	1.30
Glycine	0.21
Alanine	0.19
Cystine	0.02
Valine	0.21
Methionine	0.02
Isoleucine	0.13
Leucine	0.25
Tyrosine	0.05
Phenylalanine	0.13
Histidine	0.18
Lysine	0.05
Arginine	0.12
Total	5.26

Standardization

Traditionally, the quality and commercial value of licorice root and extract have been determined by the glycyrrhizin content. Glycyrrhizin is the portion of the licorice extract which precipitates upon mineral acid treatment.[19,20,38] This portion of the licorice extract includes the compound glycyrrhizic acid.[20] Thus, glycyrrhizin and glycyrrhizic acid have been used interchangeably in literature causing much confusion.[39-41] Glycyrrhizin content is not a good monitor for quality, since it is nonspecific and the analytical method has poor reproducibility among laboratories.

Various methods have been developed to determine the glycyrrhizic acid content of the licorice root and extract.[1,9] The currently accepted method[42] by the Association of Analytical Chemists (AOAC) is based on HPLC analysis. It was developed by MAFCO 12 years ago. The resultant chromatograph of this assay for licorice products displays well-isolated glycyrrhizic acid peaks. This was verified in a collaborative study conducted in both the U.S.[42] and in Europe.[43] Utilizing this method, licorice extract can be standardized in terms of glycyrrhizic acid content. This method is used to determine the glycyrrhizic acid in licorice products and also in finished products in the cosmetic, food, and pharmaceutical industries.

This assay method has enabled MAFCO to control some phytochemical compounds in the licorice extract. For example, when monitoring the glycyrrhizic acid during the processing of licorice extract over the period of the last eight years from 1984 to 1992, the annual average value for glycyrrhizic acid and standard deviation are fairly consistent over a long period of time. The standard deviation has actually decreased slightly.

Similar control over various other licorice extract parameters and components can be achieved through a combination of precise root blending and controlled processes' parameters. For example, the total sugar content, pH, and residue on ignition has been maintained at a very consistent level.

Applications in the Food, Cosmetic/Personal Care, and Pharmaceutical Industries

Stability of any ingredient in food products is very important. The stability of licorice products in three food applications with various processing conditions was evaluated. For example, the stability of the glycyrrhizic acid found in licorice extract was evaluated in foods when subjected to either baking, microwave, or frying conditions. Comparing the glycyrrhizic acid level of foods before and after processing yielded a 93-100% recovery rate. This indicates that glycyrrhizic acid is stable under normal food processing conditions.

Table 25.7 shows the various applications of licorice products, starting with the food industry where it is used as a flavoring agent, flavor enhancer, and sweetening enhancer.[21,44,45] Table 25.8 describes the applications of licorice products in various food items and their compatibilities. Licorice products also have a wide range of applications in the U.S. and abroad in the pharmaceutical industry.[10,30,33] They are utilized extensively in the cosmetic industry[39,46] for their anti-inflammatory, anti-irritant, and emulsifying characteristics, especially in Japan and in Europe. Additionally, they are utilized in the tobacco industry as a flavoring agent[47,48], in the metal industry to suppress sulfuric acid fumes formed during the zinc plating process, and in the gardening industry where the extracted licorice root is sold as a garden mulch.

TABLE 25.7
APPLICATIONS OF LICORICE PRODUCTS

A. FOOD INDUSTRY:	Flavoring Agent Flavor Enhancer
B. PHARMACEUTICAL INDUSTRY:	Masking/Debittering Flavor Enhancer Anti-ulcer Anti-inflammatory Anti-carie/Gingivitis Anti-tumor Detoxification of Liver
C. COSMETIC INDUSTRY:	Anti-inflammatory Anti-irritant Moisturizer Emulsifier/Stabilizer

D. TOBACCO-FLAVORING AGENT

E. METAL INDUSTRY-FOAMING AGENT

F. GARDEN MULCH

TABLE 25.8
APPLICATION OF LICORICE PRODUCTS IN VARIOUS FOOD ITEMS:
PRODUCT AND COMPATIBLE FOOD PRODUCTS

LICORICE EXTRACT

Suitable with Foods Where Strong Licorice Flavor is Desired
Licorice Confection, Licorice Pastilles, Root Beer
Stable up to pH 5.0 and Higher
Good Temperature Stability
Very Stable

AMMONIUM GLYCYRRHIZINATE

Compatible with Chocolate, Maple Brown Sugar, Caramel, Malt, and Vanilla Flavors
Stable up to 5.0 pH
Good Temperature Stability
Very Stable

MONOAMMONIUM GLYCYRRHIZINATE

Compatible with Wide Variety of Foods, Beverages, Confections, and Nutrient
    Supplements
Good pH Stability, Temperature Stability and Shelf-Life

CONCLUSION

Horticulture procedures and processing conditions play a significant role in the phytochemical composition of licorice root and its resultant products. By employing a standardized process, which is monitored by a standardized analytical technique, a final end-product of consistent quality, flavoring characteristics, and predetermined concentrations of particular phytochemicals can be manufactured on a commercial basis. Thus, it is possible to start from a natural product, standardize it, and produce it in on a consistent basis.

Standardized licorice products can be incorporated to provide unique characteristics in various applications.[49] They have known pharmacological activity; and, at the same time, have significant flavoring and sweetness characteristics. Therefore, licorice products can be combined with other food items with pharmacological activity, such as soy products, garlic, and various fruits and vegetables, and aid to develop consumer-acceptable products with positive flavoring and sweetening characteristics, while providing beneficial physiological properties.

REFERENCES

[1] Nieman, C. Licorice. *In* Advances in Food Research, Vol. 7, (Mrak, E.M. and Stewart, G.F., eds.) Academic Press, San Diego, CA, 1957:339–381.

[2] Kent, C. Licorice — More than just candy. Australian Traditional Medicine Soc. 1994:9–14.

[3] Harada, M. Pharmacological studies on herb paeony root. IV. Analysis of therapeutic effects of paeony- and licorice-containing frequent prescriptions in Chinese medicine and comparison with effects of experimental pharmacological tests. J. Pharm. Soc. Japan 1969; 89(7):899–908 (in Japanese).

[4] Wargovich, M.J. Overview of chemoprevention strategies and drug development. Clin. Chem. 1993; 39:1311.

[5] Weinberg, D.S., Mainer, M.L., Richardson, M.D. and Haibach, F.G. Identification and quantification of isoflavonoid and triterpenoid compliance markers in a licorice-root extract powder. J. Agric. Food Chem. 1993; 41:42–47.

[6] Webb, T.E., Stromberg, P.C., Abou-Issa, H., Curley, R.W. and Moeschberger, M. Effect of dietary soybean and licorice on the male F344 rat — an integrated study of some parameters relevant to cancer chemoprevention. Nutr. Cancer 1992; 18:215–230.

[7] Mirsalis, J.C., Hamilton, C.M., Schindler, J.E., Green, C.E. and Dabbs, J.E. Effects of soya bean flakes and liquorice root extraction enzyme

induction and toxicity in B6C3F1 mice. Food Chem. Toxicol. 1993; 31(5): 343–350.

8 Shou-chuan, L. and Yu-yi, T. A study on the utilization of six species of Glycyrrhiza from China. Acta Phytotaxonom. Sinica 1977; 15(2):47.

9 Molyneux, F. Licorice production and processing. Food Technol. Australia 1975; June:231–234.

10 Lutomski, J., Nieman, C. and Fenwick, G.R. Licorice, Glycyrrhiza glabra L. biological properties. Herba Polonica 1991; 37(3–4):163–178.

11 Isambaev, A.I., Kuz'min, E.V. and Saurambaev, B.N. Distribution and growth conditions of licorice in the valley of the Syr Darya River USSR. Izv. Akad. Nauk. Kaz. SSR. Ser. Biol. 1987; 0:9–13 (in Russian).

12 Hudaybergenov, E.B., Abdrakhmanov, O.K. and Isambayev, A.I. Conditions and perspectives of licorice research in Kazakhstan (glycyrrhiza, wild species, supply of natural vegetation, cultivation). Alma-Ata, Soviet Union: Central Botanical Garden of Sciences of Kazakhstan, 1984.

13 Musaev, I.F. Area graphic characteristics of Glycyrrhiza species. Arealy Rastenii Flory SSSR 1976:85–111 (in Russian).

14 Khudaibergenov, E.B., Abdrakhmanov, O.K. and Isambaev, A.I. Present state and prospects of studying licorice in Kasakhstan. Izvestiia Akademii Nauk Kazakhskoi SSR 1984:1–4 (in Russian).

15 Murav'ev, I.A. and Durdyev, D.D. Lower Amu Darya River as a new region for preparation of licorice root. Aktual Vopr Farm 1974; 2:57–61 (in Russian).

16 Shah, R.R. and Dalal, K.C. Glycyrrhiza glabra (Liquorice): From test tube to field. Plant Tiss. Cult. 1982:685–686.

17 Shah, R.R. and Dalal, K.C. In vitro multiplication of Glycyrrhiza. Curr. Sci. 1980; 49(2):69–71.

18 Yonetani, S., Himi, M., Yoshioka, S. and Fujita, K. Apparatus for high density plant tissue culture. (to Babcock-Hitachi K. K.) Application: Japan 90-86389, 30 Mar. 1990. Japan Kokai 91-285678, 16 Dec. 1991 (in Japanese).

19 Houseman, P.A. The constituents of licorice root and of licorice extract. II. Am. J. Pharm. 1912; 84:531–546.

20 Houseman, P.A. The constituents of licorice root and of licorice extract. I. Am. J. Pharm. 1910; Dec.:533–546.

21 O'Brien-Nabors, L. and Gelardi, R.C. Alternative Sweeteners. Marcel Dekker, New York, 1994.

22 Akao, T., Akao, T., Hattori, M., Kanaoka, M., Yamamoto, K., Namba, T. and Kobashi, K. Hydrolysis of glycyrrhizin to 18-beta-glycyrrhetyl monoglucuronide by lysosomal beta-D-glucoronidase of animal livers. Biochem. Pharmacol. 1991; 41(6/7):1025–1029.

[23] Bombardelli, E. and Curri, S.B. Phytosomes: New functions in cosmetics. Italy, 1987. Milan, Italy: Indena Della Beffa and Centro di Studi per la Dermocosmesi Funzionale. 1987.

[24] Inada, S., Ogasawara, J. and Takahashi, M. Extraction of a glycoside from a natural source and separation of an aglycone therefrom. (to Seitetsu Kagaku Co., Ltd.) Application: Europe 87-306642, 28 July 1987. Europe Patent 255331, 3 Feb. 1988.

[25] Yang, L., Liu, Y.L. and Lin, S.Q. HPLC analysis of flavonoids in the root of six Glycyrrhiza species. Acta Pharmacol. Sin 1990; 25(11):840–848 (in Chinese).

[26] Okada, K., Tamura, Y., Yamamoto, M., Inoue, Y., Takagaki, R., Takahashi, K., Demizu, S., Kajiyama, K., Hiraga, Y. and Kinoshita, T. Identification of antimicrobial and antioxidant constituents from licorice of Russian and Xinjiang origin. Chem. Pharm. Bull. 1989; 37:2528–2530.

[27] Lederer, M., Glomb, M., Fischer, P. and Ledl, F. Reactive Maillard intermediates leading to coloured products. Zeitschrift fur Lebensmittel-Untersuchung und -Forschung 1993; 197:413–418.

[28] Baltes, W. Maillard reactions in foods. Lebensmittelchemie 1993; 47:9–14.

[29] Kanoui, F., Guyot, P. and Marteau, J. Attempt at improving the digestive tolerance of phenylbutasone by deglycyrrhizinized licorice. RNF 1960:347–350.

[30] Fintelmann, V. Modern phytotherapy and its uses in gastrointestinal conditions. Plant Med. 1991; 57:S48–52.

[31] Nakamura, K., Sunaga, H., Ooi, I. and Ozawa, Y. Effects of FM-100, cimetidine and the combination of these two drugs on serum gastrin levels in pylorus-ligated rats. Folia Pharmacol. Jap. 1983; 81:499–505.

[32] Anon. GRAS status of licorice Glycyrrhiza, ammoniated glycyrrhizin, and monoammonium glycyrrhizinate. Fed. Regist. 1983; 48:54983–54990.

[33] Sela, M.N. and Steinberg, D. Glycyrrhizin the basic facts plus medical and dental benefits. In Progress in Sweeteners, (Grenby, T.H., ed.) Illinois:, 1989:71–96.

[34] Anon. Bassett's liquorice allsorts — The root of sweet success. Confection. Prod. 1984; Jan.

[35] Anon. Confectionery — Recent developments in the UK. Res. Studies-ICC Key Notes 1992:1–54.

[36] Gottfried, S. and Baxendale, L. Glycyrrhetinic acid derivatives. Application: United Kingdom 22556/57, 16 July 1957. United Kingdom Patent 843,133, 4 Aug. 1960.

[37] Berry, C. and Vora, P. Effect of glycyrrhizin on DMBA-induced hamster cheek pouch tumors. J. Dent. Res. 1990; 69:385.

[38] Houseman, P.A. Studies on licorice root and licorice root extract. Am. J. Pharm. 1921:1–15.

[39] Kondo, M., Minamino, H., Okuyama, G., Honda, K., Nagasawa, H. and Otani, Y. Physiochemical properties and applications of alpha- and beta-glycyrrizins, natural surface active agents in licorice root extract. J. Soc. Cosmet. Chem. 1986; 37:177–189.

[40] Baba, M. and Shigeta, S. Antiviral activity of glycyrrizin against Varicella-zoster virus in vitro. Avir. Res. 1987; 7:99–107.

[41] Kiso, Y., Tohkin, T., Hikino, H., Hattori, M., Sakamoto, T. and Namba, T. Mechanism for antihepatotoxic activity of glycyrrhizin, I: Effect on free radical generation and lipid peroxidation. Plant Med. 1984; 50:298–302.

[42] Vora, P.S. High pressure liquid chromatographic determination of glycyrrhizic acid or glycyrrhizic acid salts in various licorice products: collaborative study. J. Assoc. Off. Anal. Chem. 1982; 65:572–574.

[43] Petitbois, D., Hamann, Y., Lesgards, G. and Bozzi, R. Analytical study on licorice products. Choice of a chromatographic system for the determination of glycyrrhizic acid. Comparison with the traditional gravimetric method. Ann. Falsif. Expert. Chim. Toxicol. 1985; 78:19–31 (in French).

[44] Cook, M. and Gominger, H.B. Glycyrrhizin. In Symposium: Sweeteners (Inglett, G., ed.) Chapman Hall (AVI), New York. 1974:211–215.

[45] Schiffman, S.S. and Gatlin, C.A. Sweeteners: State of knowledge review. Neurosci. Biobehav. Rev. 1993; 17:313–345.

[46] Kondo, M., Minamino, H., Otani, Y., Miyashita, A., Okada, K. and Kuramoto, T. Alpha-glycyrrhizin solubilizers for clear liquid skin cleansers and conditioners. (to Kanebo, Ltd.; Maruzen Kasei Co., Ltd.) Application: U.S. 462,720, 1 Feb. 1983. U.S. Patent 4,481,187, 6 Nov. 1984.

[47] Tamaki, E., Morishita, I., Nishida, K., Kato, K. and Matsumoto, T. Process for preparing licorice extract-like material for tobacco flavoring. (to The Japan Monopoly Corporation) Application: U.S. 180,867, 15 Sept. 1971. U.S. Patent 3710512, 16 Jan. 1973.

[48] Vora, P. Characteristics and applications of licorice products in tobacco. Tob. Int. 1984; 4:15–20.

[49] Pierson, H. Cancer chemoprevention by design. The J. Health & Healing 1993; 16:3–9.

CHAPTER 26

PHYTOCHEMICAL CONSTITUENTS IN LICORICE

JAMES DUKE

U.S. Department of Agriculture
Agricultural Research Service
Building 003, Room 227
10300 Baltimore Avenue
Beltsville, Maryland 20705-2350
(Retired)

ABSTRACT

Some 400 phytochemicals are reported from the common licorice, Glycyrrhiza glabra L. *This is just one of an estimated 20 species of licorice in the genus* Glycyrrhiza. *Several other compounds reported from other species of* Glycyrrhiza *are not included in this report. The tabulation has been updated by use of Norman Farnsworth's NAPRALERT, and the AGRICOLA and MEDLINE databases, supplemented by entries from selected reprints provided us by the authors. These phytochemicals are tabulated in the format used in the CRC* Handbook of Phytochemical Constituents of GRAS Herbs and Other Economic Plants, *which reported closer to 300 phytochemicals in 1992. Quantitative data, in parts per million (ppm) is reported for nearly half. Biological activities are tabulated for some 25% of these phytochemicals, including dosage-related data for some 15% and lethal-dose data for nearly 10%.*

INTRODUCTION

Licorice, soy and garlic are three very important plants in this important new arena called nutraceuticals — shall we say "rootaceuticals" when we talk about *Glycyrrhiza*. *Glycyrrhiza* means "sweet root." There are 400 compounds from *Glycyrrhiza* glabra — just one of the 18 known species of licorice. Yet, you can find about 50 species' names that have been floating around out there. Most of them contain this sweet compound, glycyrrhizin, which has many biological activities. This report concentrates only on *Glycyrrhiza glabra*, the common licorice.

Like soybean, *Glycyrrhiza* is a legume, and shares many important chemopreventive compounds with soybean. It also has estrogenic isoflavones.

As with soybean and dozens of other edible legumes, licorice contains genistein. It also contains Bowman-Burke inhibitors, phytosterols, saponins and phytates, which also have their chemopreventive attributes. In 1992, I published the CRC *Handbook of Phytochemical Constituents in GRAS Herbs and Other Economic Plants* (Duke 1992). At that time, there were about 335 phytochemicals cited for licorice. When I was invited to write this chapter, I was forced to update my database. And thanks to Napralert, Medline, Agricola, and about 20 different authors who have published recently on *Glycyrrhiza*, the list is now up to 400. Surprisingly, I found published activities for about 25% of these 400 compounds. I have quantitative data, albeit trace on almost half of them. Since 1992 my colleague, Stephen Beckstrom-Sternberg, and I did a paper on spices (Beckstrom-Sternberg and Duke 1994). With our modern database, one can query which of 30 spices in the database have the most cancer-preventive compounds, which have the most anti-inflammatory compounds, which have the most pesticidal compounds, and so on.

Which of our spices have the greatest variety of pesticidal compounds? Licorice! This had good and bad implications. When I talk about the compounds in licorice, I find that I have LD_{50s} for about 10% of them. When you have the quantities of the compounds in a plant, their published activities, effective dosages and lethal dosages, you're beginning — just beginning — to get a rational basis for the interpretation of the medicinal potential of this herb. The Orientals, I think rightly, speak of licorice as a master hormonal herb. It shows up in about half of Chinese combination formulas, because it does have so many activities.

There were 40 pesticidal compounds when I published in 1992. I would wager there are 60 among the 400 today. Licorice also contains the estrogenic isoflavone, genistein, and formononetin, another estrogenic isoflavone. It contains at least 7 or 8 isoflavones — most of which I suspect are mildly estrogenic. This is bringing me to another point, there are often demonstrable synergies among closely related pesticidal compounds within a species.

If you walk from Georgetown University to the river about four blocks away, you will pass the three most important anticancer plants in the world, sources of four "hard drugs." You will pass the yew tree, the source of taxol; there's no *Taxus* out there that doesn't contain taxol in the needles. By the Security Building, the Madagascar periwinkle, *Catharanthus roseus*, is the source of vincristine and vinblastine. When you get down to the flood plain, you'll pass the mayapple, *Podophyllum peltatum*, which contains a lignan used as a starter material for the new cancer drug etoposide, approved first in 1984. Mayapple has at least four important lignans in its root: peltatin, podophyllotoxin, epipodophyllotoxin, and picropodophyllin. An interesting study several years ago reported the antiviral activities of each of these four lignans, and the antiviral activity of the whole extract. The whole extract was 2 to 5 times stronger than an equivalent amount of any one of the lignans, indicating either that there was a synergy between these four compounds, or that there was yet another unknown compound that was stronger than the four. One lignan was taken to make the drug etoposide, and we've left behind three or more, which are probably collectively much more effective. Every time I consider cancer-preventive compounds, there's not one silver bullet compound in the plant, but usually a suite of 5 or 10 or 20 closely related compounds. Evolution would favor synergy between these compounds and would not favor antagonism. I think we are missing something in this mix of lignans. Consider estrogenic isoflavones and the antiangiogenic properties of genistein — be it from licorice, peanuts, or soy — could have a flip side and a positive side, both of which need to be carefully studied. There are cases where you do not want to have new blood vessels developing. For example, if you are faced with metastasis or diabetic blindness. There are probably more cases where you do want more blood vessels.

I've seen papers which listed 23 phytoalexins in the lima bean (*Phaseolus lunatus*), including five that are genistein relatives. These are clearly plant protective: most of them had fungitoxic properties. If my speculation is correct, that these should be synergetic rather than antagonistic; then there's a powerful combination in the lima bean, with almost 10% phytoalexins — five of them based on genistein, others based on phaseollidin. Many of these same compounds occur in licorice.

Hatano *et al.* (1988) has pointed out that glycyrrhizin inhibited the formation of giant cells by the AIDS virus at a level of 500 μg/ml. Among other compounds closely related to the glycyrrhizin, the levels required for this inhibition were closer to 20 μg/ml. Searching Medline one can find where it's not one, but 5, 10, or 15 *closely related* compounds. It's also true of MAOI (monoamine oxidase inhibitors) antidepressants, at least 12 of which occur in licorice.

An article in *Science News* (Anon. 1994) relates to licorice, albeit tangentially. They reported on mice with a catheter (into the heart) which could inject themselves by certain actions. If nicotine instead of saline solution was provided, they injected 90 times a day — the equivalent of 90 cigarettes, instead of the 8 times for saline. If they injected acetaldehyde, one of the aldehydes in licorice and many if not all legumes, they injected 240 times a day — the equivalent of 240 cigarettes. But if they mixed the nicotine and the acetaldehyde, the mouse would inject 400 times — more than the sum of 240 + 90. The conclusion, as interpreted by *Science News*, was that the acetaldehyde was more important as a "hook" than the nicotine. Perhaps I should have used licorice sticks along with carrot sticks when I quit smoking — the carrots for the beta carotene and the licorice for withdrawal.

In regard to estrogens, genistein is estrogenic, and daidzin is probably estrogenic. Instead of spraying kudzu with herbicides, it might make more sense to harvest the kudzu and take out the genistein and daidzin. Daidzin hit the market in a very beautiful bottle called "Kudzu Power." If you've ever purchased "arrowroot" in an Oriental store, chances are it was kudzu. This *may* be a better source of genistein and daidzin than soybean. It may not. Until we get the licorice, soybean and kudzu all analyzed in an unbiased lab, we can't be sure which may be the best dietary source of estrogenic isoflavones.

The Hoxsey formula (Duke 1988/1989) for cancer includes red clover and is also endowed with genistein and other estrogenic isoflavones. Some clovers are so well-endowed with estrogenic isoflavones, that they cause problems on the range in Australia, namely abortions in grazing cattle.

Apigenin, possibly a precursor to glycyrrhizin and glycyrrhetinic acid, is quite active on its own merits. Widely distributed in many plants, apigenin is documented in the literature to be antimutagenic with an inhibitory dose at 10-40 nM.

I'm just a compiler and do no original research myself. I do grow plants to make sure people doing the original research know what plant they're dealing with. The computer can go through and ask which of our compounds have both cancer preventive and pesticidal activities. When we ask which licorice anticancer compounds also have pesticidal activity, the answer is about 20. Licorice, too, shows good immunomodulatory activity. Benzaldehyde probably has some reinforcing properties, but it's not very toxic, at least on an LD_{50} basis.

Formononetin is one of several estrogenic isoflavones that occur in licorice. I'm sure it occurs in many clovers and possibly in the soybean. The antiangiogenic activities of genistein are suspected of preventing the growth of newly developing tumors and metastases.

Dr. Judah Folkman at Children's Hospital in Boston is working on hemangioma. He uses petri dishes each containing a chicken egg yolk, with radiating blood vessels. He adds various test compounds to see if he can prevent the development of new blood vessels. He believes a good source of genistein (I'm not promoting soy or lupine or groundnut) would probably help a lot of these people, on whom he can't ethically operate, by slowing down the formation of new blood vessels. He also stated that he thought a good dose of genistein, or some other good antiangiogenic compound, could prevent about 10% of diabetic blindness. Genistein can prevent the growth of small tumors (1 mm) — if they don't get a blood vessel, they don't grow.

Another American Indian food, the groundnut, *Apios americana*, turns out to be a good source of genistein; it has daidzin like the kudzu, and it also has coumestrol. Are they synergetic? I don't know.

The red and black seeds of the genus *Abrus*, is a very dangerous genus but also a legume. It also contains some glycyrrhizin but at much lower levels than licorice. I'm a believer in food pharmacy and synergy among pesticidal medicinal compounds of foods. I've compiled a list of some 100 "herbal alternatives." These are alternatives to chemical drugs on the market. I think the alternatives might be as good as the drug on the market but I don't really know. In preparation of this chapter, I made a discovery in the literature that completely modified the formulas for one of my so-called food pharmaceuticals. I call it "prosnut butter." Prosnut butter is a ground up mixture of saw palmetto seeds, (saw palmetto seeds contain compounds that, if ingested, will prevent the conversion of testosterone to dihydrotestosterone), pumpkin seeds and licorice.

The seeds of the pumpkin, are very rich in three amino acids (alanine, glutamic acid, glycine) and naturopaths say alleviate benign prostatic hypertrophy (BPH). Merck's new drug Finasteride, predicted to be a billion-dollar drug, also prevents the conversion of testosterone to dihydrotestosterone. The licorice stick also contains a compound, glycyrrhetinic acid, that may prevent the conversion of testosterone to dihydrotestosterone. So I have three American Indian food plants: the weedy western licorice, *Glycyrrhiza lepidota*, Saw palmetto (eaten by the Seminole Indians), and the pumpkin seed. I grind these up into a mixture that's fairly palatable and that will do the same thing as Finasteride, at least as far as preventing conversion of testosterone to dihydrotestosterone, thereby preventing the problem of BPH.

Dr. Stephen Beckstrom-Sternberg (1994) has coaxed from our database the names of 400 licorice phytochemicals. I suspect that most of the pesticidal compounds will be synergetic in many of their activities, though there invariably will be an exceptional case of antagonism.

Synergy's the self-serving song I sing.
Why should I sing this?
A little knowledge is a dangerous thing;
Complete ignorance is bliss.

REFERENCES

[1] Beckstrom-Sternberg, S.M. and Duke, J.A. Potential for Synergistic Action of Phytochemicals in Spices. *In* Developments in Food Science, Vol. 34, Spices, Herbs and Edible Fungi, (Charalambous, G., ed.) Elsevier Science B.V., Amsterdam, 1994; 201–223.

[2] Duke, J.A. Handbook of Phytochemical Constituents in GRAS Herbs and other Economic Plants. CRC Press, Boca Raton, FL, 1992; 654.

[3]　Duke, J.A. The Synthetic Silver Bullet vs. the Herbal Shotgun Shell. HerbalGram 1988/89; 18-19: 12-13. Fall 1988/Winter 1989.

[4]　Hatano, T., Yasuhara, T., Miyamoto, K. and Okuda, T. Anti-Human Immunodeficiency Virus Phenolics from Licorice. Chem. Pharm. Bull. 1988; 36:2286–2288.

[5]　Keung, W-M. and Valley, B.L. Daidzin and daidzein suppress free-choice ethanol intake by Syrian Golden hamsters. Proc. Natl. Acad. Sci. 1993; 90:10008-10012.

CHAPTER 27

LICORICE: ABSORPTION, DISTRIBUTION, METABOLISM AND CANCER CHEMOPREVENTION

RAJENDRA G. MEHTA, VERNON STEELE*, HERBERT PIERSON**, ANDREAS CONSTANTINOU and RICHARD C. MOON

*Chemoprevention Program
Department of Surgical Oncology
University of Illinois, Chicago*

**Chemoprevention Branch
The National Cancer Institute
Bethesda, MD*

***Preventive Nutrition Consultants Inc.
Woodinville, WA 98072
Deceased*

ABSTRACT

This is a comprehensive chapter describing the bioavailability and metabolism, as well as dose tolerance of 18-β-Glcyrrhetinic acid (GA), an active component of licorice roots. In recent years, attention has been focused to consider licorice as a possible cancer chemopreventive agent. Results summarized in this report suggest that 18-β-GA and carbenoxolone are effective chemopreventive agents against carcinogen-induced preneoplastic lesions in mouse mammary gland organ culture. Moreover, both carbenoxolone and GA inhibited chemically-induced mammary carcinogenesis in rats. Further studies also have shown that while GA induces estrogen receptor modestly, it dramatically down-regulates progesterone receptors in uterus and mammary glands. Since there was a potent inhibition of progesterone receptors by GA, it may have effects towards other physiological events related to progesterone receptors.

INTRODUCTION

Licorice, the root of *Glycyrrhiza garba* (Leguminosae), has been traditionally used in Chinese herbal medicine for hundreds of years. Ethnomedical information suggests that it has been used for a variety of purposes including contraception, liver disfunction, dysmenorrhea, cough, and cancer.[1] A large number of compounds have been isolated from licorice which includes carbohydrates, coumarins, lignans, lipids, flavonoids, triterpenes, etc.[2] The biological effects include, but are not limited to antibacterial, analgesic, antimutagenic, antispasmodic, antiulcer, anti-inflammatory, anti-immunomodulatory and antifungal.[1,3,4] Such a broad scope of activity and chemical composition makes licorice a unique edible material. In order to selectively understand properties of the chemical components of licorice, considerable attention has been directed to pentacyclic triterpene β-glycyrrhetinic acid (3-hydroxy-11 oxo-18β-olean-12-en-30-oic acid) (GA) and its glucoside glycyrrhizin (GL). Glycyrrhizin is a principal component of licorice and is ingested orally as a sweetener. When administered orally to humans, R-glycyrrhetinic acid, an aglycon of GL, is detected in the serum. Since GA and its synthetic analog carbenoxolone (3-O-β-carboxypropionyl-GA) have been used clinically, and often pseudo-aldosteronism is associated as a side effect of carbenoxolone[5], attention has been focused towards understanding the absorption, distribution and dose tolerance of GL, GA and carbenoxolone in experimental models. The chemical structures of these agents are shown in Fig. 27.1.

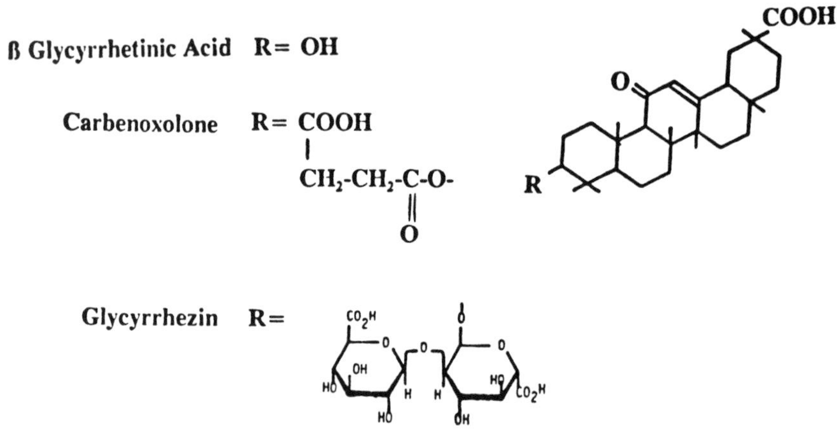

ß Glycyrrhetinic Acid   R= OH

Carbenoxolone   R= COOH
                    |
                 CH₂-CH₂-C-O-
                         ‖
                         O

Glycyrrhezin   R=

FIG. 27.1. CHEMICAL STRUCTURES OF SELECTIVE COMPONENTS FROM LICORICE ROOT

In addition to numerous activities attributed to licorice root, in recent years its possible use in cancer prevention has been explored.[1,6] The demonstration of a cancer-prevention strategy requires that the active chemopreventive agent, at a totally nontoxic concentration, either prevent initiation of transformation or should substantially inhibit the process of promotion and/or progression. Since the ultimate goal is to prevent cancer in the general population, the phytochemicals present in edible foods have been considered as prime candidates for chemoprevention research. Several major groups of foods have been identified including indole derivatives in cruciferous vegetables, limonene in citrus fruits, phenolic amides in green peppers, diallyl disulfide from garlic, etc.[6,7] GA has been studied for its chemopreventive activity in a two-stage mouse skin carcinogenesis model. Results have shown that GA effectively suppressed DMBA-induced and TPA promoted skin tumorigenesis in mice.[8,9] During the past few years, we have studied the effectiveness of several chemopreventive agents including GA and carbenoxolone against carcinogen-induced mammary carcinogenesis *in vitro* and *in vivo*. Results obtained from these studies are reviewed in this chapter.

Since multiple physiological functions are attributed to licorice root, the mechanistic studies have been less focused. However, concerted efforts have been made to understand the (anti)steroidal action of licorice root and its constituents. This was of particular interest since GA and carbenoxolone-induced pseudo-aldosteronism is of principal concern as a result of their use as therapeutic agents. It has been well-recognized that certain plant constituents, such as flavonoids function as natural phytoestrogens. These phytoestrogens are converted to equol by the intestinal flora, which in turn have both weak estrogenic and antiestrogenic activity. The anti-estrogenic activity of GA has been well-documented by Kraus and others.[10,11] Very little information is available on GA interactions with steroid receptors, which could be of significant importance in the chemoprevention of mammary carcinogenesis.

Absorption and Distribution of GL and GA

Experiments in the literature on the subject of absorption and tissue distribution of licorice components, especially GL and GA, are largely incomplete. During the past few years extraction and HPLC procedures have been developed to separate GL, GA and their metabolites in the serum, urine, bile or organ tissues.[12-14] Recently a complete distribution comparison of GL and GA has been reported. Ichikawa *et al.*[12] injected 100 mg/Kg GL or 60 mg/Kg GA in male Wistar rats. Urinary bladder, bile duct and femoral artery were cannulated for collecting samples. Bile, urine and blood were collected for a 24 h period, whereas other organs were collected 1 h after the intravenous injection with the licorice components. HPLC analysis was carried out to separate GL and GA. The half-lives of elimination for GL were 50 min, whereas for GA it was 80 min.[12] GL was largely concentrated to GA in the liver and kidney. GA on the other hand was retained in the brain, liver, kidney and skin. However, the concentration in liver was significantly lower than GL. The majority of GL was secreted in bile, whereas a negligible amount of GA was present in bile or urine. These results suggested that although GL remains intact in the rat, GA is readily metabolized.[12,13] The results from Ichikawa et al. are summarized in Table 27.1. In a separate study Sakiya et al.[14] have shown that following a bolus injection of GL, no GA was detected in the serum indicating that the conversion of GL to GA occurs in the small intestine. The results are consistent with those of Ichikawa, in that GA is rapidly metabolized, whereas GL is distributed in various organs. Recently, a comparison was made between the absorption of GL and GA in germ-free rats versus conventional rats.[15] Results showed that orally-administered GL was poorly absorbed from the gut and is hydrolyzed by the intestinal bacteria. No GA was detected for up to 17 h after the administration of GL in germ-free rats; whereas GA was detected in conventional rats after 4 h of oral treatment with GL.

TABLE 27.1
TISSUE DISTRIBUTION OF GLYCYRRHIZIN AND GLYCYRRHETINIC ACID IN RATS

Tissue	Glycyrrhizin Recovered (%)	Glycyrrhetinic acid Recovered (%)
Lung	5.0	13.8
Brain	0.0	2.4
Liver	24.5	9.0
Kidney	11.1	7.9
Skin	6.7	10.0

Adopted from Ichikawa et al. (12).

Metabolism of GL and GA

The consumption of licorice as a sweetener, or its components and analogues, such as GA and carbenoxolone, is almost always by the oral route. Therefore metabolism of GL and GA has been studied in the intestine or by the intestinal flora. Hattori and colleagues[16] reported on the incubation of intestinal bacterial mixture, prepared from human feces, with either glycyrrhizin or β glycyrrhetinic acid for 48 h. Metabolites were separated and identified using silica gel thin-layer chromatography and mass spectral analysis. Results indicated that glycyrrhizin is initially metabolized to its aglycone form 18-β-glycyrrhetinic acid and then transformed by epimerization to 3-epi-18-β-glycyrrhetinic acid. This epimerization of β-glycyrrhetinic acid was reversible via an intermediate,

3-hydro-18-β-glycyrrhetinic acid. Hattori subsequently identified, *Ruminococcus* sp., the strains of bacteria responsible for the hydrolysis of glycyrrhizin to glycyrrhetinic acid and the reduction of 3-dehydroglycyrrhetinic acid to GA. However, *Clostridium innocuum* was responsible for the conversion of 3-dehydro-GA to 3-epi-GA. A mixture of both bacterial strains completed the entire metabolic spectrum.[17]

Dose Tolerance of GA and Carbenoxolone

In addition to its therapeutic use, licorice has been extensively used as a sweetener. A high intake of licorice can cause hyper-mineralocorticoidism with potassium loss and sodium retention.[17] As a result increased blood pressure and deregulation of aldosterone effects may be noticed. This makes it imperative to evaluate the tolerance of a maximum nontoxic dose of licorice for human consumption. Stormer *et al.* in a review[19], indicated that if one assumes that 0.2% of glycyrrhizic acid is present in licorice root, then 50 g of licorice root will generate 100 mg of glycyrrhizic acid. This amount will be sufficient to produce adverse effects in a sensitive individual. Generally people consuming approximately 400 mg glycyrrhizic acid demonstrate the side effects of the agent. Since 100 mg glycyrrhizic acid is found to be the lowest concentration with measurable side effects, 10% of that concentration or 10 mg of glycyrrhizic acid (5 g of licorice roots) per day may be considered to be a safe dose.[19]

A dose tolerance study for determining the maximum tolerated dose of GA and carbenoxolone in rats was conducted in our laboratory. Rats were divided in six groups and fed increasing dose levels of GA and carbenoxolone mixed with AIN 76A diet for six weeks. The concentration of these agents in the test diet ranged from 625 mg/kg diet to 10,000 mg/kg (1%). Animals were weighed twice a week. Results showed that there was no difference in the body weights between control and the 2,500 mg/Kg GA dose. At the 5,000 or 10,000 mg/Kg dose level, there was a significant reduction in the body weight gain. These results are summarized in Table 27.2. Since 2,500 mg/kg was the maximum non-toxic tolerated dose for β glycyrrhetinic acid, carcinogenesis experiments were conducted using 80% of the maximum tolerated dose (2g/kg diet). An identical dose tolerance pattern was observed for carbenoxolone. Carbenoxolone also affected body weight at 5,000 mg/kg diet.

In a separate toxicity study with glycamil (ammonium glycyrrhitinate) it was observed that the LD_{50} for rats and mice was >5,000 mg/kg of oral dose, whereas for guinea pigs it was >3,000 mg/kg in an acute toxicity study. In a subchronic study of eight weeks with 700 mg/kg of glycamil, it was also found to be non-toxic in terms of its effects of body weight, hepatic function, electrolyte balance and organ weights. Similar results have been obtained in a chronic study with rats and mice with a 90 mg/kg dose level.

TABLE 27.2
DOSE TOLERANCE OF DIET SUPPLEMENTED WITH β
GLYCYRRHETINIC ACID (GA) AND CARBENOXOLONE IN RATS

β GA in Diet (mg/kg diet)	Initial Body Weight	Terminal Body Weight (grams)	% Increase in Body Weight	% Survival
None	135	220	63.0	100
625	135	218	61.5	100
1,250	135	221	63.7	100
2,500	135	222	64.4	100
5,000	135	213	57.8	100
10,000	135	210	55.6	100

Female Sprague/Dawley rats were fed diets supplemented with β glycyrrhetinic acid for 47 days. Animals were weighed twice weekly and observed for any signs of toxicity twice a day.

Chemoprevention of Mammary Carcinogenesis by GA and Carbenoxolone

Several reports in the literature have appeared, which strongly suggest chemopreventive activity of components of licorice root in a two-stage skin carcinogenesis model. Nishino et al.[8,9] reported that 18-β-GA inhibited 7,12 dimethylbenz[a]anthracen (DMBA)-induced and 12-O-tetradecanoylphorbol-13-acetate promoted skin cancer incidence by more than 50 %. This inhibitory effect was more prominent when teleocydin was used as a promoter. We have been evaluating potential chemopreventive agents in a variety of in vivo and in vitro carcinogenesis models. Chemically-induced in vivo carcinogenesis in urinary bladder, lung, pancreas, prostate, skin and mammary gland have been extensively employed.[20] On the other hand, we have utilized an in vitro approach in which the effectiveness of a suspect chemopreventive agent is identified using a mouse mammary gland organ culture model.[21] Once the compound is considered effective in the organ cultures, then the potential chemopreventive agent is further evaluated in a carcinogen-induced in vivo rat mammary carcinogenesis model. This approach was employed for the evaluation of β-GA and carbenoxolone as chemopreventive agents.

Mammary Gland Organ Culture (MMOC)

Based upon the concept that the mammary gland responds to growth promoting protein and steroid hormones to alter morphology and functional state of the organ in animals and humans, an organ culture model was established.[22,23]

It was found that organ tissue not only responds to hormones in culture to induce differentiation, but also responds to the carcinogen. Under appropriate hormonal conditions, the organ culture tissue develops preneoplastic mammary lesions (ML), which would form adenocarcinoma upon transplantation in syngeneic mice.[24] We have further modified this model system to determine if the chemopreventive agents would suppress the development of these ML (Fig. 27.2). Results have consistently shown that the compounds that suppress the development of ML also inhibit carcinogen-induced incidence or multiplicity of mammary tumors.[21,25] We examined effects of 18-β-GA and carbenoxolone on the development of DMBA-induced ML in organ culture.

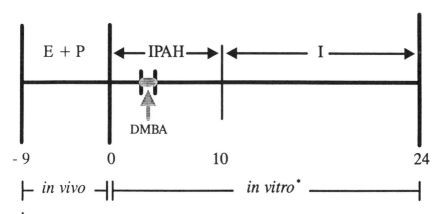

*Chempreventive Agent Present Days 0 - 10

FIG 27.2. EXPERIMENTAL DESIGN FOR THE INDUCTION OF MAMMARY LESIONS IN ORGAN CULTURES

The detailed procedure of MMOC has been described elsewhere.[22-25] Briefly, the glands were incubated with growth-promoting hormones for 10 days followed by the withdrawal of hormones for an additional 14 days. The carcinogen, DMBA (2μg/ml) was included in the medium for 24 h between 72 and 96 h of the incubation. Fifteen glands per group were incubated with five concentrations of β-GA or carbenoxolone during the growth-promoting phase of the first 10 days of culture. At the end of the study, glands were fixed and evaluated for the presence or absence of the lesions. Results are shown in Fig. 27.3. Both 18-β-GA and carbenoxolone inhibited ML development in a dose-responsive manner. GA at 10^{-5}M showed reduced effectiveness. This was attributed to the toxic effect of GA at that concentration. Carbenoxolone was less effective than GA, however, no toxicity was associated with carbenoxolone in MMOC. These results suggest a possible chemopreventive activity of the licorice-derived agents β-GA and carbenoxolone in a mammary carcinogenesis model.

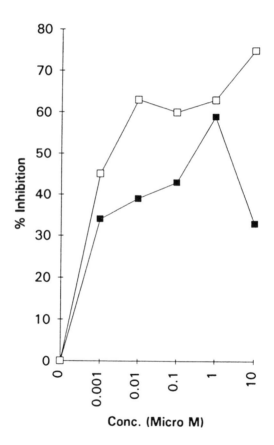

FIG. 27.3. INHIBITION OF DMBA-INDUCED MAMMARY LESIONS BY
β GLYCYRRHETINIC ACID AND CARBENOXOLONE
Mammary lesions were induced in the mammary glands in organ cultures as described in the text. β Glycyrrhetinic acid (---■---) and carbenoxolone (---□---) were included in the medium between days 0-10 of the growth promoting phase. Percent inhibition was calculated by normalizing the data for the incidence in the control group of glands.

Effects of GA and Carbenoxolone in Carcinogen-Induced Mammary Carcinogenesis Model

There are two widely used *in vivo* mammary carcinogenesis models: one utilizes DMBA as a carcinogen, whereas the other one utilizes N-methyl-N-nitrosourea (MNU) as a carcinogen. We have extensively used both of these models in our laboratory.[20,26] For the present study, the MNU-induced mammary carcinogenesis model was employed. The detailed procedure has been described previously.[26] Typically, 50-day-old female Sprague Dawley rats are treated with a single intravenous injection of 50 mg/Kg body weight of MNU. Animals are fed the AIN 76A semi-purified diet as a basal diet. The chemopreventive agents are included in the diet by thoroughly mixing them with the basal diet. Normally the chemopreventive agents are mixed at 40 and 80% of the maximum tolerated dose level, to ensure that the effect observed with the test agent is not related to the toxicity of the agent. As discussed above, the maximum tolerated non-toxic dose for both GA and carbenoxolone was 2,500 mg/Kg diet. Thus, the carcinogenesis experiments were carried out at 1,000 and 2,000 mg/kg dose levels. Both GA and carbenoxolone did not affect the tumor incidence during a 100-day experiment. At the 2,000 mg/kg of GA dose level, the tumor multiplicity was significantly reduced by 59% (6.2 tumors/rat in control vs 2.6 tumors/rat). At the reduced dose level of GA, 1 g/kg, the effect was 48% inhibition. At 2,000 mg/kg, carbenoxolone also inhibited multiplicity by 49% (6.9 control vs 3.5 tumors/rat). Both GA and carbenoxolone increased tumor latency and the tumor weights were reduced (unpublished).

The effect of GA and carbenoxolone has been evaluated in a variety of experimental carcinogenesis systems. In addition to skin and mammary carcinogenesis models, carbenoxolone was found effective against liver and lung carcinogenesis; however, carbenoxolone was ineffective against colon[27,28] and buccal pouch (unpublished) carcinogenesis.

Effects of Licorice Root Extract (LRE) on Steroid Receptors

Although the effect of licorice in mineralocorticoid, metabolism, has been recognized for a long time, only recently has the mechanism of its action been understood. Briefly it has been postulated that licorice, and specifically GA, inhibits 11β hydroxysteroid dehydrogenase.[29] This enzyme inactivates corticosterone and cortisol and converts them to dehydrocortisone and cortisone. Aldosterone is not metabolized by this enzyme. Thus, in cells with high concentration of 11β hydroxysteroid dehydrogenase, the native glucocorticoids are inactivated leaving aldosterone as the steroid to regulate electrolyte balance. Inhibition of this enzyme activity by licorice would elevate the levels of corticosterone and cortisol resulting in increased mineralocorticoid activity.

The other major functional property of licorice has been reported to be its antiestrogenic activity.[10,11] Although the steroid receptor modulation by licorice is highly relevant to cancer prevention research, it has not been investigated. We investigated the effects of dietary licorice root extract on the steroid receptors in uterus and mammary glands. Uteri of mice and rats contain relatively high amounts of estrogen and progesterone receptors.[30] Similarly, the mammary gland is the target organ for hormone dependent and steroid receptors play a significant role in the modulation and management of breast cancer. Therefore, the effects of licorice root extract on the modulation of steroid receptors were evaluated in these two tissues.

Young Sprague Dawley female rats were maintained on modified AIN 76A diet. Diets were supplemented with 0.5, 5 and 10% licorice root extract (LRE). Animals consumed the licorice supplemented diet for 21 days. There was no effect of LRE on the body weight gain of the animals. Estrogen and progesterone receptors were measured by receptor titration assays. Typically, aliquots of cytosol prepared from the tissues were incubated with increasing concentration of [H³]estradiol for estrogen receptors or [H³]R5020 for progesterone receptors, either alone or in the presence of 100-fold excess unlabeled diethylstilbestrol or R5020 for 16 h at 0-4°C. Reactions were terminated by dextran coated charcoal treatment. Specific binding was calculated and Scatchard plots were generated. Results are shown in Table 27.3. As expected, the uterus contained high concentrations of estrogen and progesterone receptors. LRE increased estrogen receptors in a dose-related manner. There was a two-fold increase in the estrogen receptors with 10% LRE treatment. However, there was a dramatic down regulation of progesterone receptors by the licorice components. The binding was reduced from 413 fmoles per mg protein to 18 fmoles per mg as also observed for the protein. Such total "shut-down" of progesterone receptors was also observed for the mammary gland tissue. It is realized that both estrogen and progesterone receptor concentrations are very low in the mammary gland, nonetheless the trend consistent with that observed with uterus is quite evident.

To verify the effect of LRE on progesterone receptors a second experiment was carried out. Animals were either kept intact or ovariectomized. One week following the surgery, animals received a 30-day release 0.5 mg estrogen 17β pellets s.c, and either received a basal diet or a diet supplemented with 5% LRE for one month. Uteri were removed and steroid receptors were measured. Results are shown in Table 27.4. As expected, estrogen treatment enhanced both estrogen and progesterone receptors in ovariectomized animals. Ovariectomy almost totally inhibited available protein receptors, however, estrogen treatment increased it from 24 to 268 fmole per mg protein. LRE, however, inhibited estrogen induction of progesterone receptors. Again these results clearly suggest that LRE down regulates progesterone receptors in rats.

In intact rats, the results were consistent to that described in the previous experiment. Although estrogen modestly increased estrogen and progesterone receptors. LRE up regulated estrogen receptors and down regulated progesterone receptors.

TABLE 27.3
EFFECT OF LICORICE ROOT EXTRACT (LRE) ON STEROID RECEPTORS

Tissue	Treatment	ER (fmole/mg protein)	PR (fmole/mg protein
Uterus	Control	243 ± 174	413 ± 40
	0.5% LRE	373 ± 142	10 ± 1
	5.0% LRE	418 ± 101	25 ± 9
	10.0% LRE	373 ± 142	18 ± 9
Mammary Gland	Control	4.0	25 ± 1
	0.5% LRE	4.0	No Binding
	5.0% LRE	2.0	No Binding
	10.0% LRE	13.0	No Binding

Rats were fed either modified AIN 76A diet supplemented with LRE. LRE was obtained from the Arthur D. Little and Co. Steroid receptor assays represent mean of three experiments, each in duplicate. Scathard plots were generated and the number of binding sites and binding affinity were calculated.

TABLE 27.4
EFFECT OF OVARIECTOMY AND LICORICE ROOT EXTRACT ON STEROID
RECEPTORS IN RAT UTERUS

Treatment	ER (fmole/mg protein)	PR (fmole/mg protein)
Intact	64	108
Intact + E_2	70	173
Intact + E_2 + 5% LRE	162	No Binding
Ovariectomized	133	24
Ovariectomized + E_2	198	268
Ovariectomized + E_2 + 5 % LRE	139	158

Rats were either left intact or ovariectomized bilaterally. Animals were fed either modified AIN 76A diet or AIN 76A diet supplemented with LRE. Estrogen treatment was employed by transplanting an estradiol pellet dorsally in the animal as described in the text. LRE was obtained form the Arthur D. Little and Co. Steroid receptor assays represent mean of three experiments, each in duplicate. Scatchard plots were generated and the number of binding sites and binding affinity were calculated.

REFERENCES

[1] Davis, E.A. and Morris, D.J. Medicinal uses of licorice through the millennia: The good and plenty of it. Mol. Cell Endocrinol. 1991; 78:1-6.

[2] Hatono, T., Kagawa, H., Yasuhara, T. and Okuda, T. Two new flavonoids and other constituents in licorice root: Their relative astringency and radical scavenging effects. Chem. Pharm. Bull. 1988; 36:2090-2097.

[3] Zhang, H.Q., Liu, F., Sun, B. and Li, G.H. Antiallergic action of glycyrrhizin. Chung Kuo Yao Li Hseun Pao 1986; 7:175-177.

[4] Pompei, R., Flore, O., Marccealis, M.A. and Loddo, B. Glycyrrhetinic acid inhibits virus growth and inactivates virus particles. Nature 1979; 281:689-690.

[5] Clavenna, G., Musci, R. and Dorrigoti, L. Antiulcer and mineralocorticoid activities of carbenoxolone and desoxycorticosterone in rats. J. Pharm. Pharmacol. 1982; 34:517-519.

[6] Hocman, G. Prevention of cancer: Vegetables and Plants Comp. Biochem. Physiol. 1989; 93B:201-212.

[7] Wattenberg, L.W. Inhibition of carcinogenesis by minor dietary components Cancer Res. 1992; 52:2085s-2091s.

[8] Nishino, H., Kitagawa, K. and Iwashima, A. Antitumor promoting activity of glycyrrhetinic acid in mouse skin tumor formation by DMBA plus teleocidin. Carcinogenesis 1984; 5:1529-1530.

[9] Agrawal, R., Wang, Z.Y. and Mukhtar, H. Inhibition of skin tumor initiating activity of 7,12, dimethylbenzo[a]anthracene by chronic feeding of glycyrrhizin in drinking water to SENCAR mice. Nutr. Cancer 1991; 15:187- 192.

[10] Kraus, S.D. Estrogenic properties of licorice root. Nature 1962; 193:1082-1084.

[11] Sharaf, A. and Goma, N. Phytoestrogen and their antagonism to progesterone and testosterone. J. Endocrinol. 1982; 31:289-290.

[12] Ichikawa, T., Ishida, S., Sakiya, Y. and Henano, M. Biliary excretion and enterohepatic cycling of glycyrrhizin in rats. J. Pharmacol. Sci. 1986; 75: 672-675.

[13] Ichikawa, T., Ishida, S., Sakiya, Y. and Akada, Y. High performance liquid chromatographic determination of glycyrrhizin and glycyrrhetinic acid in biological material. Chem. Biol. Bull. 1984; 32:3734-3738.

[14] Sakiya, Y., Akada, Y., Kawano, S. and Miyauchi, Y. Rapid estimation of glycyrrhizin and glycyrrhetinic acid in plasma by high-speed liquid chromatography. Chem. Pharm. Bull. 1979; 27:1125-1129.

[15] Akao, T., Hayashi, T., Kobashi, K. and Kanaoka, M. Intestinal bacterial hydrolysis is indispensable to absorption of 18 β glycyrrhetinic acid after oral

administration of glycyrrhizin in rats. J. Pharm. Pharmacol. 1994; 46:135–137.

[16] Hattori, M., Sakamoto, T., Yamagishi, T., Sakamoto, K., Konichi, K., Kobashi, K. and Namba, T. Metabolism of glycyrrhizin by human intestinal flora II. Isolation and characterization of human intestinal bacteria capable of metabolizing glycyrrhizin and related compounds. Chem. Pharm. Bull. 1985; 33:210–217.

[17] Krahenbuhl, S., Hasler, F. and Krapf, R. Analysis and pharmacokinetics of glycyrrhizic acid and glycyrrhetinic acid in humans and experimental animals. Steroids 1994; 59:121–126.

[18] Doll, R., Hill, I.D., Hutton, C. and Underwood, D.J. Clinical trial of triterpenoid licorice compounds in gastric and duodenal ulcer. Lancet 1963; 11:793–796.

[19] Stormer, F.C, Reistad, R. and Alexander, J. Glycyrrhizic acid in liquorice-evaluation of health hazard. Food Chem. Toxicol. 1991; 31: 303–312.

[20] Moon, R.C., Mehta, R.G. and Rao, K.V.N. Retinoids and cancer. *In* The Retinoids, 2nd Ed., (Spom, M.B., Roberts, A.B. and Goodman, D.S., eds.) Raven Press 1994:597–630.

[21] Mehta, R.G., Moon, R.C., Steele, V., Kelloff, G.J. Influence of thiols and inhibitors of prostaglandin biosynthesis on the carcinogen-induced development of mammary lesions *in vitro*. Anticancer Res. 1991; 11:587.

[22] Lin, F.K., Banerjee, M.R. and Cump, L.R. Cell cycle related hormone carcinogen interaction during chemical carcinogen induction of nodules like lesions in organ culture. Cancer Res. 1976; 36:1607–1614.

[23] Mehta, R.G. and Banerjee, M.R. Progesterone receptors and possible role of prolactin for its regulation in the mammary gland. J. Cell Biol. 1974;63: 220.

[24] Telang, N.T., Banerjee, M.R., Iyer, A.P. and Kundu, A.B. Neoplastic transformation of epithelial cells in whole mammary gland *in vitro*. Proc. Natl. Acad. Sci. USA 1979; 76:5886–5890.

[25] Mehta, R.G., Liu, J., Constantinou, A., Hawthome, M., Pezzuto, J.M., Moon, R.C. and Moriarty, R.M. Structure-activity relationships of brassinin in preventing the development of carcinogen-induced mammary lesions in organ culture. Anticancer Res. 1994; 14(3A):1209–1213.

[26] Moon, R.C. and Mehta, R.G. Cancer chemoprevention by retinoids: animal models. Methods Enzymol. 1990;190:395.

[27] Reddy, B.S., Tokumo, K., Kulkami, N., Aligi, C. and Kelloff, G. Inhibition of colon carcinogenesis by prostaglandin synthesis inhibitors and related compounds. Carcinogenesis 1992; 13:1019–1023.

[28] Wang, Z.Y., Agrawal, R., Khan, W.A. and Mukhtar, H. Protection against benzo[a]pyrene and N-nitrosodiethylamine induced lung and forestomach

tumorigenesis in A/J mice by water extracts of green tea and licorice. Carcinogenesis 1992; 13:1491–1494.

[29] Baker, M.E. and Fanestil, D.D. Licorice, computer based analyses of dehydrogenase sequences and the regulation of steroid and prostaglandin action. Mole. Cellular Endocrinol. 1991; 78:C99–C102.

[30] Wittliff, J.L., Mehta, R.G., Boyd, P.A., Rwko, W.L. and Park, D.C. Steroid receptor interaction in normal and neoplastic mammary tissues. *In* Biochemical Markers of Cancer, (T.M. Chu, ed.) Marcel Decker, New York, 1982:183.

CHAPTER 28

CANCER PREVENTION BY LICORICE

HOYOKU NISHINO

Cancer Prevention Division
National Cancer Center Research Institute
1-1, Tsukiji 5-chome, Chuo-ku, Tokyo 104, Japan

ABSTRACT

Licorice (sweetening material) has been demonstrated to have various pharmacological activities, which are widely applied in Chinese medicine. In the present study, a new aspect in the disease preventive activity of licorice, i.e., a preventive effect on carcinogenesis, is demonstrated. Glycyrrhizin and its aglycon glycyrrhetinic acid showed antitumorigenic activity in in vivo experiments. Licochalcone A, a constituent in licorice, was also demonstrated to suppress tumorigenesis.

Besides licorice, Allium vegetables, such as garlic, are also commonly used as medicinal materials in Asia, and some of their extracts have potent anticarcinogenic activity. Thus, the use in combination of constituents prepared from licorice and Allium vegetables might be valuable for the creation of new designer foods for cancer prevention.

INTRODUCTION

In Asia, various kinds of foodstuffs are used as medicinal materials. Typical examples are licorice (sweetening material) and *Allium* vegetables such as garlic (spice). Licorice especially is an important material used in Chinese medicine. In old Chinese medical books, licorice was described as being an antispasmodic agent, as well as being a harmonizer to soften various drug actions. Therefore, licorice has frequently been combined with other herbal drugs in numerous prescriptions. And in Dioscorides' De Materia Medica, licorice was described as an anti-inflammatory drug. Recently, licorice was reported to be effective as an antihepatitis drug.

In this chapter, the new aspect of licorice in disease prevention, i.e., a cancer-preventive activity of licorice constituents, is presented as the main topic. In addition, antitumorigenic agents in *Allium* vegetables are also introduced. For example, the extract or its constituents prepared from aged garlic has been shown to prevent tumorigenesis. Thus, the possible advantage of the combination use of the constituents obtained from licorice and garlic for cancer prevention is discussed.

MATERIALS AND METHODS

Spontaneous Liver Carcinogenesis in Mice

Male C3H/He mice were treated with or without test compounds for 44 weeks, and then killed for the measurement of the number of tumor nodules developed in the liver.

Two-stage Skin Carcinogenesis in Mice

Female ICR mice had their backs shaved, and were treated with a single topical painting of 7,12-dimethylbenz[a]anthracene (DMBA, 100 μg) as the tumor initiator. 12-O-tetradecanoylphorbol-13-acetate (TPA, 0.5 μg) with or without the test compound was applied twice a week for 18 weeks. The number of skin tumors was measured once a week.

^{32}Pi-Incorporation into Phospholipids of Cultured Cells

Confluent HeLa cells were treated with or without test compound for 1 h, and then ^{32}Pi (370 kBq/culture) was added into the culture medium with or without TPA (50 nM). Incubation was continued for 4 h, and then the radioactivity incorporated into phospholipid fraction was measured.

RESULTS

Glycyrrhizin (Fig. 28.1), sweetening material in licorice, had anti-tumorigenic activity *in vivo*. It suppressed the spontaneous liver carcinogenesis in male C3H/He mice, as shown in Table 28.1. In addition, glycyrrhetinic acid, the aglycon of glycyrrhizin, showed antitumor-promoting activity in DMBA-initiated and TPA-promoted skin tumorigenesis in female ICR mice (Fig. 28.2).

Licochalcone A (Fig. 28.1), a constituent of licorice, has also been proven to possess antitumor-promoting activity in a mouse skin two-stage carcinogenesis experiment (data not shown).[1]

Since the crude extracts and purified constituents [for example, allixin[2]] prepared from garlic and onion have been shown to have anticarcinogenic activity, we expanded the screening test to survey other types of anticarcinogenic compounds in various kinds of *Allium* vegetables. Among *Allium* spp. tested, the extract of the tuber of *Allium bakeri* Rgl. showed the highest antitumor-promoter activity (Table 28.2). It is worth noting that this *Allium* plant is widely consumed as a vegetable, as well as a medicinal material in Japan, China and other Asian countries. As antitumor-promoter agents, several compounds were purified from the crude extract, and the antitumorigenic activities of these compounds are now going to be evaluated.

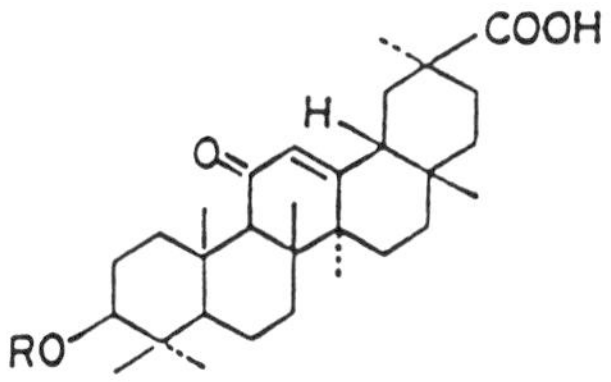

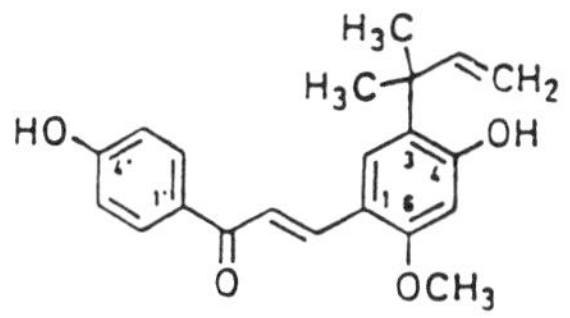

R=GlcuA-GlcuA- : Glycyrrhizin   Licochalcone A

R=H- : Glycyrrhetinic acid

FIG. 28.1. STRUCTURES OF GLYCYRRHIZIN AND LICOCHALCONE A,
CONSTITUENTS OF LICORICE

TABLE 28.1
EFFECT OF GLYCYRRHIZIN ON LIVER CARCINOGENESIS IN C3H/HE MALE MICE

Group	Number of Mice	Number of Tumor-bearing mice (%)		Tumors per Mouse
I. Control	15	12	(80.0)	3.07
II. +Glycyrrhizin	15	11	(73.3)	1.40

C3H/He male mice at age of 8 weeks were divided into two groups of 15 each: group I was the control group, and group II received glycyrrhizin dissolved in the drinking water at the concentration of 5 mg/100 ml for 44 weeks.

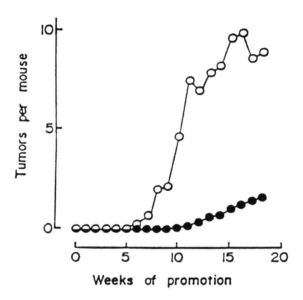

Weeks of promotion

FIG. 28.2. EFFECT OF GLYCYRRHETINIC ACID ON DMBA INITIATED AND
TPA PROMOTED SKIN CANCER IN MICE
o = Control, ● = Glycyrrhetinic acid 10 μ moles 2X week

TABLE 28.2
INHIBITORY EFFECT OF EXTRACTS FROM ALLIUM PLANTS ON
TPA-ENHANCED PHOSPHOLIPID SYNTHESIS IN CULTURED HELA CELLS

Allium spp.	Name as Drugs	Inhibition %
A. *bakeri* Rgl.	Xiebai	68.2
A. *sativum* L.	Dasuan	15.3
A. *fistulosum* L.	Congbai	13.4
A. *tuberosum* Rottl.	Jiucai	4.7
A. *cepa* L.	Hucong	2.2

HeLa cells were incubated with one of the test samples at the concentration of 100 ug/ml, and after
1 h ^{32}Pi (370 kBq per culture) was added with or without TPA (50 nM). Incubation was continued
for 4 h, and then the radioactivity incorporated into phospholipid fraction was measured.

DISCUSSION

It has been suggested that a large number of cancers develop due to the consumption of certain foods. On the other hand, various constituents in foodstuffs have also been shown to suppress carcinogenesis. In this context, designer foods for cancer prevention, in which carcinogenic factors are reduced and anticarcinogenic factors are fortified, makes sense. As anticarcinogenic agents for preparing designer foods, the constituents in plants, which are commonly used as foods, as well as medicinal materials (for example, licorice and garlic), seem especially promising, since the potency of some ingredients is relatively high and their safety has been well-documented.

A combination use of these constituents may result in synergistic effects in cancer prevention, and study of this aspect is ongoing.

Analysis of the action mechanism of these constituents is also being currently carried out; at present, anti-inflammatory and antiproliferative effects, accompanied with the suppressive effect on tumor-promoter enhanced phospholipid metabolism, seem to be important.

REFERENCES

1   Shibata, S., Inoue, H., Iwata, S. *et al.* Inhibitory effects of licochalcone A isolated from *Glycyrrhiza inflata* root on inflammatory ear edema and tumor promotion in mice. Planta Med. 1991; 57:221–224.
2   Nishino, H., Nishino, A., Takayasu, J. *et al.* Antitumor-promoting activity of allixin, a stress compound produced by garlic. Cancer J. 1990; 3:20–21.

CHAPTER 29

CLINICAL EVALUATION AND SAFETY OF LICORICE-CONTAINING FORMULATIONS

HERBERT F. PIERSON*

Vice President for Research and Development
Preventive Nutrition Consultants, Inc.
19508 189TH Place NE
Woodinville, WA 98072-8269

ABSTRACT

Licorice root extract has been used in folkloric medicine for thousands of years. The variety of phytochemicals present in the root not only impart intense sweetness and mask bitterness, but also possess anti-inflammatory and immune modulating activity. Some of the major classes of phytochemicals found in licorice root extract include triterpenoids, chalcones, isoflavones, coumarins, acetophenones, flavonoids, and phenolic acids. Many of the classes of phytochemicals in licorice root extract are finding therapeutic use in HIV treatment. HIV-positive patients are benefiting from treatment with licorice compounds and survivability in many cases is significantly enhanced. From a daily intake purpose, much literature suggests that products delivering under 200 mg of pure glycyrrhizin may be safe and minimally influence mineral balance. There is enormous potential to formulate now food products with safe and effective phytochemicals in licorice root extract.

INTRODUCTION

The purpose of this presentation is to summarize evidence on the misunderstood safety of licorice extract and glycyrrhizin, and highlight recent clinical applications of this food ingredient in the treatment of HIV disease. Despite the biological activities reported for licorice root extract and constituents such as glycyrrhizin, the major use of licorice extract and glycyrrhizin entails sweetness and palatability enhancement of bitterness and flavors in foods and medicines. Over the last decade licorice extract and licorice phytochemicals have become important ingredients in "designer" food development, because they improve food flavor and texture and impart pharmacological value to rationally developed foods.

*Deceased

Aqueous extracts of the roots of *Glycyrrhiza* species have been used therapeutically for over 4,000 years.[1] One major phytochemical constituent of medicinal importance in licorice root extracts is a triterpenoid diglycoside called glycyrrhizin (also called glycyrrhetinic acid glycoside and glycyrrhizic acid). Along with triterpenoids, other major phytochemicals isolated from licorice extract include coumarins (licopyranocoumarin, glabrene, glabridin, and glycycoumarin), flavanones (liquiritigenin and isoliquiritigenin), an isoflavone (glycyrhisoflavone), an aurone (licocoumarone), and chalcones (licochalcones A and B).[2] Current evidence suggests that the aglycone of glycyrrhizin (glycyrrhetinic acid) is one major pharmacologically active phytochemical in *Glycyrrhiza* species.[3,4]

Glycyrrhizin is hydrolyzed by intestinal floral enzymes to release the aglycone glycyrrhetinic acid, which is absorbed and conjugated in the liver to a more polar constituent prior to urinary excretion.[5,6] Glycyrrhetinic acid binds extensively to serum albumins and undergoes enterohepatic recycling indicating that the biliary route also is used for excretion.[7] Both glucuronide and sulfate conjugates of glycyrrhetinic acid have been isolated from biological fluids after ingestion of licorice extracts containing glycyrrhizin.[8]

Medicinal activities reported for licorice extract and glycyrrhizin include suppression of inflammation[9], prevention of diabetic cataract[10], reduction of serum lipids[11], treatment of hepatitis[12], cancer prevention[13], and treatment of gastric ulcer[14]. Apthous lesions in the oral cavity also appear to be healed by licorice extract constituents.[15] Phytochemicals in licorice extract modulate key regulatory enzymes in animal models such as ornithine decarboxylase[16], cyclo-oxygenase[17], cytochrome P450 isozymes[18], and oncogene protein product synthesis.[19] These metabolic modulations may account in part for the cancer therapeutic, anti-inflammatory, drug detoxification, and cancer preventive activities, respectively, of licorice extracts.

Regulatory

In the United Kingdom, glycyrrhizin is permitted at a level of 50 mg/kg food product as a flavor.[20] In the Netherlands, licorice is considered to be a natural flavor, and in both the Netherlands and in Belgium, glycyrrhizin is regulated in beer at a level of 100 mg/L with 55-60% of the glycyrrhizin allowed as the intensely sweet ammonium glycyrrhizinate.[21] In the People's Republic of China, licorice is used as a sweetener in candy, biscuits, and canned food under Good Manufacturing Practices (GMP) conditions.[2] Glycyrrhizin is used without limit in soy sauce and miso as a nonnutritive sweetener in Japan.[2] Licorice extract and glycyrrhizin are GRAS constituents and their use in foods as sweetness enhancers and flavors is determined by GMP in the U.S.[22]

The triterpenoid diglycosides in licorice extract impart a sweetness to foods assessed at 150- to 200-times greater than table sugar.[23] Based on glycyrrhizin as a marker compound in a licorice extracts, 0.1% of the triterpenoid diglycoside is permitted in beverages, 16% in hard candies, 3.1% in soft candies, and 1.1% in chewing gum in the U.S. The extract and its contained phytochemicals impart few calories to foods, and selectively destroy oral *Streptococcus mutans*, the bacteria causing tooth decay.[24]

Safety of Licorice Extract

Large dietary levels encompassing grams of glycyrrhizin per kilogram of body weight for various licorice products including ammonium glycyrrhizinate were subjected to chronic toxicology tests (carcinogenicity and reproductive toxicity) and were found to be negative.[25-28] Moreover, during drug development of glycyrrhetinic acid for human cancer chemoprevention trials, the safety of dietary levels up to several percent of the diet was confirmed in laboratory animals.[29]

Elevation of blood pressure in susceptible individuals who abuse licorice products has been the major safety concern for the clinical use of licorice extract. Numerous retrospective case histories have been published linking licorice overindulgence with sodium retention, water accumulation, blood volume expansion, hypertension, hypokalemia, and resulting myocardial and renal sequelae. It is important to note from Table 29.1 that levels of glycyrrhizin exceeding 200 mg daily are usually reported for chronic imbalance of salt and water in the human body. Most reports of adverse effects to licorice products associate the chronic intake of gram quantities of glycyrrhizin with hypokalemia and hypertension (Table 29.1). Adverse effects to glycyrrhizin or licorice products rarely occurred for dietary levels less than 200 mg glycyrrhizin consumption.

In susceptible individuals and in a dose-related manner, the triterpenoids in licorice extract may inhibit the renal enzyme 11-beta-hydroxysteroid dehydrogenase.[30] This enzyme converts cortisol into cortisone, thereby protecting the kidney mineralcorticoid receptors from stimulation by cortisol accumulation and corticosterone binding.[31] If the 11-beta-hydroxysteroid dehydrogenase is inhibited competitively by structural analogs of cortisol such as glycyrrhetinic acid (typically through licorice product abuse), mineralcorticoid activity may be potentiated. This invariably results in sodium reabsorption, subsequent water accumulation in tissues (edema), hypokalemia and hypertension.[32] In all cases reported, glycyrrhetinic acid-induced hypokalemia and hypertension appear to be reversible after several days of discontinued licorice ingestion.[33]

TABLE 29.1
RETROSPECTIVE CASES OF ADVERSE CLINICAL EFFECTS ATTRIBUTABLE
TO LICORICE ABUSE

Case 1.	One man; 62 years old; consumed 100 g extract while fasting for several weeks; developed hypokalemia[64]
Case 2.	Six men and women; consumed 814 mg of glycyrrhizin for 4 weeks; developed hypokalemia[65]
Case 3.	One man; unknown age; consumed 8-12 three ounce bags of chewing tobacco over several days containing 0.15% glycyrrhizin (8.3% extract); developed hypokalemia[66]
Case 4.	One person; consumed 3-4 licorice bars daily for unknown time period; developed hypertension[67]
Case 5.	One woman; young adult with anorexia nervosa; consumed 600 g licorice extract daily for unknown time period as laxative; developed hypokalemia[68]
Case 6.	One man; 72 years old; chronic glycyrrhizin consumption; developed hypokalemia and acute renal failure[69]
Case 7.	One man; 58 years old; consumed three 36-g-licorice bars daily for 6-7 years; developed pseudoaldosteronism[70]
Case 8.	Fourteen people; mixed sexes; consumed up to 1.4 g glycyrrhizin daily for 4 weeks; 11 out of 14 developed hypokalemia; 4 out of 11 dropped-out of study [71]
Case 9.	Four women; ages 38-55 years; consumed 25-200 g extract daily for unknown time period; developed pseudoaldosteronism[72]
Case 10.	Unspecified number of people; consumed 100-200 g extract daily for 6 months to 5 years; urinary cortisol increased 2-fold[73]
Case 11.	One man; 70 years old; consumed 60-100 g licorice candy daily; developed hypertension and hypokalemia[74]
Case 12.	Twelve people; unspecified ages; consumed 100 g of extract daily for 8 weeks; developed about 80% increased atrial natriuretic peptide[75]
Case 13.	One woman; 50 years old; second injection of Neo-minophagen C (0.2% glycyrrhizin) caused skin rash[76]
Case 14.	One man; 35 years old; consumed Chinese herbal drugs containing licorice extract for 6 months; developed hypokalemia[77]
Case 15.	One man; 55 years old; ingested glycyrrhizin for hepatitis treatment for one year; developed hypokalemia and hypertension[78]
Case 16.	Twelve people; ages unspecified; consumed 546 mg glycyrrhizin daily for 4 weeks to treat chronic hepatitis; plasma renin activity greatly increased[79]
Case 17.	One man; ingested unreported amount of 18b-glycyrrhetinic acid for hepatitis treatment; 3-o-glucuronide of glycyrrhetinic acid found in urine and linked with pseudoaldosteronism[80]
Case 18.	Fifty-four people with peptic ulcer; unspecified ages; treated with 300 mg oral carbenoxolone for 1 week; no adverse reports[81]
Case 19.	One man; 66 years old; used licorice containing cough drops for pulmonary expectoration (150 mg glycyrrhizin daily for 5 weeks); developed pseudoaldosteronism[82]
Case 20.	One man; former alcoholic; consumed 1.0-1.5 g glycyrrhizin per month for 11 months; developed hypokalemia[83]
Case 21.	One woman; 53 years old; consumed 2 tablets containing 10% enoxolone; developed facial angioderma; 10 controls consuming 20% enoxolone experienced no such reaction[84]

TABLE 29.1 (Cont.)

Case 22.	One woman; 47 years old; reported to chronic use of licorice-chewing gum; developed hypokalemia and hypertension[85]
Case 23.	One woman; 22 years old; unrestrained use of glycyrrhizin laxative and diuretics (furosemide); developed hypokalemia[86]
Case 24.	Unspecified person; consumed 55 g licorice extract daily for unknown time period; developed hypokalemia[87]
Case 25.	Fifty-nine people; unspecified ages; reportedly abused glycyrrhizin for an unspecified period of time; developed hypokalemia and myopathy[88]
Case 26.	Two women; ages 71 and 68; consumed between 273-546 mg glycyrrhizin daily for 6 months for liver disease treatment; both developed hypokalemia and hypertension[89]
Case 27.	Two people; unspecified ages; showed serum glycyrrhetinic acid values between 70-80 ng/ml during pseudoaldosteronism[90]
Case 28.	One male; 15 years old; consumed 0.5 kg licorice candy daily for 5 months; developed hypertension encephalopathy[91]
Case 29.	One man; 51 years old; reported to be an excessive licorice abuser; developed hypertension and hypokalemia[92]
Case 30.	One woman; 22 years old; consumed excessive amounts of glycyrrhizin for several years; developed hyperprolactinemia and amenorrhea[93]

A review of licorice safety suggested that in susceptible individuals, symptoms and biochemical signs of adverse effects may occur following regular daily intakes of 100 mg glycyrrhizin, and that it would be unlikely that adverse effects could be observed at lower dosages.[34] Additionally it was suggested that most individuals who express adverse effects to glycyrrhizin were likely do so at a minimum dietary level of 400 mg/day.[35] When the lowest-observed-adverse-effect level is set at 100 mg glycyrrhetinic acid daily (over several weeks) and then subjected to the safety factor of 10, the safe intake level for healthy subjects becomes 10 mg glycyrrhizin daily or about 5 g of licorice extract-containing confectionaries.[36]

The Information Bureau for Nutrition in The Netherlands claimed that 20 g per day of licorice candy was safe for adults.[37] This figure was based upon a report that 20 g of licorice candy (containing 5% licorice extract with 15% glycyrrhizin content) would deliver a safe glycyrrhizin dose of 150 mg.[2] Further investigation showed that the glycyrrhizin content of most Dutch candies was 3- or 4-fold lower than expected, supporting a safer daily glycyrrhizin dose of 60 mg.[2]

Most peer-reviewed reports suggest that between 100-200 mg glycyrrhizin daily may be safely consumed. One systematic study confirmed that 150 mg glycyrrhizin probably was safe, as 350 mg daily was linked with intoxication.[38] The governments of both Japan and Holland recommend that the consumption of glycyrrhizin be less than 200 mg/day as a sweetener.[37] A safe dietary level of 150 mg glycyrrhizin daily corresponds to about 60 g of licorice candy containing various "licorice root species extracts."[2]

Absorption and Metabolism

Rodents completely absorb glycyrrhetinic acid and humans absorb most of an oral dose.[39,40] When 80 mg of glycyrrhizin was administered intravenously to three patients, the maximum serum concentration of glycyrrhizin ranged between 10-15 μg/ml with a half-life of about 5 h.[41]

The aglycone, glycyrrhetinic acid, appeared at 6 h after administration of glycyrrhizin with a maximum serum level of 150 ng/ml at 15 h.[42] When five participants were administered 100 mg glycyrrhizin orally, a glycyrrhetinic acid level of 50-500 ng/ml (100-1,000 nm) was measured in serum.[42] Administering either a single 4-gram dosage of 18-beta-glycyrrhetinic acid orally or 12 g divided into six treatments in one day showed that at days 9 and 13 more than 50% of the compound could be isolated from feces as conjugated metabolites.[43] Additionally, incubation with human feces directly did not alter the structure of 18-beta-glycyrrhetinic acid.[44] Metabolic studies have shown that glycyrrhetinic acid is isomerized by human intestinal bacteria into 3-epi-18-beta-glycyrrhetinic acid.[45,46]

One study showed that 15 individuals who ate 2-4 twists of licorice candy for 2-4 weeks had plasma glycyrrhetinic acid levels of 0-480 ng/ml.[47] Four licorice-addicted patients exhibited between 0-84 ng/ml of 18-beta-glycyrrhetinic acid in their plasma.[47] A 70-year-old man hospitalized for licorice intoxication (consumed 60-100 g licorice candy containing 0.3% glycyrrhizin for 4-5 years) showed urinary glycyrrhetinic acid levels decline from 320 to 0 nmol/day over an 18-day-discontinuation period.[32] Consuming 200 g of licorice candy (580 mg of glycyrrhizin/day) for 10 days did not elevate 18-beta-glycyrrhetinic acid levels in plasma of 7 participants.[48]

Treatment of Viral Diseases

In vitro studies suggest that there are qualitiative and structure-activity relationships for antiviral activity among phytochemicals in licorice extract. For example, glycyrrhizin inhibited the replication of both DNA- and RNA-viruses *in vitro*[49] including herpes.[50] Potential activity *in vivo* focused on interferon induction by glycyrrhetinic acid.[51] However, both glycyrrhizin and glycyrrhizin sulfate counteracted the deleterious effects of HIV, inhibited infection of cells, and blocked viral replication *in vitro*.[52] Using an MT-4 cell line after HIV-1 infection, 0.25 μg/ml glycyrrhizin sulfate inhibited viral-induced cytolysis, suppressed expression of viral antigen (LC 100 = 1.0 mg/ml), and was four-times more potent than glycyrrhizin as a reverse transcriptase inhibitor.[53-55] At about 20 μg/ml various licorice extract phytochemicals (glycyrrhizin, licochalcone A, isolicoflavonal, glycyrrhisoflavone, and licopyranocoumarin) selectively inhibited HIV-induced giant cell formation in MT-4 cells without cytotoxicity, whereas licochalcone B was cytotoxic and isoliquiritigenin and isoliquiritin were inactive.[56,57] Many of the above *in vitro* observations were extended to clinical investigations on HIV disease.

In one clinical trial three HIV-positive hemophiliacs were treated intravenously with 400-1,600 mg glycyrrhizin (7.2-30.8 mg/kg body weight/day) per day for a total of 30 days using 3 treatment cycles. This treatment significantly decreased P24 antigen detection which was interpreted as inhibition of HIV-1 replication *in vivo*.[58] In another study, 9 HIV positive hemophiliacs were administered 800 mg of glycyrrhizin intravenously and then subsequently maintained daily on 150 mg glycyrrhizin orally for over 10 years. There was an increased absolute lymphocyte count, increased CD4/CD8 ratio due to increased CD4 cells, increased NK cell activity, and increased white cell thymidine uptake by PHA, Con A, and PWM stimulation. Five patients survived with normal CD4 counts, and glycyrrhizin prevented AZT-induced bone marrow suppression. Results of this study suggested that glycyrrhizin prevented the conversion of AC/ARC to AIDS.[59,60]

In another study 42 HIV-infected hemophiliacs with liver impairment were treated intravenously with "Stronger Neominophagen" (a liquid glycyrrhiz-in-containing preparation) for 3 weeks, then every second day for 8 additional weeks (21 cases received 100-200 ml, and 21 cases received 400-800 ml containing 0.2% glycyrrhizin, i.e., up to 160 mg glycyrrhizin in a single treatment). There were no changes in the CD4/CD8 ratio, no change in CD4 counts, increased mitogenic response of white cells to PHA, ConA, and PWM with best results observed in AC carriers given 400 ml. The latter group showed complete liver improvement, regression of lymphoadenopathy, resistance to *Candida* infection, suppression of skin rashes, amelioration of anorexia, and abrogation of night sweats. The study concluded that prophylactic high doses of glycyrrhizin were effective for preventing immune deterioration and for preserving and restoring liver function.[61]

Additionally, 16 asymptomatic HIV positive hemophiliacs were treated with oral glycyrrhizin (150-225 mg) for 3-7 years. There was no progression toward AIDS in any of the patients. Glycyrrhizin was converted into glycyrrheti-nic acid in the body as measured in urine. No side effects including hypertension were observed during treatment. This treatment delayed immunological deterioration according to elevated CD4 counts and maintained CD4/CD8 T-lymphocyte ratios. There was a decreased level of both P24 and P17 antigen and Nef antibody titers. It was concluded that oral glycyrrhizin is an effective and safe "prodrug" for preventing progression of HIV disease toward AIDS.[62,63]

CONCLUSION

Licorice extract and its contained phytochemicals have been used for centuries as a therapeutic agent. The anti-inflammatory, enzyme modulatory,

TABLE 29.2
REPORTS SHOWING MODULATION OF IMMUNE FUNCTION BY
LICORICE PHYTOCHEMICALS

- glycyrrhizin and glycyrrhetinic acid induce interferon in mice[94]
- glycyrrhizin suppresses inflammatory free-radicals produced by neutrophils[95]
- glycyrrhizin inhibits inflammatory prostaglandin biosynthesis[96]
- licorice saponins enhance cell mediated immunity *in vitro*[97]
- glycyrrhizin patented as immunostimulant for fish[98]
- glycyrrhizin stimulates gamma-interferon production by human spleen[99]
- glycyrrhizin modulates rat peritoneal cell platelet activating factor *in vitro*[100]
- licorice chalcones inhibit human PMN leukotriene biosynthesis[101]
- glycyrrhizin stimulates human natural killer cell (NK) interferon production[102]
- glycyrrhizin stimulates interferon-gamma production in human mononuclear cells[103]
- glycyrrhizin and monoammonium glycyrrhizinate stimulate NK cell activity[104]
- glycyrrhizin stimulates leukocyte interferon-gamma production[105]
- glycyrrhizin induces tumor-necrosis factor in diabetic mouse model[106]
- licorice extract neutral polysaccharides stimulate immune response[107]
- glycyrrhizin stimulates human lymphocyte response to hepatitis antigen[108]
- glycyrrhizin activates suppressor macrophages in mouse tumors[109]
- glycyrrhizin stimulated human natural killer cell activity[110]
- ammonium glycyrrhizinate stimulates mouse immunity[111]
- glycyrrhetinic acid modulates antioxidant activity in neutrophils[112]
- glycyrrhizin inhibits two-stage skin carcinogenesis in mice[113]
- glycyrrhizin prevented chemotherapy side-effect in breast cancer patients[114]
- licorice phenolics demonstrated anti-HIV activity *in vitro*[115]
- licorice flavonoids patented as treatment for AIDS[116]

detoxification, and immune potentiating properties of this food ingredient have been verified by modern research methodologies. Medicinal uses of licorice extract have survived the test of time and it is used by cultures around the world. It is important to note that adverse effects of any food item may appear after chronic overindulgence. Licorice is no exception. With licorice products this may be related to potentiation of endogenous mineralocorticoid activity and subsequent retention of sodium and water and loss of potassium. Abnormal mineral balance in addicted people is reversible and normal sodium and potassium values may be resumed in a matter of days or weeks after discontinuation of licorice. Recent reports suggest that one promising area of future research for licorice phytochemicals will be in the area of designer food development for adjunct HIV treatment and prevention. An abundance of reports on licorice pharmacology suggest that a daily consumption between 100-200 mg glycyrrhizin appears to be a safe dietary level.

REFERENCES

[1] Gibson, M.R. Glycyrrhiza in old and new perspectives. Lloydia 1978; 41: 348–354.

[2] Lutomski, J., Nieman, C. and Fenwick, G.R. Liquorice, *Glycyrrhiza glabra* L. Biological Properties. Herba Polonica 1991; 37(3-4):163–178.

[3] Fenwick, G.R., Lutomski, J. and Nieman, C. Liquorice, *Glycyrrhiza glabra* L. — Composition, uses and analysis. Food Chem. 1990; 38: 119–143.

[4] Ikram, M. and Zirvi, A.K. Chemistry and pharmacology of licorice. Herba Polonica 1976; 3-4:312–320.

[5] Hattori, M., Sakamoto, T., Kobashi, K. and Namba, T. Metabolism of glycyrrhizin by human intestinal flora. Planta Med. 1983; 48:38–42.

[6] Akao, T., Akao, T. and Kobashi, K. Glycyrrhizin beta-D-glucuronidase of *Eubacterium* sp. from human intestinal flora. Chem. Pharm. Bull. 1987; 35:705–710.

[7] Helbing, A. A new method for the determination of 18-beta-glycyrrhetinic acid in biological material. Clinica Chimica Acta; 8:752–756.

[8] Iveson, P., Lindup, W.E., Parke, P.D. and Williams, R.T. The metabolism of carbenoxolone in the rat. Xenobiotica 1971; 1:79–95.

[9] Finney, R.S.H., Somers, G.F. and Wilkinson, J.H. Pharmacological properties of glycyrrhetinic acid — a new anti-inflammatory drug. J. Pharm. Pharmacol. 1958; 10:687–695.

[10] Aida, K., Tawata, M., Shindo, H., Onaya, T., Sasaki, H., Yamaguchi, T., Chin, M. and Mitsuhashi, H. Isoliquiritigenin: a new aldose reductase inhibitor from glycyrrhizae radix. Planta Med. 1990; 56(3):254–258.

[11] Mezenowa, T.D. Hypolipidemic effect of the licorice root extract. Chim. Farm Zb. 1983; 4:432–435.

[12] Fujisawa, K., Watanabe, Y. and Kimura, K. Therapeutic approach to chronic active hepatitis with glycyrrhizin. Asian Med. J. 1980; 23(10): 745–756.

[13] Li, J., Xie, Y., Shi, G., Jin, M. and Lu, G. Experimental and clinical studies of inhibitory effects of *Glycyrrhiza uralensis fisch.* and *Chelidonium majus* L. on carcinogenesis and progression of stomach carcinoma. Gastroenterology 1991; 100(5 part 2):A380.

[14] Revers, F.E. Heeft succus liquirititae een genezende werking op de maagzweer. Nederlands Tijdschrift voor Geneeskunde 1946; 90:135–137.

[15] Poswillo, D. and Partridge, M. Management of recurrent apthous ulcers. A trial of carbenoxolone sodium mouthwash. Brit. Dent. J. 1984; 157: 55–57.

[16] Inoue, H., Saito, H., Koshihara, Y. and Murata, S. Inhibitory effect of glycrrhetinic acid derivatives on lipoxygenase and prostaglandin synthesis.

Chem. Pharm. Bull. 1986; 2:847–901.

[17]   Okamoto, H., Yoshida, D., Saito, Y. and Mizusaki, S. Inhibition of 12-O-tetradecanoyl phorbol-13-acetate induced ornithine decarboxylase activity in mouse epidermis by sweetening agents and related compounds. Cancer Lett. 1983; 21(1):29–36.

[18]   Nishino, H., Saito, Y., Kanamura, N. and Hasegawa, T. Effect of glycyrrhetinic acid on arachidonic acid metabolism. J. Kyoto. Pref. Univer. Med. 1987; 96(9):765–769.

[19]   Martin, W., Zempel, G., Hulser, D. and Willecke, K. Growth inhibition of oncogene-transformed rat fibroblasts by cocultured normal cells: relevance of metabolic cooperative mediated by gap junctions. Cancer Res. 51(19):5348–5351.

[20]   Weedon, B.C.L. (Chairman). Report on the review of flavorings in food. FAC/REP/22 of Food Additives and Contaminants Committee. H.M. Stat. Office, London, 1976.

[21]   Spinks, E.A. and Fenwick, G.R. The determination of glycyrrhizin in selected UK liquorice products. Food Add. Contam. 1990; 7:769–778.

[22]   Life Sciences Research Office: Tentative evaluation of the health aspects of licorice, glycyrrhiza and ammoniated glycyrrhizin as food ingredients. Bureau of Foods, Food and Drug Administration, Department of Health, Education and Welfare, Washington, D.C., 1974.

[23]   Anon. Shuri Shokuhin Tokei Geppo Mar. 1991; 33(1):70–74.

[24]   Steinberg, D., Sgan-Cohen, H.D., Stabholz, A., Pisanty, S., Segal, R. and Sela, M.N. The anti-cariogenic activity of glycyrrhizin: preliminary clinical trials. Isr. J. Dent. Sci. 1989; 2(3):153–157.

[25]   Poulsen, E. Report of the scientific committee for food sweeteners. Rep. Sci. Committee Food (Brussels). 1986; 16:1–20.

[26]   GRAS status of licorice (Glycyrrhiza), ammoniated glycyrrhizin and monoammonium glycyrrhizinate. Federal Reg. 1983; 48: 54983-54990.

[27]   Dusemund, B. and Grunow, W. Glycyrrhizin in liquorice confectionery. Safety aspects. Max-von-Pettenkofer-Institut des Bundegesundheitsamtes, Berlin, 1991.

[28]   Savelkoul, T.J.F. Glycyrrhizin in liquorice confectionery. Safety aspects. Max-von-Pettenkofer-Insitiut des Bundesgesund-heitsamtes, Berlin, 1990.

[29]   Akimoto, M., Kimura, M., Sawano, A., Iwasaki, H., Nakajima, Y. et al. Prevention of cancer chemotherapeutic agent-induced toxicity in postoperative breast cancer patients with glycyrrhizin. Gan No Rinsho 1986; 32:869–872.

[30]   Armanini, D., Karbowiak, I. and Funder, J.W. Affinity of liquorice derivatives for mineralocoritcoid and glucocorticoid receptors. Clin. Endocrinol. 1983; 19:609–612.

[31]   Edwards, C.R.W. Renal 11-beta-hydroxysteroid dehydrogenase: a

mechanism ensuring mineralocorticoid specificity. Hormone Res. 1990; 47: 114–117.

[32] Farese, R.V., Biglieri, E.G., Shackelton, C.H.L., Irony, I. and Gomez-Fontes, R. Licorice-induced hypermineralocorticoidism. New Engl. J. Med. 1991; 325:1223–1227.

[33] Epstein, M.T., Espiner, E.A., Donald, R.A., Hughes, H., Cowles, R.J. and Lun, S. Licorice raises urinary cortisol in man. J. Clin. Endocrinol. Metab. 1978; 47:397–400.

[34] Simpson, F.O. and Currie, I.J. Licorice consumption among high school students. New Zealand Med. J. 1982; 95:31–33.

[35] Stormer, F.C., Reistad, R. and Alexander, J. Glycyrrhizic acid in liquorice-evaluation of health hazard. Food Chem. Toxicol. 1993; 31(4):303–312.

[36] Ibsen, K.K. Liquorice consumption and its influence on blood pressure in Danish schoolchildren. Danish Med. Bull. 1977; 28: 124–126.

[37] Anon. Suikerwerk(Glycyrrhizinezuur). Voedingsinform 1988; 11:22–23.

[38] Cereda, J.M., Trono, D. and Schifferli, J. Liquorice intoxication caused by alcohol-free pastis. Lancet 1983; i (8339):1442.

[39] Iveson, P., Lindup, W.E., Parke, D.V. and Williams, R.T. The metabolism of carbenoxolone in the rat. Xenobiotica 1971; 1:79–95.

[40] Parke, D.V., Pollock, S. and Williams, R.T. The fate of trifium-labelled beta-glycyrrhetinic acid in the rat. Pharmacol 1963; 15:500–506.

[41] Ishida, S., Sakiya, Y., Ichikawa, T., Taira, Z. and Awazu, S. Prediction of glycyrrhizin disposition in rat and man by a physiologically based pharmacokinetic model. Chem. Pharm. Bull. 1990; 38:212–218.

[42] Nakono, N., Kato, H., Suzuki, H., Nakao, N., Yano, S. and Kanaoka, M. Enzyme immunoassay of glycyrrhetinic acid and glycyrrhizin II. Measurement of glycyrrhetinic acid and glycyrrhizin in serum. 1980; Jap. Pharmacol. Ther. 8:4171–4174.

[43] Helbing, A. A new method for the determination of 18-beta-glycyrrhetinic acid in biological material. Clinica Chimica Acta 1962; 8:752–756.

[44] Carlat, L.E., Magraf, H.W., Weathers, H.H. et al. Human metabolism of orally ingested glycrrhetinic acid and monoammonium glycyrrhizinate. Proc. Soc. Exp. Biol. Med. 1959; 102:245–248.

[45] Akao, T., Akao, T. and Kobashi, K. Glycyrrhizin beta-D-glucuronidase of Eubacterium sp. from human intestinal flora. Chem. Pharm. Bull. 1987; 35:705–710.

[46] Hattori, M., Sakamoto, T., Kobashi, K. and Namba, T. Metabolism of glycyrrhizin by human intestinal flora. Planta Medica 1983; 48:38–42.

[47] Hughes, H. and Cowles, R. Estimation of plasma glycyrrhetinic acid. New Zealand Med. J. 1977; 85:298.

[48] Stewart, P.M., Wallace, A.M., Valentino, R., Birt, D., Shackleton,

C.H.L. and Edwards, C.R.W. Mineralocoritcoid activity of liquorice: 11-beta-hydroxysteroid dehydrogenase comes of age. Lancet 1987; ii:821-823.

[49] Pompei, R., Pani, A., Flore, O., Marcialis, M.A. and Loddo, B. Antiviral activity of glycyrrhizic acid. Experientia 1980; 3:304.

[50] Segal, R. and Pisanty, S. Glycyrrhizin gel as a vehicle for idoxuridine. Part 1. Clinical investigations. J. Clin. Pharm. Therap. 1987; 12:165-171.

[51] Abe, N., Ebina, T., Ishida, N. Interferon induction by glycyrrhizin and glycyrrhetinic acid in mice. Microbiol. Immunol. 1982; 26(6):535-539.

[52] Nakashima, H., Matsui, I., Yoshida, O., Isowa, Y., Kido, Y., Motoki, Y., Ito, M., Shigeta, S., Mori, T. and Yamamoto, N. A new anti-human immuno-deficiency virus substance; endowment of glycyrrhizin with reverse transcriptase-inhibitory activity by chemical modification. Jap. J. Cancer Res. 1987; 78:767-771.

[53] Nakashinma, H., Matsui, I., Yoshida, O., Isowa, Y., Kido, Y., Motoki, Y., Ito, M., Shigeta, S., Mori, T. and Yamamoto, N. A new anti-human immunodeficiency virus substance; endowment of glycyrrhizin with reverse transcriptase-inhibitory activity by chemical modification. Jap. J. Cancer Res. 78(8):767-771 (1987).

[54] De Clercq, E. New promising inhibitors of the human immunodeficiency virus. Curr. Opin. Infect. Dis. 1989; 2(3):401-410.

[55] Harada, S. and Yamamoto, N. AIDS studies in Japan. Jap. J. Cancer Res. 78(5):415-427 (1987).

[56] Hatano, T., Yashuhara, T., Miyamoto, K. and Okuda, T. Anti-human immunodeficiency virus phenolics from licorice. Chem. Pharm. Bull. 1988; 36:2286-2288.

[57] Ito, M., Nakashima, H., Baba, M., Pauwels, R., DeClercq, E., Shigeta, S. and Yamamoto, N. Inhibitory effect of glycyrrhizin on the in vitro infectivity and cytopathic activity of the human immunodeficiency virus [HIV(HTLV-III/LAV)]. Antiviral Res. 7(3):127-137 (1987).

[58] Hattori, T., Ikematsu, S., Koito, A., Matsushita, S., Maeda, Y., Hada, M., Fujimaki, M. and Takatsuki, K. Preliminary evidence for inhibitory effect of glycyrrhizin on HIV replication in patients with AIDS. Antiviral Res. 1989; 11(5-6):255-261.

[59] Endo, Y. The immunotherapy for AIDS with glycyrrhizin and/or neurotropin. Int. Conf. Aids 1993; 9(1):492.

[60] Endo, Y., Mamiya, S., Iwamoto, K., Niitsu, H., Itoh, T. and Miura, A. The frequency of patients with positive HIV-antibody in the various groups and the prognosis of these patients treated with SNMC. Rinsho Byori 1990; 38(2):188-192.

[61] Mori, K., Sakai, H., Suzuki, S., Akutsu, Y., Ishikawa, M., Aihara, M., Yokoyama, M., Sato, Y., Sawada, Y., Endo, Y. et al. Effects of high-dose

glycyrrhizin (SNMC: stronger neominophagen C) on hemophilia patients with HIV infection. Int. Conf. Aids 1990; 6(2):394.

[62] Ikegami, N., Akatani, K., Imai, M., Yoshioka, K. and Yano, S. Prophylactic effect of long-term oral administration of glycyrrhizin on AIDS development of asymptomatic patients. Int. Conf. Aids 1993; 9(1):234.

[63] Ikegami, N., Yoshioka, K. and Akatani, K. Clinical evaluation of glycyrrhizin on HIV-infected asymptomatic hemophiliac patients in Japan. Int. Conf Aids 1989; 5:401.

[64] Achar, K.N., Abduo, T.J. and Menon, N.K. Severe hypokalemic rhabdomyolysis due to ingestion of licorice during Ramadan. Aust. N.Z. J. Med. 1989; 19(4):365–367.

[65] Bernardi, M., D'Intino, P.E., Trevisani, F., Cantelli-Forti, G., Raggi, M.A., Turchetto, E. and Gasbarrini, G. Effects of prolonged ingestion of degraded doses of licorice by healthy volunteers. Life Sci. 1994; 55(11): 863–872.

[66] Blachley, J. and Knochel, J. Tobacco chewer's hypokalemia: licorice revisited. New Engl. J. Med. 1980; 302(14):784–785.

[67] Brandon, S. Liquorice and blood pressure. Lancet 1991; 337(8740):557.

[68] Brayley, J. and Jones, J. Life-threatening hypokalemia associated with excessive licorice ingestion. Am. J. Psychiatry 1994;151(4):617–618.

[69] Chubachi, A., Wakui, H., Asakura, K., Nishimura, S., Nakamoto, S. and Miura, A.B. Acute renal failure following hypokalemic rhabdomyolysis due to chronic glycyrrhizic acid administration. Intern. Med. 1992; 31(5): 708–711.

[70] Conn, J.W., Rowner, D.R. and Cohen, E.L. Licorice-induced pseudoaldosteronism. J. Am. Med. Assoc. 1968; 205(7):80–84.

[71] Epstein, M.T., Espiner, E.A., Donald, R.A. and Hughes, H. Effect of eating liquorice on the reninangiotensin-aldosterone axis in normal subjects. Brit. Med. J. 1977; 1:488–490.

[72] Epstein, M.T., Espiner, E.A., Donald, R.A. and Hughes, H. Liquorice toxicity and the reninangiotensin-aldosterone axis in man. Brit. Med. J. 1977; 1(1):209–210.

[73] Epstein, M.T., Espiner, E.A., Donald, R.A., Hughes, H., Cowles, R.J. and Lun, S. Licorice raises urinary cortisol in man. J. Clin. Endocrinol. Metab. 1978;47(2):397–400.

[74] Farese, R., Biglieri, E., Shackleton, C., Irony, I. and Gomez-Fontes, R. Licorice-induced hypermineralocorticoidism. New Engl. J. Med. 1991; 325(17):1223–1227.

[75] Forslund, T., Fyhrquist, F., Froseth, B. and Tikkanen, I. Plasma atrial natriuretic peptide concentration (P-ANP-C) during licorice ingestion in healthy volunteers. J. Intern. Med. 1989; 225:95–99.

[76] Ishii, M., Miyazaki, Y., Yamamoto, T., Miura, M., Ueno, Y., Takahashi,

T. and Toyota, T. A case of drug-induced ductopenia resulting in fatal biliary cirrhosis. Liver 1993; 13(4):227–231.

77 Izumotani, T., Ishimura, E., Tsumura, K., Goto, K., Nishizawa, Y. and Morii, H. An adult case of Fanconi syndrome due to a mixture of Chinese crude drugs. Nephron 1993; 65(1):137–140.

78 Kageyama, Y. A case of pseudoaldosteronism induced by glycyrrhizin. Nippon Jinzo Gakkai Shi 1992; 34(1):99–102.

79 Kageyama, Y., Suzuki, H. and Saruta, T. Renin-dependency of glycyrrhizin-induced pseudoaldosteronism. Endocrinology 1991; 38(1):103–108.

80 Kanaoka, M., Yano, S., Kato, H. and Nakada, T. Synthesis and separation of 18-beta-glycyrrhetyl monoglucuronide from a patient with glycyrrhizin-induced pseudoaldosteronism. Chem. Pharm. Bull. 1986; 34(12):4978–4983.

81 La Brooy, S.J., Taylor, R.H., Hunt, R.H., Golding, P.L., Laidlaw, J.M., Chapman, R.G., Pounder, R.E., Vincent, S.H., Colin-Jones, D.G., Milton-Thompson, G.J. and Misiewicz, J.J. Controlled comparison of cimetidine and carbonoxolone sodium in gastric ulcer. Brit. Med. J. 1979; 19(1):1308-1309.

82 Levesque, H., Cailleux, N., Poutrain, R.J., Noblet, C., Moore, N., Courtois, H., et al. Pseudohyperaldosteronism by glycyrrhizin overdose from Pulmoll expectorant lozenges. Therapie 1992; 47(5):439–440.

83 Luchon, L., Meyrier, A. and Paillard, F. Hypokalemia without arterial hypertension by licorice poisoning. Nephrologie 1993; 14(4):177–181.

84 Martinez, F.V., Badas, A.J., Goitia, J.F.G. and Aguirre, I. Sensitization to oral enoxolone. Contact Dermat. 1994; 30:124.

85 Michaux, L., Lefebvre, C. and Coche, E. Perverse effects of an apparent harmless habit. Rev. Med. Interne 1993; 14(2):121–122.

86 Paccalin, J. and Traissac, F.J. Polydrug overmedication and potassium. Therapie 1973; 28(6):1143–1152.

87 Salassa, R.M., Mattox, V.R. and Rosevear, J.W. Inhibition of the "mineralocortocoid" activity of licorice by spironolactone. J. Clin. Endocrinol. Metab. 1962; 22:1156–1159.

88 Shintani, S., Murase, H., Tsukagoshi, H. and Shiigai, T. Glycyrrhizin licorice-induced hypokalemic myopathy report of two cases and review of the literature. Euro. Neurol. 1992; 32(1):44–51.

89 Takeda, R., Morimoto, S., Uchida, K., Nakai, T., Miyamoto, M., Hashiba, T., Yoshimitsu, H., Kim, K.S. and Miwa, U. Prolonged pseudoaldosteronism induced by glycyrrhizin. Endocrinology Japan 1979; 26(5):541–548.

90 Terasawa, K., Bandoh, M., Tosa, H. and Hirate, J. Disposition of glycyrrhetic acid and its glycosides in healthy subjects and patients with pseudoaldosteronism. J. Pharmacodyn. 1986; 9(1):95–100.

[91] Van Der Zwan, A. Hypertension encephalopathy after liquorice ingestion. Clin. Neurol. Neurosurg. 1993; 95(1):35–37.

[92] Wash, L.K. and Bernard, J.D. Licorice-induced pseudoaldosteronism. Am. J. Hosp. Pharm. 1975; 32(1):73–74.

[93] Werner, S., Brismar, K. and Olsson, S. Hyperprolactinemia and liquorice. Lancet 1979; 1(2):319.

[94] Abe, N., Ebina, T. and Ishida, N. Interferon induction by glycyrrhizin and glycyrrhetinic acid in mice. Microbiol. Immunol. 1982; 26(6):535–539.

[95] Akamatsu, H., Komura, J., Adada, Y. and Niwa, Y. Mechanisms of anti-inflammatory action of glycyrrhizin effect on neutrophil functions including reactive oxygen species generation. Planta Med. 1991; 57(2): 119–121.

[96] Capasso, F., Mascolo, N., Autore, G. and Duraccio, M.R. Glycyrrhetinic acid leukocytes and prostaglandins. J. Pharm. Pharmacol. 1983; 35(5): 332–335.

[97] Chavali, S.R., Francis, T. and Campbell, J.B. An in vitro study of immunomodulatory effects of some saponins. Int. J. Immunopharmacol. 1987; 9(6):675–683.

[98] Edahiro, T. Glycyrrhizin as immunostimulants for fish and feeds containing it. Japanese Patent No. 90250832, Oct. 8, 1990.

[99] Feng, Y. and Zhu, T. Effect of glycyrrhizin on the production of gamma-interferon (IFN) from human spleen. Zhonghua Weishengwuxue He Mianyixue Zazhi 1986; 6(2):99–102.

[100] Ichikawa, Y., Mizoguchi, Y., Kioka, K., Kobayashi, K., Tomekawa, K., Morosawa, S. and Yamamoto, S. Effects of glycyrrhizin on the production of platelet-activating factor from rat peritoneal exudate cells. Jap. J. Allergol. 1989; 38(4):365–369.

[101] Kimura, Y., Okuda, H., Okuda, T. and Arichi, S. Effects of chalcones isolated from licorice roots on leukotriene biosynthesis in human polymorphonuclear neutrophils. Phytother. Res. 1988; 2(3):140–145.

[102] Kuwano, K., Munakata, T., Semba, U., Shibuya, Y. and Arai, S. Interferon production from human natural killer cells stimulated by glycyrrhizin. Igaku No Ayumi 1984; 129(3):173–175.

[103] Li, J. and Guo, L. Effect of glycyrrhizin on the interferon-gamma production of human neonatal cord-mononuclear cells. Acat. Acad. Med. Shanghai 1991; 18(2):104–108.

[104] Li, T. The enhancing effect of glycyrrhizin and monoammonium glycyrrhizinate on NK activity augmented by IL2. J. China Med. Univer. 1990; 19 (6):430–432.

[105] Sakuma, T., Yoshida, T., Shinada, M. and Azuma, M. Effect of glycyrrhizin on the production of gamma-interferon in human leukocytes. Igaku No Ayumi 1983; 127(6):669–671.

[106]   Satoh, J., Shintani, S., Tanaka, S., Tamura, K., Seino, H., Ohta, S., Nobunaga, T., Kumagai, K. and Toyota, T. Inhibition of development of insulin-dependent (type 1) diabetes mellitus in non-obese diabetic mice by TNF and TNF inducers. Igaku No Ayumi 1988; 147(1):63–64.

[107]   Shimizu, N., Tomada, M., Kanari, M., Gonda, R., Satoh, A and Satoh, N. A novel natural neutral polysaccharide having activity on the reticuloendothelial system from the root of Glycyrrhiza uralensis. Chem. Pharm. Bull. 1990; 38(11):3069–3071.

[108]   Shinada, M., Azuma, M., Kawai, H., Sazaki, K., Yoshida, I., Yoshida, T., Suzutani, T. and Sakuma, T. Enhancement of interferon production in glycyrrhizin-treated human peripheral lymphocytes in response to Concanavalin A and to surface antigen of hepatitis B virus. Proc. Soc. Exp. Biol. Med. 1986; 181:205–210.

[109]   Suzuki, F. and Maeda, H. Suppressor macrophages: a role on the growth of transplanted tumors and regulation by an extract of licorice glycyrrhizin. Oncologia (Tokyo) 1987; 20(5):124–133.

[110]   Wada, T., Arima, T. and Nagashima, H. Natural killer activity in patients with chronic hepatitis treated with OK432, interferon, adenine arabinoside and glcyrrhizin. Gastroenterol. Japan 1987; 3(June 22):312–321.

[111]   Wang, Z.G., Ren, J. and Zhang, R. Immunoregulatory studies on ammonium glycyrrhizinate in mice. Zhongguo Yu Dulixur Zazhi 1990; 4(1):36–38.

[112]   Wu, Y. and Li, X. Effects of sodium 18-beta glycyrrhetinic acid on the production of active oxygen species and the concentration in intracellular free calcium in rat neutrophils. Acta Pharmacol. Sin. 1991; 12(3):280–284.

[113]   Yasukawa, K., Takido, M., Takeuchi, M. and Nakagawa, S. Inhibitory effect of glycyrrhizin and caffeine on two-stage carcinogenesis in mice. J. Pharm. Soc. Jap. 1988; 108(8):794–796.

[114]   Akimoto, M., Kimura, M., Sawano, A., Iwasaki, H., Nakajima, Y., Matano, S. and Kasai, M. Prevention of cancer chemotherapeutic agent-induced toxicity in postoperative breast cancer patients with glycyrrhizin (SNMC). Gan No Rinsho 1986; 32(8):869–872.

[115]   Hatano, T., Yashuhara, T., Miyamoto, K. and Okuda, T. Anti-human immunodeficiency virus phenolics from licorice. Chem. Pharm. Bull. 1988; 36(6):2286–2288.

[116]   Watanabe, K., Yashiro, J. and Makoto, M. Licorice flavonoids for treatment of AIDS. Japanese Patent No. 89175942, Dec. 28, 1987.

BRIDGING THE GAPS IN KNOWLEDGE FOR DESIGNER FOOD APPLICATIONS

CHAPTER 30

STUDIES OF PLANT FOODS FOR DISEASE PROTECTION

BRIAN D. SIEBERT

CSIRO
Division of Human Nutrition
Adelaide, SA, Australia

ABSTRACT

A synopsis of one human and several animal studies is provided. Each study demonstrates significant decreases and adverse pre-cancerous signs (e.g., micronuclei). Occurrence of micronuclei is diminished in the presence of antioxidant compounds, in particular ascorbic acid and vitamin E.

INTRODUCTION

It is estimated that approximately 50% of cancers have a dietary link. This is not surprising when one considers that the human diet is a major environmental source of both natural and synthetic chemicals, many of which have the potential to be mutagenic. There is a large body of evidence, however, that shows that dietary supplementation with micronutrients has positive health benefits in this regard, many of which micronutrients can be derived from plants. In addition, some plant macro components can be useful in reducing tumor incidence.

Fruits, vegetables, some plant oils and cereals appear to be protective because the phytochemicals they contain have antioxidant properties. These defenses against free radical damage include enzymes, such as glutathione peroxidase, catalase, superoxide dismutase, as well as vitamins, such as C and E. Selenium-containing glutathione peroxidase destroys lipid peroxides in membranes, while vitamin E protects polyunsaturated fatty acids in cell membranes via free-radical quenching. Antioxidants may have beneficial effects as anti-carcinogens above the recommended RDI levels. In addition, a plant macro component such as fiber may reduce cancer of the colon and other tissues.

As consumers become aware of the anticancer potential of some foodstuffs, the demand will grow for natural and processed foods to contain these health benefits.

METHODS

A battery of genotoxicity tests enables the antimutagenic properties of specific plant micronutrients to be tested by using assays, which are primarily designed to assess chromosomal damage or gene mutation. The induction of micronuclei in bone marrow, splenocytes or blood detects breakages in the DNA strands, whereas the P^{32} post-labelling technique detects the formation of adducts on the strands that result from damage. Another test can determine oxidative damage to deoxy guanosine bases, whereas a more specific assay can measure the frequency of somatic mutation at the hypoxanthine phosphoribosyl transferase locus (HPRT).

Many of the above tests can be used to assess antimutagenic food components. For instance, rather than measure the number of micronuclei found in splenic or bone marrow cells following the injection of say an agricultural chemical, the protective effects of a potential plant antioxidant can be assessed in long-term dietary studies in the presence of a known mutagen using the same final end-point. A number of experiments in progress are using these techniques.

Other methods of assessing the protective effects of plant foods include examination of the colon by the continual feeding of anticarcinogen after an initial insult by a known carcinogen. In these experiments microscopic examination is carried out to detect the frequency of aberrant crypt foci (ACF) prior to the macroscopic appearance of tumors.

RESEARCH

A number of research projects using the above methods are under investigation using extracts of green and orange vegetables, of green tea containing polyphenolics and of genistein from soy, all being sources of antioxidants. These extracts have been investigated following studies where it has been shown that antioxidants were able to significantly reduce chromosome damage. One instance of this is the use of a supplement of beta-carotene, ascorbic acid and alpha tocopherol in human volunteers and the response in terms of micronucleus frequency in circulating lymphocytes (Fig. 30.1).[1] When a total of 187 human volunteers were examined before receiving supplementation, it was found that women had higher micronucleus counts than men and that the frequency increased with age in both sexes. Following supplementation for 6 months, male and female controls, and females taking antioxidants exhibited average micronucleus frequencies about 20% less than that at baseline, but supplemented males showed a further 18.6% decrease that was statistically lower than in controls.

Present research is aimed at determining the most sensitive conditions for a rodent model in terms of the level of antioxidant fed and of the animal's age (Siebert, unpublished data). Since cancer tends to be more frequent in the elderly, bone marrow micronucleated erythrocytes are being determined in young and old animals fed various levels of antioxidants. Similar measurements are being carried out on rodents being fed wheat produced by standard agricultural practices and without the use of pesticides.

Apart from the orange and green vegetables, some plant oils are rich in the tocopherols and tocotrienols. Palm oil can contain approximately 500 $\mu g\%$ each of the tocopherols and the tocotrienols. When fed to mice treated with a known carcinogen, fewer mammary tumors were present compared to mice fed other vegetable oils of lesser tocotrienol content.[2]

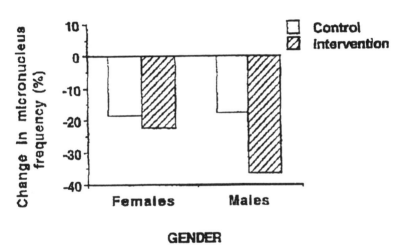

FIG. 30.1. MICRONUCLEUS FREQUENCY IN LYMPHOCYTES OF MEN AND WOMEN FED A CONTROL DIET AND AN INTERVENTION DIET HIGH IN ANTIOXIDANTS

In the area of skin cancer, topical and dietary use of vitamin E has been studied in relation to damage caused by UV irradiation of hairless mice.[3] Restriction of dietary vitamin E had little effect on the degree of epidermal lipid peroxidation or DNA synthesis, but high levels of dietary vitamin E did return DNA synthesis to levels in unirradiated animals. Topical application of the vitamin similarly restored levels to unirradiated values and reduced lipid peroxidation. In a separate study, the effects of vitamins E and C were studied in relation to irradiation by exposing liposomes to UVA and UVB emissions.[4] Alpha tocopherol was more effective in reducing radiation reaction products than its acetate ester, and protected more potently when in the presence of ascorbic acid.

Other research has examined the benefits of cereal fractions on colon cancer. Some fractions of barley bran and wheat bran reduced the incidence of tumors induced in rats by the use of a procarcinogen.[5] These reductions were correlated with butyrate concentration in the distal colon. Butyrate formed from microbiological fermentation of polysaccharides has been proposed as an antineoplastic agent. In a separate study, using similar techniques, whey protein was found to be more effective than red meat or soya protein in reducing the incidence of tumors.[6]

The importance of butyrate has been emphasized in another study which examined the frequency of micronuclei in the distal section of the colon of pigs after they received a diet similar to that of the Australian human population. Together with negative correlation with fecal mass and micronuclei frequency, was the positive correlation between butyrate concentration and micronuclei in this section of the colon.[7] Studies of this nature emphasize the opportunities to design foods with increased non-starch polysaccharide content to improve the colon environment.

REFERENCES

[1] Baghurst, P.A. and Luderer, W. An intervention study of the effects of antioxidant supplements on chromosome damage, 1994. Proc. American Association for Cancer Research, Vol. 35, p. 630, Abstract No. 3759.

[2] Sundrum, K., Hkor, H.T., Ong, A.S.H. and Pathmanathan, R. Effect of dietary palm oils on mammary carcinogenesis in female rats induced by 7,12-dimethylbenz(a)anthracene. Cancer Res. 1989; 49:1447–1451.

[3] Record, I.R., Dreosti, I.E., Konstatinopoulas, M. and Buckley, R.A. The influence of topical and systemic vitamin E on ultraviolet light-induced damage in hairless mice. 1991; 16:220–225.

[4] Record, I.R., Cavallo, L. and Dreosti, I.E. Protection against UVA and UVB induced lipid peroxidation in liposomes by different forms of vitamins E and C. (Yagi et al. eds.) Intern. Congress Series no. 998, 5th ICOR, Kyoto, Japan 1992:621–624, Elsevier, Amsterdam.

[5] McIntosh, G.H., Regester, G.O., Le Leu, R. and Royal, P. Dietary protein and colon cancer risk. Proc. XV Intern. Congr. Nutr. 1993:504.

[6] McIntosh, G.H., Le Leu, R.K., Royal, P.J. and Young, G.P. A comparative study of the influence of differing barley brans on DMH-Induced intestinal tumors in male S-D rats. J. of Gastroenterology & Hepatology. 1996; 11:113–119.

[7] Topping, D.L., Illman, R.J., Clarke, J.M., Trimble, R.P. and Marsono, Y. Human foods and large bowel volatile fatty acids in a suitable animal model — the pig. Proc. XV Intern. Congr. Nutr. Adelaide, 1993.

CHAPTER 31

ONGOING AND FUTURE CLINICAL NUTRITION RESEARCH NEEDS

DANIEL W. NIXON

Associate Director, Cancer Prevention and Control
Hollings Cancer Center and Folk Professor of Experimental Oncology
Medical University of South Carolina
Charleston, South Carolina

ABSTRACT

Clinical nutrition research is moving into new areas. Appropriate prevention and treatment methods for most human undernutrition syndromes have been developed. Basic research and epidemiologic studies now link other chronic diseases to nutrition (atherosclerotic cardiovascular disease, diabetes, cancer), but effective nutritional intervention strategies are lacking, especially for cancer. A crucial need in clinical nutrition research is to develop and complete nutrition intervention clinical trials, so that basic and epidemiologic research results can be transmitted into public health practice.

INTRODUCTION

The greatest challenge facing every living organism, plant or animal, single celled or multicelled, is proper nutrition. Normal biologic functions depend on an adequate and sufficient supply of energy and nutrients in the correct balance. An unbalanced energy and nutrient supply, either too little or too much, results in disease. Undernutrition syndromes in humans (marasmus, kwashiorkor, beriberi, scurvy, etc.) are well-studied but unfortunately still common even though methods of prevention and cure are known. Overnutrition likewise is associated with disease: obesity, arteriosclerotic cardiovascular disease, diabetes, certain cancers. These diseases also remain a major health burden throughout the world.

The syndromes caused specifically by undernutrition, while still subject to research in some aspects, are today largely a social and political problem. On the other hand, the mechanisms of chronic diseases associated with, but not necessarily caused entirely by nutrition, especially cancer, remain largely unknown. Discovery of these mechanisms and ways to overcome them is a major clinical nutrition research need now and in the future. This chapter focuses on cancer as a model for clinical nutrition research.

BACKGROUND

We now know that certain vitamins, major minerals, micronutrients, fat, fiber and calories are all related in varying degrees to cancer, as are hundreds of non-nutrient components of plants (including the so-called "chemopreventive" compounds). These relationships have long been suspected[1], but only in the last few decades have mechanisms for the relationships begun to emerge. Early in the 20th century, it was suggested that excessive weight and calorie intake might be the cause of the observed increase in cancer in the developed world.[2] Since then, abundant basic and epidemiologic evidence has accumulated that greatly strengthens the diet cancer association. This association is most clear with several common human cancers, including breast and colon cancer.

In breast cancer, a consistent survival advantage, stage by stage, has been seen for Japanese compared to American patients[3,4]; this advantage has been related to the traditionally low intake of dietary fat in Japan.[5,6] A number of other studies show that increased body weight is associated with shorter disease-free intervals and shorter overall survival in breast cancer patients.[7,8,9] High weight and elevated serum cholesterol together was associated with a substantial decrease in five-year survival compared to patients in whom both parameters were low.[10]

In the laboratory, transplanted breast cancer is enhanced by increased fat intake; both primary tumor growth and metastatic growth are increased.[11,12,13] Decreased fat intake has a tumor-inhibiting effect.[14,15] Certain fatty acids, especially linoleic acid, are known to stimulate mammary tumor growth in animals[16], but the relative importance of total fat, fatty acids and fat calories in breast cancer remains uncertain.

In colorectal cancer, dietary fat, protein and fiber all appear to be involved. Higher incidence and mortality of this cancer is associated with increased consumption of animal fat and animal protein[16,17]; the association is perhaps stronger with beef intake rather than chicken, pork or fish.[18] Animal experiments also support the fat-colon cancer connection.[19,20] Dietary fiber is strongly associated with colorectal cancer in a preventive role rather than an enhancing one. Many epidemiologic studies show that high fiber intake correlates with low colorectal cancer risk and vice versa.[6,21,22,23] Possible mechanisms for fiber's antineoplastic effects include decreased gut transit time, binding of carcinogens in the gut lumen, increased bacterial degradation of intraluminal carcinogens and changing the concentration of bile acids in the gut.

Diet is also linked to cancers other than breast and colon cancer. Risk of prostate, endometrial, pancreas and other cancers is influenced by diet.[24] An interesting recent study showed that a low-fat diet reduced the incidence of premalignant actinic keratoses of the skin[25], so now squamous carcinomas may have joined the ranks of dietary related tumors, most of which have been adenocarcinomas to date (breast, colon, etc.).

CLINICAL TRIALS

Despite the strong associations between diet and cancer, conclusive proof of the cancer preventive benefit to be gained by dietary changes or supplements requires prospective, randomized clinical trials. Approximately 40 such trials are now being sponsored by the National Cancer Institute, and some have now been completed.[26] Completed trials have shown that increased dietary fiber decreased the formation of colonic polyps in patients with familial adenomatous polyposis[27], that supplemental vitamins A and E with selenium reduced stomach cancer mortality in an area of China[28], that 13-cis-retinoic acid caused regression of oral leukoplakia and a lower rate of second primary head and neck cancers[29,30] and that direct application of all-*trans* retinoic acid to the cervix caused regression of early dysplastic lesions of the cervix.[31] Negative results include the finding that 13-cis-retinoic acid and beta carotene did not prevent basal and squamous cell skin cancers[32,33], and that beta carotene and vitamin E did not prevent lung cancer in smokers.[34] The latter study did reveal a lower lung cancer incidence in the placebo group of smokers who had higher blood levels of beta carotene and vitamin E, and these individuals have higher intakes of foods rich in beta carotene and vitamin E.

Future clinical trials in nutrition and cancer will have to overcome the barriers of cost, large subject numbers, and long study duration. Some trials have faltered in the past because of these obstacles. The Women's Health Trial was designed to evaluate a low-fat diet as a breast cancer preventive in high-risk women but was not completed in part because of the high costs involved (estimates as high as $150,000,000) to complete the study.

Clearly such costly trials are difficult. Strategies to decrease expense (and the related large subject number and long study duration) include the use of "intermediate markers" as surrogates for cancer end points and enrollment of subjects at very high risk for the disease in question. Intermediate markers, such as colon polyps or leukoplakia, shorten trial duration because they occur or reoccur faster than the cancers they proceed. High risk subjects develop more end points during the study duration than subjects at low risk.

The American Cancer Society is sponsoring two cancer prevention trials designed for cost-effectiveness through the use of intermediate markers and enrollment of high-risk subjects and by using non-paid trained personnel to help conduct the studies. The first trial, being piloted in the Virginia Division of the American Cancer Society, focuses on prevention of adenomatous colon polyp recurrence. This two-arm, randomized prospective study enrolls subjects who have had previous polyps removed and therefore are at increased risk for polyp recurrence. Volunteer physicians identify appropriate patients and perform routine follow-up colonoscopy as for any polyp patient. Trained volunteers (volunteer adjunct researchers or VARS) evaluate potential subjects for protocol eligibility, collect baseline nutritional information and counsel subjects on the requirements of the study. After randomization, VARS are assigned to help monitor compliance to the required cereal intake and to report any problems to the study office. A small paid staff operates this office in Richmond, Virginia. The pilot (200 patient) phase is scheduled for completion in FY94-95; performance of the VARS, as well as subject compliance, will be evaluated. If satisfactory, the study will be expanded to 1,000 patients in each arm.

The second study evaluates a 15% fat calorie diet versus a 30% fat calorie diet in the prevention of breast cancer recurrence. It is being piloted in New York State and Long Island. Patients who have had breast cancer treated for cure and are postmenopausal are eligible; this group is at high risk for recurrence and second breast primaries, so the study requires fewer subjects and can be done sooner (approximately 5 years) than a study enrolling women at risk, but who have not already had the disease (10-15 years). As in the colon polyp study, volunteer physicians and trained VARS identify, evaluate, counsel and help follow compliance. This study's pilot phase is scheduled for completion in FY94-95 and, if successful, will expand to an implementation phase of 1,000 patients in each arm. It is hoped that these two ACS projects will serve as time-efficient and cost-effective models for future nutritional and chemoprevention clinical trials. Such trials will complete the TRIAD of basic research, epidemiologic research and clinical research, so that appropriate preventive nutritional advice can be given to those at risk for cancer.

REFERENCES

[1]   Kritchevsky, D. Nutrition and breast cancer. Cancer 1990; 66:1321–1325.

[2]   Hoffman, F.L. The menace of cancer. Am. J. Obstet. 1913; 68:88–96.

[3]   Wynder, E.L., Kajitani, T. et al.: A comparison of survival rates between American and Japanese patients with breast cancer. Surg. Gynecol. Obstet. 1963; 117:196–200.

[4]   Sakamoto, G., Sugono, H. and Hartman, W.H. Stage by stage survival from breast cancer in the U.S. and Japan. Jap. J. Cancer 1979; 25: 161–170.

5   Carroll, K.K., Gammal, E.B. and Plunkett E.R. Dietary fat and mammary cancer. Can. Med. Assn. J. 1968; 98:590–594.

6   Armstrong, B. and Doll, R. Environmental factors and cancer incidence and mortality in different countries, with special reference to dietary practices. Intern. J. Cancer 1975; 15:617–631.

7   Donegan, W.L., Hartz, A.J. and Rimm, A.A. The association of body weight with recurrent cancer of the breast. Cancer 1978; 41:1590–1594.

8   Newman, S.C., Miller, A.B. and Howe, G.R. A study of the effect of weight and dietary fat on breast cancer survival time. Am. J. Epidemiol. 1986; 123:767–774.

9   Tretli, S., Haldorsen, T. and Ottestad, L. The effect of pre-morbid height and weight on the survival of breast cancer patients. Brit. J. Cancer 1990; 62:299–304.

10 Tartter, P.I., Papatestos, A.E. et al. Cholesterol and obesity as prognostic factors in breast cancer. Cancer 1981; 47:2222-2227.

11 Gabor, H., Hillayed, L.A. and Abraham, S. Effect of dietary fat on growth kinetics of transplantable mammary adenocarcinoma in BALB/c mice. J. Nat. Cancer Inst. 1985; 74:1299–1305.

12 Katz, E.B. and Boylon, E.S. Stimulatory effect of a high polyunsaturated fat diet on lung metastasis from the 13762 mammary adenocarcinoma on female retired breeder rats. J. Nat. Cancer Inst. 1987; 79:351–358.

13 Katz, E.B. and Boylon, E.S. Effect of the quality of dietary fat on tumor growth and metastasis from a rat mammary adenocarcinoma. Nutr. Cancer 1989; 12:343–350.

14 Davidson, M.B. and Carroll, K.K. Inhibitory effect of a fat free diet on mammary carcinogenesis in rats. Nutr. Cancer 1982; 3:207–215.

15 Kalamegham, R. and Carroll, K.K. Reversal of the promotional effect of a high fat diet on mammary tumorigenesis by subsequent lowering of dietary fat. Nutr. Cancer 1984; 6:22–31.

16 Rose, D.P., Hatala, M.A., Connolly, J.M. and Raybum, J. Effect of diets containing different levels of linoleic acid on human breast cancer growth and lung metastasis in nude mice. Cancer Res. 1993; 53:4686–4690.

17 Drasar, B.S. and Irving, D. Environmental factors and cancer of the colon and breast. Brit. J. Cancer 1973; 27:167–172.

18 Howell, M.A. Diet as an etiological factor in the development of cancers of the colon and rectum. J. Chron. Dis. 1975; 28:67–80.

19 Nigro, N., Campbell, R., Gantt, J., Lin, Y. and Singh, D. A comparison of the effect of the hypocholesteric agents cholestyramine and candicidin on the induction of intestinal tumors in rats by azoxymethane. Cancer Res. 1977; 37:3198–3203.

[20]   Broitman, S., Vitale, J., Varousek-Jakuba, E. and Gottleib, L. Polyunsaturated fat, cholesterol and large bowel tumorigenesis. Cancer 1977; 40: 2455–2463.

[21]   Zaridze, D.G. Environmental etiology of large bowel cancer. J. Nat. Cancer Inst. 1983; 70:389–400.

[22]   Kritchevsky, D. Dietary fiber and cancer. Nutr. Cancer 1985; 6:213–219.

[23]   Greenwald, P., Lanza, E. and Eddy, G.A. Dietary fiber in the reduction of colon cancer risk. J. Am. Dietet. Assoc. 1987; 87:1178–1188.

[24]   Cancer Facts and Figures, American Cancer Society, 1994.

[25]   Black, H.S., Herd, J.A., Goldberg, L.H. et al. Effect of a low fat diet on the incidence of actinic keratosis. N. Engl. J. Med. 1994; 330:1272–1275.

[26]   Nixon, D.W. Chemoprevention clinical trials. In Cancer Chemoprevention, Chapter 9. CRC Press, Boca Raton, FL (1994).

[27]   De Cosse, J.J., Miller, H.H. and Lesser, M.L. Effect of wheat fiber and vitamins C and E on rectal polyps in patients with familial adenomatous polyposis. J. Natl. Cancer Inst. 1989; 81:1290–1297.

[28]   Blot, W.J., Li, J.Y., Taylor, P.R. et al. Nutrition intervention trials in Linxian, China: supplementation with specific vitamin/mineral combinations, cancer incidence, and disease-specific mortality in the general population. J. Natl. Cancer Inst. 1993; 85:1483–1492.

[29]   Hong, W.K., Endicott, J. and Itri, L.M. 13-cis-retinoic acid in the treatment of oral leukoplakia. N. Engl. J. Med. 1986; 315:1501–1505.

[30]   Hong, W.K., Lippman, S.M., Itri, L.M. et al. Prevention of second primary tumors with isotretinoin in squamous and cell carcinoma of the head and neck. N. Engl. J. Med. 1990; 323:795–801.

[31]   Meyskens, F.L., Serwit, E. and Moon, T.E. Enhancement of regression of cervical intraepithelial neoplasia II (moderate dysplasia) with topically applied all trans-retinoic acid: a randomized trial. J. Natl. Cancer Inst. 1994; 86:539–543.

[32]   Tangrea, J.A., Edwards, B.R., Taylor, P.R. et al. Long-term therapy with low dose isotretinoin for prevention of basal cell carcinoma: a multicenter clinical trial. J. Natl. Cancer Inst. 1992; 84:328–332.

[33]   Greenberg, E.R., Bacon, J.A., Stukel, T.A. et al. A clinical trial of beta-carotene to prevent basal cell cancers of the skin. N. Engl. J. Med. 1990; 323:789–795.

[34]   The Alpha-Tocopherol, Beta Carotene Prevention Study Group. The effect of vitamin E and beta carotene on the incidence of lung cancer and other cancers in male smokers. N. Engl. J. Med. 1994; 330:1029–1035.

CHAPTER 32

ROLE OF PHYTOCHEMICALS IN CHRONIC DISEASE PREVENTION

GERDA GUHR
PAUL A. LACHANCE

Department of Food Science
Rutgers - The State University
New Brunswick, New Jersey 08903-0231

ABSTRACT

There is an abundance of scientific evidence which indicates that certain naturally-occurring non-nutritive, and some nutritive, chemical components of fruits, vegetables, grains, nuts, tea and seeds may prevent or reduce the risk of some chronic diseases, such as various cancers and cardiovascular disease. These natural compounds are commonly referred to as "phytochemicals" or "active compounds."

Phytochemicals have various disease-fighting properties. They may function via one or multiple biochemical mechanisms to interfere with or prevent carcinogenesis. Phytochemicals with antioxidant properties may help inhibit cardiovascular disease. Certain vitamins and the trace mineral selenium also function as antioxidants and play a role in disease prevention.

The past few years have shown an abundance of interest in the link between diet and cancer; however, this concept is not a novel one. A large number of researchers have focused their attention on cancer prevention and the potential of phytochemicals; therefore, many more studies are in progress. The goal of this research was to review the current literature and report on the progress of the latest studies regarding phytochemicals and their role in the prevention of carcinogenesis and cardiovascular disease, as well as to categorize various phytochemicals according to their functional groups and disease-preventive mechanisms.

Some of the benefits of phytochemicals are their low toxicity, low cost, easy availability and their ability to prevent some chronic diseases. However, due to the complex mechanisms of cancer and cardiovascular disease, prevention will rely on more than a single compound. Researchers have just scratched the surface of the identification of various compounds, their disease preventive properties, and their efficacy against certain diseases. However, many questions still remain, such as the bioavailability of the compounds, how exactly humans metabolize the compounds, the relative toxicity of phytochemicals, their storage stability, doses required for effective outcomes, and the use of acceptable universal methods of analyses for each phytochemical.

INTRODUCTION

There is an abundance of scientific evidence which indicates that certain naturally-occurring non-nutritive chemical components of fruits, vegetables, grains, nuts, tea and seeds may prevent or reduce the risk of some chronic diseases, such as various cancers and cardiovascular diseases. These natural compounds are commonly referred to as phytochemicals or active compounds. Other terminology used in reference to these compounds and products which contain them is provided in Table 32.1.

Phytochemicals have various disease-fighting properties. They may function via one or multiple biochemical mechanisms to interfere with or prevent carcinogenesis. Phytochemicals with antioxidant properties may help inhibit cardiovascular disease. Certain vitamins and the trace mineral selenium also function as antioxidants and play a role in disease prevention.

The past few years have shown an abundance of interest in the link between diet and cancer; however, this concept is not a novel one. A large number of researchers have been focusing their attention on cancer prevention and the potential of phytochemicals, and many more studies are in progress. The main scientific approaches used to gather data on these compounds are epidemiological studies, the use of animal model studies to observe the effects of the compounds *in vitro* and *in vivo*, and the study of the effects of the compounds on human cells both *in vitro* and *in vivo*. Many of the studies conducted to date have shown a link between diet and cancer risk. Epidemiological data from 1933 to the present have been quite consistent in regard to indicating a link between cancer and the diet.[62] There is also a great amount of evidence indicating that high meat and fat consumption increases the risk of developing cancer, while the consumption of diets high in cereals, fruits and vegetables decreases the risk of certain cancers.[21] Studies also suggest that an increased intake of fruits and vegetables, and eating a wide variety of foods, decreases chronic disease risk factors.

Phytochemicals have been reported to be one of the contributors towards cancer prevention. Some of the benefits of these compounds are their low toxicity, low cost, easy availability and their ability to prevent some chronic diseases. However, due to the complex mechanisms of cancer and cardiovascular disease, prevention will rely on more than a single compound. Researchers have just scratched the surface of the identification of various compounds, their

TABLE 32.1

DEFINITIONS OF TERMS USED IN REFERENCE TO NATURAL COMPOUNDS AND
THEIR PRODUCTS WHICH HELP FIGHT CHRONIC DISEASE

Term	Definition
Chemopreventors	Non-nutritional components in our regular diet that possess antimutagenic and anticarcinogenic properties. These components are classified as food entities that can prevent the appearance of some long term diseases like cancer or cardio-vascular disorders. *
Designer Foods	Processed foods that are supplemented with food ingredients naturally rich in disease-preventing substances. This may involve genetic engineering of food.**
Functional Food	Any modified food or food ingredient that may provide a health benefit beyond the traditional nutrients it contains.**
Pharmafood	A food or nutrient that claims medical or health benefits, including the prevention and treatment of disease.**
Phytochemical (Greek: "plant chemical")	Non-nutritive or nutritive, biologically active compounds present in edible natural foods including fruits, vegetables, grains, nuts, seeds and tea, which, when ingested, have the potential to prevent or delay the onset or continuation of chronic diseases in humans and animals.
Nutraceutical	Any substance that may be considered a food or part of a food and provides medical or health benefits, including the prevention and treatment of disease.**

Sources:
* Stavric, B.[82]
** American Dietetic Association[4]

disease preventative properties, and their efficacy against certain diseases. Many questions still remain. Although the mechanisms of action of a variety of phytochemicals have been uncovered, many more need to be investigated. Acceptable universal methods of analysis of the compounds still must be established and the stability of many phytochemicals during processing and storage is still unknown. Storage conditions and levels for optimal benefit and also levels of toxicity (if any) still need to be determined. Also, the possibility of synergism between the compounds is yet to be explored in great detail. Despite all of these gaps in our knowledge of phytochemicals, there is an interest in isolating the compounds and fortifying other foods with them to produce "designer foods" or marketing them as dietary supplements. The marketing of these products is ongoing and far ahead of the evidence needed. Before these compounds can be formed into a dietary supplement beneficial to health, or before designer foods should be produced, more precise and more targeted research must be conducted to shed light on the unknown variables.

Facts About the Diets of Various Nations

Overview of the American Diet. Diet is believed to play a role in the morbidity and mortality associated with four of the major diseases in the United States, cardiovascular disease, cancer, hypertension and obesity. Cardiovascular disease and cancer combined account for 70% of all U.S. deaths. Cardiovascular disease accounts for 51% of all deaths in the world. Scientists estimate that one-third of all cancer cases and one-half of cardiovascular diseases and hypertension can be attributed to diet.[14,31] The National Cancer Institute further estimates that eight of ten cancers have a nutrition/diet component.[4] Emerging theory holds that cardiovascular diseases can be attributable to oxidized cholesterol and oxidized fats coupled with not enough fiber, fruits and vegetables in the diet.[31] Tobacco and alcohol are the leading two avoidable causes of cancer. Tobacco is responsible for approximately 30% of all cancer deaths and alcohol for 3%.[14]

The American Dietetic Association conducted a nation-wide Nutrition Trends Survey[3] between April 20 and May 4, 1995, in which a randomly selected group of 807 adults aged 25 or older were asked various questions about nutrition. Diet and nutrition were rated by 79% of the respondents to be important to him/her personally, and 94% felt that foods they eat have a strong/moderate impact on their personal health in the long term. Despite their feelings towards good nutrition and its effect on their health, only 35% of the participants said that they were doing all they could to achieve balanced nutrition and a healthful diet, and only 35% of the respondents claimed to be careful in food selection to achieve balanced nutrition and health. These results indicate that a considerable gap exists between how the public feels about the benefits of proper nutrition and what choices they actually make when it comes to their diets.

Awareness of the USDA Food Guide Pyramid, which indicates, among other things, that at least 2-4 servings of fruits and 3-5 servings of vegetables should be consumed per day was quite low, with 58% of the respondents being aware of it. Only 13% of the 58% were "very familiar" with the Pyramid. Based on the results, a change toward more healthful diet habits, including eating at least five to nine servings of fruits and vegetables a day, would correlate positively with a decrease in certain chronic diseases. Consumers will need not only to be more educated about proper diet choices, but they need to be motivated to alter their current dietary status.

Epidemiological Studies of Diets in Various Countries Versus the U.S. Although some biases[1] exist in epidemiological studies, there is overwhelming evidence from these studies that there is a link between diet and cancer which cannot be ignored. Although there are certainly exceptions, different countries, regions, and certain religious groups, have their own culinary traditions with an emphasis on certain ingredients which tend to be consumed more often than others. Due to this generalization, correlations between a certain group's typical diet and the absence or presence of certain cancers have been made. For example, the typical high fat, omnivorous diet of Western countries has been implicated in a high incidence of coronary heart diseases, breast cancer, prostate cancer, colon cancer, ovarian cancer, and endometrium cancer, while in Japan and China a high incidence of hypertension, stroke, stomach cancer and esophageal cancer has been linked to high intake of salt and smoked foods in their diets associated with their choices of preservation techniques.[89]

The following table lists some nations, regions or religious groups and the implications their typical diets have on cancer.

TABLE 32.2

A SAMPLE OF EPIDEMIOLOGICAL STUDIES EXHIBITING THE DIET-CANCER LINK

Nation/Group	Major Food Included in the Diet/Excluded from Diet	Food's Proposed Effect on Cancer
Mediterranean Countries (e.g. Italy, Greece)	High tomato consumption Low saturated fat consumption	Low prostate cancer[30]
China: Gangshan Province Quixia Province	Average 20g garlic/day/capita Little garlic eaten	Low gastric cancer death rate (3.5/100,000) High gastric cancer death rate (40/100,000)[70]
Southeast Asian Countries	Soy Foods Seaweed	Low breast cancer, colon cancer, and prostate cancer[37,69]
U.S. Adventist Men	Beans, lentils, peas, dried fruit	Low prostate cancer[37]
South Pacific Islands (e.g., Fiji, Cook, Tahiti)	Many fruits and vegetables high in carotenoids	Low incidence of several types of cancers[51]

---

[1] Biases include the possibility that the people in epidemiological studies which observe healthful diets may be healthier and more health conscious to begin with.[40] The sequence in which the foods are consumed is not accounted for in epidemiological studies, which may have an impact on the diet-cancer study depending on the mechanism of the compound.[88]

Epidemiological studies show that people living in Western countries are at a higher risk for hormone-dependent cancers, such as breast and prostate cancer, versus countries such as Japan and Singapore. These epidemiological studies emphasize certain foods as being protective. Researchers have broken down these foods to determine which of their components are eliciting their protective effects. The Classification of Active Compounds section of this chapter classifies the biologically active compounds implicated in protecting against chronic diseases.

Pathogenesis of Chronic Diseases

Cancer. Carcinogenesis has been described as "a progressive series of disorders in the function of signal transduction pathways — signal transduction pathways being the means by which the hormones and growth factors that regulate cell growth, proliferation, and differentiation communicate across cell membranes via receptors and receptor-associated enzymes, then through the cytoplasm and into the nucleus via a network of intermediary molecules known as second messengers."[49] Important general mechanisms of carcinogenesis are mutagenesis and uncontrolled proliferation[49] The five most common cancer sites are lung, colon, breast, prostate and pancreas. Widespread cancers at these five sites are all associated with low cure rates[2]; thus, prevention of these cancers is the more desirable route. Chemoprevention of cancer is a "means of cancer control in which the occurrence of the diseases, as a consequence of exposure to carcinogenic agents, can be slowed, completely blocked, or reversed by the administration of one or several naturally-occurring or synthetic compounds."[85]

Cardiovascular Disease. The term cardiovascular disease (CVD) encompasses a variety of diseases which affect the heart and arteries. Occlusive cardiovascular disease occurs when blood circulation is blocked. This type of cardiovascular disease includes atherosclerosis, thrombosis and/or embolism, and vasoconstriction/vasospasm and research has shown its development to be linked to diet.[31] Atherosclerosis is a condition in which fatty deposits accumulate in and on the lining of the artery walls and is precursory to many cardiovascular diseases which include angina (chest pain), congestive heart failure, and myocardial infarction (heart attack). Saturated fat was once believed to be the culprit in cardiovascular diseases, mainly atherosclerosis; however, only one study (the Seven Countries study) found a strong link between the two.[31] There are also many contradictions, based on epidemiological studies, to the idea that saturated fat is linked to cardiovascular disease. A few of these contradictions are listed in Table 32.3.

TABLE 32.3
CONTRADICTIONS TO THE BELIEF THAT SATURATED FAT IS RELATED
TO CARDIOVASCULAR DISEASES[31]

Country/Group	Diet and Health Trends
France	High saturated fat intake: Low CVD
United Kingdom, upper class	Highest saturated fat intake in the UK: Lowest CVD in UK
Asian immigrants to U.S.	Lower blood pressure and lower plasma cholesterol levels than U.S. Caucasians: higher CVD than U.S. Caucasians
Providence, R.I. Veterans Hospital, U.S.	Of 194 consecutive autopsies of patients with fatal atherosclerosis, only 10% had elevated serum cholesterol. In most cases, no evidence of serum cholesterol, diabetes or hypertension.

Despite these contradictions, saturated fat, especially laurate and palmitate, have been reported to raise LDL cholesterol[13], which has been reported to increase the risk of cardiovascular disease.[60] Fernandez, et al.[27], reported that levels of plasma low density lipoprotein (LDL) cholesterol significantly increased with increasing concentrations of saturated fat in guinea pigs. Other dietary factors that have been reported to increase plasma LDL cholesterol are dietary cholesterol and excess calories leading to obesity.[84]

The emerging hypothesis for the cause of cardiovascular disease is the "Antioxidant Hypothesis," which states free radical mediated oxidation of cholesterol's low density lipoproteins (LDL) as the main step in atherogenesis.[31,52,82] The phospholipid layer of LDL contains a large number of polyunsaturated fatty acids (PUFA). These fatty acids are highly susceptible to oxidation due to their methylene interrupted double bond structure; thus, they make LDL cholesterol highly susceptible to free radical mediated peroxidation. Studies show that oxidized LDL is more atherogenic than non-oxidized LDL. Macrophages which create foam cells, the precursors of plaque that ultimately clog the artery, have a much greater affinity towards oxidized LDL and take these compounds up more readily than the non-oxidized form of LDL. Thus, the oxidation of LDL is considered to be the early phase of the development of atherosclerosis.

Oxidation of LDL is thought to be initiated by smoking, ingestion of PUFAs and low nutritional intake of antioxidants and folic acid. Studies have shown that high levels of homocysteine in the blood, which is due to insufficient intake of folic acid, promotes the formation of oxidized LDL. Antioxidants (e.g., vitamin E, carotenoids, vitamin C, flavonoids) prevent the oxidation of LDL from occurring; therefore, have been found to prevent atherosclerosis. Vitamin E, especially, has been found to be a powerful antioxidant against LDL oxidation. Due to its lipophilic nature, vitamin E positions itself directly at the core of the atherogenic process in the phospholipid layer near the PUFAs, and prevents or reduces the formation of the disease. Vitamin C spares vitamin E and is associated with a lower mortality from cardiovascular disease.[24,45]

Defense Mechanisms Working Against Chronic Diseases. Two defense systems that work against chronic diseases, the deactivation of carcinogens and antioxidant activities, either exist endogenously in human cells or are obtained from exogenous sources, such as fruits or vegetables.

The Body's Defense: Hepatic Enzymes. The liver contains two classes of detoxification reactions, which function to metabolize carcinogenic compounds into non-toxic compounds which can be excreted from the body. These two reactions are referred to as Phase I and Phase II reactions. Phase I reactions are responsible for the oxidation, reduction and hydrolysis of foreign compounds, resulting in more polar compounds. These reactions are catalyzed by the cytochrome oxidase system (COS). The COS consists of about eight different enzymes which work in different substrates. These enzymes include cytochrome P450 enzymes and cytochrome P448 enzymes. Enzymes in the COS function by bringing the carcinogen in very close proximity with oxygen and forcing an atom of oxygen into the foreign molecule. The other molecule of oxygen goes to water. Most Phase I reactions reduce the toxicity of carcinogens; however, sometimes a carcinogen is actually more toxic than the original material after it has been oxidized, reduced or hydrolyzed.[76]

During Phase II reactions, conjugates are formed with Phase I products, or carcinogens directly, between endogenous chemicals to give more polar materials that are more readily excreted from the body. The endogenous chemical uridine diphosphate-α glucuronic acid (UDP-αGA), an active form of glucuronic acid, forms conjugates with compounds containing hydroxy, carboxy-acid, free amino, or thiol groups especially when catalyzed by UDP-αGA transferase. Glutathione, another Phase II reaction chemical, forms polar conjugates with foreign compounds when catalyzed by glutathione-S-transferase enzymes. Glutathione is the most highly reactive Phase II detoxifying chemical. Other endogenous chemicals involved in Phase II reactions include S-adenosyl-methionine (SAM), 3'-phosphoadenosine-5'-phosphosulfate (PAPS), and the amino acids glycine and glutamine. Most Phase II reactions detoxify a carcinogen; however, although not as frequently as in Phase I reactions, sometimes a non-toxic compound is activated into a toxic form after undergoing Phase II conjugation.[76]

Antioxidants. Free radicals are reactive chemical species that contain one or more unpaired electrons and are formed in the ordinary process of oxygen metabolism.[34] They are an omnipresent group of carcinogenic compounds which must constantly be battled and are detrimental to human health, having been linked to cardiovascular disease and carcinogenesis. They can also be generated during tissue injury, infections, excess exercise, food preparation and from environmental sources (i.e., radiation, heat, air pollution, cigarette smoking.[34,75] For example, each puff of a cigarette contains $\sim 10^{14}$ free radical species in the tar phase and $\sim 10^{15}$ in the gas phase.[31])

A group of compounds called the Reactive Oxygen Species (ROS) includes oxygen-derived free radicals and compounds that have great potential for activation *in vivo* to form free radicals. Oxygen derived free radicals include superoxide, hydroperoxyl, hydroxyl, peroxyl, and alkoxyl radicals. Hydrogen peroxide, hypohalous acids and N-chlorinated amines are all ROS compounds that, although they exist in a single state and are not free radicals, are considered ROS compounds due to their ability to form highly reactive free radicals.[34] Oxygen binds readily with metals containing unpaired electrons; thus, iron and copper are also implicated in free-radical-mediated reactions, acting mainly as catalysts.[34] The detrimental effects of these free radicals include their reaction with DNA, lipids, and cell membranes. These reactions lead to aging, organ injuries, and greater susceptibility to cancer and cardiovascular disease. They participate in the activation of certain carcinogens and cause DNA strand breaks and chromosome deletions and rearrangements.[49] Certain compounds which exist in the body or are ingested with food utilize certain defense mechanisms against free radicals as outlined in Table 32.4.

TABLE 32.4
FREE RADICAL DEFENSE METHODS[75]

Defense	Compounds
Enzymes	Superoxide dismutase, glutathione-peroxidases, catalase
Iron- and copper-binding extracellular proteins	Albumin, transferrin, lactoferrin, haptoglobin, ceruloplasmin
Antioxidants	Vitamin C, vitamin E, quinones, glutathione, uric acid, bilirubin, carotenoids, flavonoids, lignans
Constantly target oxidized molecules for catabolism	Nuclear repair enzymes, proteases

Activated oxygen species will react with the nearest fat, protein, carbohydrate, RNA or DNA molecule altering its structure and function if not quenched by an antioxidant.[75] Polyunsaturated fatty acids may be a host to additional free radicals which can generate in them; therefore, it is possible to consume too high of a quantity of polyunsaturated fatty acids, especially in the presence of too little antioxidant.

Oxidative stress occurs when more free radicals are generated than quenched. Taking into consideration that approximately 10,000 oxidative "hits" target the DNA of human cells per day[5], and the damage that free radicals incur in the body, the need for an efficient antioxidant system is clear.

Compounds that are present in fruits, vegetables, grains and nuts, possess chemopreventive properties including antioxidative functions and/or mechanisms which induce protective hepatic enzymes. The remainder of this chapter will focus on the reported evidence of the protective abilities of these phytochemicals.

CLASSIFICATION OF ACTIVE COMPOUNDS

Active compounds could be classified into many different groups to elucidate their protective effects against diseases and other functions, to indicate where they can be found and to provide information on other facts necessary for understanding these compounds. One important property of active compounds is their mode of action in the prevention of carcinogenesis. The need to classify and reclassify these active compounds arises from current discoveries of new compounds and more specific mechanisms. Knowledge of the source of the compounds is important as well, so the benefits due to increased consumption of these food sources can take place and that further research can be done on these sources. Other information, such as intake levels, amounts of the compounds found in foods, and the effects of processing on the stability and bioavailability of the compounds can also help increase our understanding of these compounds and how they can be employed to benefit humans against certain diseases.

Two classifications of the potential mechanisms of the chemopreventive activity of active compounds can be applied to phytochemicals. A classification system originated by L.W. Wattenberg[88] categorizes the compounds into three groups, Blocking Agents, Suppressing Agents and Compounds preventing the formation of carcinogens from precursor compounds. Blocking Agents are defined as compounds which "block the activity of some enzymes which aid in metabolism of genotoxic carcinogens into carcinogenic species"; Suppressing Agents are defined as compounds that "suppress different steps in metabolic pathways required for the development of tumors."[82] Carcinogenic species are usually electrophilic and react with DNA, a nucleophile. Wattenberg's third category of chemopreventive compounds, which includes those compounds which prevent the formation of carcinogens from precursor compounds, focuses mainly on the prevention of the formation of nitroso compounds from reactions of precursor amines or amides with nitrite. Wattenberg's classifications can be found in Fig. 32.1, which also illustrates where each category mechanistically interferes with carcinogenesis.

Another classification of chemopreventive mechanisms developed by Kelloff *et al.*[49] also breaks the potential mechanisms down into three groups; however, this classification is an expansion of Wattenberg's and the categories are broken down into more specific sub-categories. The categories are Carcinogen Blocking Activities, Antioxidant Activities, and Antiproliferation/Antiprogression Activities. These categories and their subsets are provided in Table 32.5.

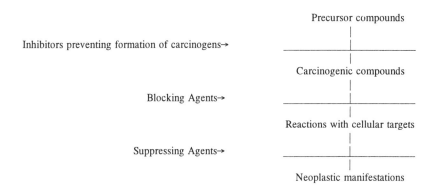

FIG. 32.1. WATTENBERG'S CLASSIFICATION OF CHEMOPREVENTIVE MECHANISMS AND DIAGRAM ILLUSTRATING THE POINT AT WHICH THE ACTIVE COMPOUNDS INTERFERE WITH CARCINOGENESIS[88]

The sources of phytochemicals are all natural, including fruits, vegetables, nuts, grains, legumes, tea and seeds. The phytochemicals from each of these sources varies greatly. Some sources contain a number of phytochemicals covering a wide range of anticarcinogenic mechanisms, and some contain mainly one or two active compounds beneficial to health. Classification of the specific phytochemicals found in each source, the levels of the phytochemicals present in each source, as well as the optimum intake levels for protection against diseases, assist in proper food selection decisions.

The levels of active compounds in each source and also the protective intake levels of these compounds need to be further explored. Many of the completed animal studies give some indication of the amounts of certain active compounds that are necessary for protection against certain diseases; however, the American Dietetic Association explains that "these levels are difficult to extrapolate into human dietary intake requirements" and that "continued *in vivo* and *in vitro* research must be completed before specific recommendations can be formulated."[4] Individual recommendations may be most likely once the intake values are established due to an individual's specific condition and the extent of the disease.

TABLE 32.5
CLASSIFICATION OF CHEMOPREVENTIVE MECHANISMS AND SUBSETS OF THESE GROUPS DESIGNATED BY KELLOFF *ET AL.*[49]

Mechanism	Subset
Carcinogen Blocking Activities	1. Inhibit Carcinogen Uptake 2. Inhibit Formation or Activation of Carcinogen 3. Deactivate/Detoxify Carcinogen 4. Prevent Carcinogen Binding to DNA 5. Increase Level of Fidelity of DNA Repair
Antioxidant/Free Radical Scavenging Activities[a]	1. Scavenge Reactive Electrophiles 2. Scavenge Oxygen Radicals 3. Inhibit Arachidonic Acid Metabolism
Antiproliferation/Antiprogression Activities	1. Modulate Signal Transduction 2. Modulate Hormonal/Growth Factor Activity 3. Inhibit Oncogene Activity 4. Inhibit Polyamine Metabolism 5. Induce Terminal Differentiation 6. Restore Immune Response 7. Increase Intercellular Communication 8. Restore Tumor Suppressor Function 9. Induce Programmed Cell Death (Apoptosis) 10. Correct DNA Methylation Imbalances 11. Inhibit Angiogenesis 12. Inhibit Basement Membrane Degradation 13. Activate Antimetastasis Genes

[a] Further mention of these activities will be understood under "antioxidant activities" or "antioxidant."

The effect of various processing conditions on active compounds still needs to be studied. In some instances, such as in sulforophane in broccoli, it has been observed that cooking increases the amount and availability of the active compound. In other cases, processing has a detrimental effect on the phytochemicals. The tolerance of various phytochemicals to different processing techniques requires further research.

Active compounds contain a variety of different chemical structures which help dictate their function. The main categories of phytochemicals include Phenols, Indoles, Isothiocyanates, Allylic Sulfur Compounds, Monoterpenes, Carotenoids, and Antioxidant Vitamins. The compounds which are categorized into these different groups have been found to participate in some form of disease protection and/or prevention.

Phenols

Phenolic compounds are "substances which possess an aromatic ring bearing one or more hydroxy substitutes, including functional derivatives (esters, methyl esters, glycosides, etc.). Most phenolics have two or more hydroxyl groups and are bioactive substances occurring widely in food plants that are eaten regularly by substantial numbers of people."[39]

There are three major groups of phenols: simple phenols and phenolic acids, hydroxycinnamic acid derivatives, and flavonoids. Phenols can be found in almost all fresh fruits and vegetables, cereal grains, tea, nuts, and seeds. This category of phytochemicals includes a wide range of compounds. A list of some phenols, their sources and their organoleptic and/or aesthetic properties may be found in Table 32.6.

The compounds comprising the large phenolic category of phytochemicals all exert their protective effects against disease via antioxidative activity. Some of these individual compounds have been studied extensively due to strong evidence of an inverse correlation with cancer and/or cardiovascular disease and also due to the interests of the individual researchers. The various mechanisms of some of these highlighted phenols on cancer and cardiovascular disease are depicted in Table 32.7.

Simple Phenols and Phenolic Acids.

Curcumin. Curcumin is the yellow pigment in turmeric, a spice used for its yellow color. Turmeric is contained in the rhizomes of the root of the plant *Curcuma longa* Linn. Curcumin can also be found in curry and mustard. The structure of this active compound contains two phenol groups and a diketone group. Curcumin has been found to possess various chemopreventive properties. The classification of anticarcinogenic mechanisms by Kelloff *et al.* from Table 32.5 has been applied to curcumin in Table 32.8.

TABLE 32.6
LIST OF SOME PHENOLIC COMPOUNDS, THEIR SOURCES AND
ORGANOLEPTIC/AESTHETIC PROPERTIES

Phytochemical (s)	Source	Organoleptic/Aesthetic Property
PHENOLS	Almost all fresh fruits and vege-tables, cereal grains, tea (black and green), nuts and seeds	
A. Simple Phenols and Phenolic Acids		
1. 3-ethylphenol	Cocoa bean	Smoky flavor
2. 3,4-dimethlylphenol	Cocoa bean	Smoky flavor
3. Capsaicin	Chili peppers	Hot pepper warming/burning sensation
4. Carnosol	Rosemary, sage	Major flavor compound
5. Coumarin	Vanilla, cloves, nutmeg	Aids in imparting flavor characteristic of the spice. Principle aroma and pungency element of cloves.
6. Curcumin, Demeth-oxycurcumin, Bisdemth-oxycurcumin	Curcumin (from turmeric)	Yellow pigment
7. p-cresol	Raspberry, blackberry, butter	Flavor compound
8. Diphenol: Isoflavonoid	Soybeans	
9. Diphenol: Lignan	Whole grains, whole soybeans, seeds of berries	Participates in skikimic acid pathway to produce aromas and flavors
10. Ellagic Acid	Grains, nuts (esp. walnuts), straw-berry, cranberry, blackberry, raspberry	Polymers produce tannins which contribute to astringent flavor and enzymic browning reactions.
11. Eugenol	Vanilla, cloves, nutmeg	Aids in imparting flavor characteristic of the spice.
12. Gallic Acid	Coffee, strawberry	Polymers produce tannins which contribute to astringent flavor and enzymic browning reactions.
13. Gingerol	Ginger	Pungent flavor
14. Hydroquinone (in-cludes Sesamol derivatives)	Sesame seeds, sesame oil	Flavor compound
15. Myristicin	Nutmeg, mace	Major flavor compound
16 Piperine	Black pepper	Organoleptic 'bite'
17. Safrole	Nutmeg, mace, black pepper	Major flavor compound
18. Shogaol	Ginger	Pungent flavor
19. Vanillic Acid	Soybeans, peanuts, strawberry	Flavor compound
20. Vanillin	Vanilla	Major flavor compound
B. Hydroxycinnamic Acid Derivatives		
1. Caffeic Acid	Green pepper, olive oil, soy-beans, rosemary, grains, apple, strawberry, peach, grapes, orange, grapefruit, lemon	Flavor compound

2. Chlorogenic Acid	Coffee beans, soybeans, pea-nuts, tea, instant coffee, pear, quince, cherry, plum/prune, peach, blackberry, strawberry, raspberry, current, gooseberry, grapes	Contributes to enzymic and nonenzymic browning reactions. Flavor Compound.
3. *p*-coumaric Acid	Soybeans, peanuts, rice hull, quince, cherry, plum/prune, strawberry, current, gooseberry, orange, grapefruit, lemon	Flavor compound
4. Ferulic Acid	Soybeans, peanuts, grains, peach, quince, plum/prune, blackberry, raspberry, current, gooseberry, orange, grapefruit, lemon	Flavor precursor
C. Flavonoids		
1. Anthocyanins	Red wine, eggplant, radish, red cabbage, onion	Plant colorants (orange, pink scarlet, red, mauve, violet, blue)
2. Catechins	Green tea, black tea, red wine, apple, pear, peach, apricot, plum/prune, cherry, blackberry, strawberry, black current, raspberry, grape, peach	Astringent flavor
3. Flavones	Orange, soy, pineapple, celery, carrot, tomato	Flavor compounds

4. Flavonols
   a. Quercetin
 	
   b. Naringen
	a. Red wine, green onion, green pepper, potato peel, apples, pear, apricot, plum,/prune, cherry, blackberry, black current, raspberry, grapes, peach, yellow onion, red onion
	
	 b. Grapefruit and other citrus fruits
		a. Flavor compound
		
		 b. Bitter flavor compound

5. Glycosides of flavons and flavonols	Same fruits, vegetable, spices as flavones & flavonols.	Flavor compounds
6. Proanthocyanidins	Red wine, apple, grape strawberry, plum, sorghum, barley	• Contribute to astringent flavor when react with a protein • Contribute to enzymic browning reaction • Haze formation in beer and wine • Colorless, but produce a colorful anthocyanin when heated in presence of mineral acid.

TABLE 32.7
SOME REPORTED MECHANISMS OF CERTAIN PHENOLS IN DISEASE PREVENTION

Phytochemical (s)	Functions in Disease Prevention
PHENOLS	
A. Simple Phenols and Phenolic Acids	
1. Curcumin, Demethoxy-curcumin, Bisdemethoxy-curcumin	1. Reverse liver damage induced by Aflatoxin B1 [82] 2. Reduce urinary mutagens in smokers[82] 3. Promote detoxification mechanisms (Phase II enzymes)[82] 4. Inhibit cancer at initiation, promotion, and progression stages[82] 5. Inhibited nitrosamine induced skin, colon, and stomach cancers[16] 6. Anti-inflammatory[16] 7. Inhibited cancers of the skin, forestomach, mammary gland (in vitro), colon, and duodenum in mice[85] 8. Inhibit epidermal arachidonic acid metabolism, cell proliferation, tumor promotion and inflammatory action of arachidonic acid in mice[85] 9. Reduce nitrite production; thus, protect against nitrosation[16]
2. Diphenol: Isoflavonoid	1. Antitumor effect in mammary cancer[9,37,49,73] 2. Protective in hormone-dependent cancers.[9,37,73] 3. Influence sex hormone metabolism, intracellular enzymes, protein syntheses, growth factor action, malignant cell proliferation and angiogenesis[37]
3. Diphenol: Lignan	1. Antitumor, antimitotic, antiviral[82] 2. Protective in hormone-dependent cancers[37] 3. Inhibited mitogen-induced proliferation of malignant cells[37] 4. Influence sex hormone metabolism, intracellular enzymes, protein synthesis, growth factor action and angiogenesis[37]
3. Ellagic Acid	1. Inhibit formation of carcinogen metabolites[49,82,85] 2. Inhibits microsomal P-450 enzymes[49,82,85] 3. Stimulates glutathione-S-transferase[49,82,85] 4. Directly binds to DNA, making sites on guanine less accessible for methylating agents[49,82,85] 5. Antioxidant[49,82,85]
B. Hydroxycinnamic Acid Derivatives	
1. Caffeic Acid	1. Inhibit formation of carcinogen metabolites[82]
2. Chlorogenic Acid	1. Reduce colon cancer in humans[39,82] 2. Inhibit nitrosation[82] 3. Exhibited chemopreventive properties[82] 4. Inhibited lipid peroxidation in rat liver induced by the liver carcinogen carbon tetrachloride[82]
3. Ferulic Acid	1. Inhibit formation of carcinogen metabolites (82)
C. Flavonoids	1. Reduce incidence of myocardial infarction and risk of death from coronary heart disease[82] 2. Lower platelet aggregation 3. Reduce thrombic tendencies 4. Modify various enzymatic functions

1. Catechins	1.	Anticarcinogen in lung, forestomach, esophagus, colon, duodenum, liver, pancreas, mammary gland[62]
	2.	Prevent chronic inflammation associated with carcinogenesis[16]
	3.	Inhibit oxidation of LDL; thus, reduce atherosclerosis and cardiovascular disease[82]
	4.	Inhibited nitrosamine induced skin, colon, stomach cancers in mice[16]
	5.	Anti-inflammatory[16,85]
	6.	Enhance PII enzymes[85]
	7.	Chemoprotective effects against each stage of carcinogenesis leading to skin cancer[85]
	8.	Antimutagenic and anticarcinogenic[82]
	9.	Reduce nitrite production; thus, protect against nitrosation[37]
2. Flavonols Quercetin	1.	Inhibit oxidation of LDL; thus, reduce atherosclerosis and cardiovascular disease[38,68,82]
	2.	Inhibit colon cancer *in vitro*[37]
	3.	Antitumor effects in colon cancer[37]

TABLE 32.8
CLASSIFICATION OF CHEMOPREVENTIVE MECHANISMS BY KELLOFF *ET AL.*
APPLIED TO CURCUMIN[49]

Mechanism	Subset
Carcinogen Blocking Activities	Deactivate/Detoxify Carcinogens[82,85]
Antioxidant Activities	Scavenge Oxygen Radicals
Antioxidant Activities	Inhibit Arachidonic Acid Metabolism[49,85]
Antiproliferation/Antiprogression	Inhibit Polyamine Metabolism[85]

Curcumin is a potent antioxidant which scavenges peroxyradicals, superanions and hydroxyl radicals, thereby inhibiting lipid peroxidation. This compound also chelates iron; thus, preventing iron oxidation. Hepatic Phase I and Phase II enzymes have been shown to increase with the presence of curcumin, enhancing hepatic detoxification activity in scavenging reactive electrophiles.

Arachidonic acid (AA) metabolism has been strongly implicated in carcinogenesis due to the high number of activated oxygen species formed, as well as an increase in AA metabolism during inflammation. Curcumin inhibits both procarcinogenic effects of AA metabolism; it is an antioxidant as well as an anti-inflammatory agent.

An abundance of evidence exists that suggests that the enzyme ornithine decarboxylase (ODC) participates in carcinogenesis. This enzyme is necessary for polyamine metabolism which, although the mechanism is unknown, is significant in cell proliferation, differentiation, and malignant transformation. The tumor promoter 12-O-tetradecanoyl-phorbol-13-acetate (TPA) increases ODC activity in the skin, colon, bladder and liver.[49] Curcumin has been shown to inhibit TPA-induced ODC activity, cell proliferation and tumor promotion in the post-initiation phase of carcinogenesis in mouse epidermis.[85] In mice, curcumin has prevented cancers of the skin (topical application), forestomach, mammary gland, colon and duodenum.[85] Micromolar concentrations of the compound reduced nitrite production significantly *in vitro* when stimulated by lipopolysaccharide and interferon-γ, two common inflammation inducers.[16] Inflammation produces nitric oxide, a damaging free radical, which reacts with superoxides to produce N-nitrosating agents, carcinogens that cause nitrosative deamination of DNA.[16]

Curcumin has been found to inhibit cancer at the initiation, promotion and progression stages of the disease.[16,82] This phytochemical has reduced urinary mutagens in smokers and reversed liver damage induced by aflatoxin B_1.[82] Studies suggest that greater than 50% of curcumin is absorbed by the body when ingested.[85]

Diphenols. The diphenol group of phenols contains two group of compounds, isoflavonoids and lignans, which have exhibited hormonal activity and a consequent reduction in some hormonal-dependent cancers when consumed. Isoflavonoids and lignans are commonly referred to as phyto-estrogens.

Isoflavonoids. Epidemiological studies indicate that Japanese citizens, and other South East Asian populations, have a four-to ten-times lower incidence of and death from breast and prostate cancers than Americans.[10] These studies also show that the incidence of these cancers in Japanese descendants increases when they immigrate to America, growing with each new generation until the second generation of Japanese descendants in the U.S. (Nisei) are at the same risk as U.S. citizens of non-Japanese descent. Investigations of diet changes in these populations have indicated that Japanese (and most Asian residents in general) consume a much larger quantity of soy and soy products than U.S. citizens. The first generation of Japanese immigrants to the U.S. (Issei) maintain a diet that contains some traditional Japanese foods; however, subsequent generations have been shown to eventually adapt to the typical Western diet, which is virtually devoid of soy and soy products. The lower risk of breast and prostate cancer is attributed to soy consumption and to the isoflavonoids.

Isoflavonoids in soy consist mainly of the compounds genistein and daidzein in glycosidic forms and small amounts in free form. The isoflavonoid precursors Biochanin A and formonetin, and the isoflavone glycetin are minor components of soy isoflavonoids. Intestinal bacteria convert isoflavonoid glycosides into hormone-like compounds with weak estrogenic and antioxidative activity.

Genistein

Chemopreventive Mechanisms. Genistein, the principal isoflavone of soy, is the main compound of interest to researchers studying soy isoflavones due to its wide range of chemopreventive activities. Kelloff et al.[49] designed a classification system of chemopreventive mechanisms which has been outlined in Table 32.5. This outline can be applied to the chemopreventive activities of genistein, which are depicted in Table 32.9.

TABLE 32.9
CLASSIFICATION OF CHEMOPREVENTIVE MECHANISMS BY KELLOFF ET AL.
APPLIED TO GENISTEIN[49]

Mechanism	Subset
1. Antiproliferation/Antiprogression	Modulate Hormonal/Growth Factor Activity
2. Antiproliferation/Antiprogression	Inhibit Oncogene Activity
3. Antiproliferation/Antiprogression	Inhibit Angiogenesis
4. Antiproliferation/Antiprogression	Induce Terminal Differentiation[37]
5. Antioxidant Activities	Inhibit Arachidonic Acid metabolism[10]

Genistein exhibits various antiproliferative and antiprogressive properties. This phytoestrogen modulates hormonal or growth factor activity. Genistein modulates hormonal activity due to its anti-estrogenic effect. With respect to modulating its growth factor activity, genistein is a specific inhibitor of tyrosine kinase, an epidermal growth factor receptor-linked enzyme. Tyrosine kinase plays an important role in cell proliferation and transformation; it catalyzes phosphorylation which activates the phosphoprotein receptors involved in signal transduction. Tyrosine kinases are associated with cellular receptors for epidermal growth factor, insulin, insulin-like growth factor- I, platelet-derived growth factor, and mononuclear phagocyte growth factor.[37] About half of the protein products of known oncogenes are either membrane-bound receptors with tyrosine kinase activity or intracellular proteins undergoing or catalyzing tyrosine phosphorylation.[9]

Genistein inhibits ras oncogene activity, specifically the ras oncogenes which are involved in mammary gland carcinogenesis (ras oncogenes are implicated in other cancers, such as colon cancer). There is evidence indicating that chemoprevention occurs when ras is inhibited.[49] Tyrosine kinase activates ras, since genistein inhibits tyrosine kinase it is therefore an inhibitor of ras oncogenes.

Angiogenesis, a highly-regulated process leading to the formation of new blood vessels, is essential for reproduction, development, and wound repair and is only necessary in adults in the case of a implantation of an embryo into the uterus, severe injury, or heart attack. Some tumors require capillaries to supply them with oxygen in order for them to grow; thus, angiogenesis may occur before tumors are formed. Genistein is one of the compounds shown to inhibit angiogenesis which in turn may inhibit tumor formation.[49]

Proliferative cancer cells do not have the ability to differentiate that normal proliferating cells do. There is extensive data supporting the fact that carcinogenesis can be inhibited by restoring the ability to differentiate in abnormally proliferating cells[49] It has been demonstrated *in vitro* that genistein gives abnormal cells with various malignancies the ability to differentiate. A sample of the diseases prevented by genistein *in vitro* and the mechanism of preventive action is depicted in Table 32.10.

TABLE 32.10
SOME DISEASES PREVENTED BY GENISTEIN AND THE
MECHANISMS INVOLVED[37]

Disease	Effect/Result
Breast cancer	Competition with estradiol
Melanoma	Differentiation (animal cells *in vitro*)
Myebid leukemia	Differentiation and inhibition of proliferation (animal and human cells *in vitro*)
Leukemia	Differentiation and inhibition of cell cycle progression and growth (human cells *in vitro*)
Embryonal carcinogenesis	Differentiation (mouse cells *in vitro*)
Endothelial cells	Inhibition of angiogenesis (many different cells *in vitro*)
Gastric cancer	Growth inhibition (human cells *in vitro*)
Liver cancer	Inhibition of proliferation (Hep G2 cells)
Monoblastic leukemia	Differentiation (U937 cells)

Apoptosis is programmed cell death, a process which is a "well-regulated function of the normal cell cycle" in which damaged and excessive cells are eliminated.[49] Genistein is thought to induce apoptosis; however, there is no published evidence to date to support this theory.

Several studies demonstrate that genistein prevents carcinogenesis via the various aforementioned modes of action; however, some studies show that rather than preventing carcinogenesis, genistein only delays onset of the disease.[9] If there is strong evidence pointing towards genistein as a delaying agent of carcinogenesis, rather than a chemopreventor, the increase in latency of the disease will still give afflicted humans a few more years before malignancy occurs. Either way, genistein has a positive potential effect on human health.

Breast Cancer. Estradiol, the precursor to estrogen, may follow two pathways to produce the hormone. One pathway leads to 16-alpha-hydroxyestrone and the other leads to 2-hydroxyestrone. The former of the two estrogens is more effective and binds more efficiently to the estrogen receptor than the latter estrogen; thus implicating it in breast cancer and deeming 16-alpha-hydroxyestrone as dangerous compared to the less effective estrogen, 2-hydroxyestrone.

There is strong evidence that breast cancer is affected by a number of factors including gender (females are at a higher risk than males), age at first menarche (higher risk at 11 years old or younger), age at first birth (higher risk at 30 years old or above), age at natural menopause (55 years old or older are at higher risk). Genistein has been demonstrated to compete with estradiol[37]; thus, inhibiting the onset of breast cancer. Also, the chemopreventive mechanisms of genistein (Table 32.8) may play a role in breast cancer prevention. An epidemiological study conducted by Persky and Van Horn[73] indicates that factors in soy products may affect the risk of breast cancer by alteration in steroid hormone levels.

Persky and Van Horn[73] compared serum levels of endogenous hormones, including testosterone, estradiol, % free estradiol and dehydroepiandrosterone sulfate (DHS) in vegetarian girls versus non-vegan girls (girls in both groups were 14-18 years old, an age range thought to be vulnerable to environmental influences on breast cancer risk). The prominent difference found between the diets of the two groups was the abundance of soy consumed by the vegetarians and lack of soy in non-vegans. The researchers found no significant differences in levels of testosterone, estradiol, or % free estradiol; however, a significant difference in the levels of DHS was found. The vegetarian group had higher levels of DHS than the non-vegan subjects. The researchers attribute the elevated levels of DHS in the vegetarian girls to soy consumption. Lower levels of the adrenal androgens DHS and dehyrdoepiandrosterone (DHEA) are reported in women with breast cancer and populations at high risk for breast cancer. The protective effect of DHS and DHEA against breast cancer has been postulated to be due to the inhibition of estrogen receptor binding, inhibition of glucose-6-phosphate dehydrogenase, inhibition of estrone sulfatase, and inhibition of interleukin 6.[73] Incontrovertible data regarding the risk of breast cancer and the effects of elevated serum levels of hormones has yet to be presented; however, factors such as adrenal androgens may be implicated with educed breast cancer risk.

Prostate Cancer. Findings from epidemiological studies, *in vitro* cell cultures and *in vivo* animal studies suggest that the risk of prostate cancer is substantially decreased by the consumption of isoflavonoids.[37] In cell culture, genistein and Biochanin A (the precursor of genistein) have inhibited the growth of androgen-dependent and -independent prostatic cancer cells. Studies suggest that "phytoestrogens may inhibit prostatic cancer cell growth during the promotional phase of the disease or they may influence differentiation as shown for genistein with other cancer cells."[37]

These findings are extremely important because prostate cancer has a very low treatment success rate in its advanced stages. By the year 2000, 40,000 men in the U.S. are predicted to be terminally afflicted annually with prostate cancer if treatment does not improve.[30]

Dietary Levels. Soybeans contain 1-3 mg per gram of the glycosides of genistein and daidzein.[1,10,37] People living in southeast Asian countries (e.g., Japan, China, Indonesia, Korea, Singapore, Taiwan) consume 20-80 mg of genistein per day ($\sim$20-80 grams of soy per day).[10] In comparison, citizens of Western countries consume only 1-3 mg of genistein per day ($\sim$1-3 grams of soy per day).[10] Until further research is completed to determine doses effective for chemoprevention, 50 mg of genistein seems to be enough to exert a protective effect in southeast Asian populations, and may be considered beneficial.[10]

An extended half-life of genistein in the body has not been observed; thus, there are minimal day-to-day carry over effects, indicating that soy must be incorporated into the diet daily to replenish the supply of genistein in the body. The addition of soy to the typical Western diet has increased urinary isoflavone levels by as much as 1,000 times.[64]

Soy can be incorporated into the diet via various processed foods, such as tofu, soy milk, miso soup, and tempeh (a fermented soy product). Soy flours and soy protein isolates are also sources of isoflavonoids. The isoflavones of whole soybeans are resistant to the heat utilized during the processing of soy flour. The total isoflavone content of soy flour is the same as in whole soybeans.[9] Full fat soy milk and tofu have isoflavone concentrations similar to soy flour and whole soybeans, while lower isoflavone contents are found in "lite" soy milk and tofu. Soy protein concentrates and isolates are prepared from soy flour and their isoflavone content depends on the washing medium used during manufacturing. Water or acid washing does not alter the isoflavone content, while alcohol-water mixtures significantly reduce it due to the high solubility of isoflavonones in aqueous alcohol. Supro™ isolated soy protein contains 50 mg of genistein in 40 g of product.[10] Soy sauce does not contain any isoflavonoids.

"Americanized" versions of soy foods have recently been introduced to the market. Some of these products are tofu hot dogs, tofu ice-cream, veggie burgers, tempeh burgers, soy milk yogurt, soy milk cheeses and soy flour pancakes. The isoflavone content of these products is lower than that of typical soybean products eaten by southeast Asians (roasted soybeans, tofu, tempeh, miso), because the soy ingredient is diluted.[6] A comparison of total isoflavone content (genistein, daidzein, and glycetein) in traditional Asian foods and Americanized foods made with soybeans is depicted in Table 32.11.

TABLE 32.11
TOTAL ISOFLAVONE CONTENT OF TRADITIONAL ASIAN FOODS VERSUS AMERICANIZED SOY FOODS

	Product	Total Isoflavone Content Mg/g
Asian Foods	Roasted soybeans	2661
	Tofu	531
	Tempeh	865
	Bean paste (miso)	647
Americanized Foods	Soy hot dog	236
	Soy bacon	144
	Tempeh Burger	386
	Tofu yogurt	282
	Soy parmesan	88
	Mozzarella cheese	123
	Cheddar cheese A	43
	Cheddar cheese B	197

Total Isoflavones = Genistein, Daidzein, Glycetein
Source: Table adapted from Anderson, R.L. and Wolf, W.J.[6]

Lignans. Lignans are diphenolic compounds derived from intestinal bacterial digestion of polymeric precursors in plants.[31] Hormone-like compounds are formed via intestinal bacteria when lignans and other diphenolic compounds are consumed. These metabolically formed phytoestrogens bind weakly to estrogen receptors and also exhibit antioxidative properties.[1] Lignans and isoflavonoids share some functional similarities in which they influence sex hormone metabolism, intracellular enzymes, protein synthesis, growth factor action, malignant cell proliferation, and angiogenesis.[37] Lignans prevent the growth of tumors and have antitumor, antimitotic and antiviral properties.[82]

Studies suggest that lignans, as well as isoflavonoids, may reduce the risk of prostate cancer during the promotional phase of the diseases. Other possible anticarcinogenic effects of lignans are inhibition of the proliferation of mitogen-induced proliferative cancer and inhibition of aromatase to inhibit cancer of placental microsomes.[37]

Seeds have high amounts of lignan precursors while unprocessed soybeans are rich in these compounds as well. Lignans cannot withstand the processing of soybeans (e.g., to produce tofu). A lignan precursor, coniferyl alcohol, can be found in soy sauce, even though isoflavonoids are not part of this product. Lignans are components of whole grain bread; however, modern milling techniques substantially eliminate the aleurone layer of the grain to produce flour. Only whole grain flour retains the fiber-containing aleurone layer where lignans are localized.

Ellagic Acid. Ellagic acid is a phenolic lactone. This compound, found mainly in blackberries, raspberries, strawberries, blueberries, cranberries, walnuts and pecans, is very stable and relatively insoluble in water. Ellagic acid is naturally found in various forms, as well as in its free form. It may be present in plants in the form of the hydrolyzable tannins, also known as ellagitannins, which yield ellagic acid.

Ellagic Acid prevents the metabolic activation of procarcinogens, especially nitroso compounds, aflatoxin B_1, and polycyclic aromatic hydrocarbons such as benzo(a)pyrene and 3-methylcholanthrene, via non-selective destruction of cytochrome P-450 enzymes, which are responsible for activating carcinogens.[48,85] Carcinogen detoxification is also promoted by ellagic acid via the stimulation of various iso-forms of the carcinogen-detoxifying Phase II glutathione-S-transferase enzymes. Ellagic acid also prevents carcinogens from binding to DNA by binding with DNA itself at the O-6 and N-7 positions of guanine, making this site unavailable to carcinogens and their metabolites for methylation.[18,49,85] Antioxidant activities are also exhibited by this bioactive compound. Electrophilic carcinogens are trapped by ellagic acid, which forms conjugates with these compounds and detoxifies them[18,49,82,85] The chemopreventive mechanisms of ellagic acid are listed in Table 32.12, categorized according to the classification system of Kelloff *et al.*[49] (Table 32.5).

Ellagic acid has been demonstrated in *in vivo* rat studies as an effective inhibitor only when consistently administered before, during, *and* after the carcinogen has been introduced.[85] Inhibitory effects against nitroso compounds have been observed in rats at dietary levels of 0.4-4.0 mg per kg body weight.[85]

TABLE 32.12
CLASSIFICATION OF CHEMOPREVENTIVE MECHANISMS BY KELLOFF *ET AL.*
APPLIED TO ELLAGIC ACID[49]

Mechanism	Subset
Carcinogen Blocking Activities	Inhibit Formation or Activation of Carcinogen[18,49,82,85]
Carcinogen Blocking Activities	Deactivate/Detoxify Carcinogen[85]
Carcinogen Blocking Activities	Prevent Carcinogen Binding to DNA[18,49]
Antioxidant Activities	Scavenge Reactive Electrophiles[18,49,82,85]

Hydroxycinnamic Acid Derivatives. Hydroxycinnamic acid derivatives can be found in a wide range of fruits and vegetables (Table 32.6). The main compounds in this group are *p*-coumaric, ferulic acid, chlorogenic acid and its derivative caffeic acid. Chlorogenic acid is found in coffee beans and instant coffee, and is a major component of this beverage. Coffee drinking has been linked with reduction in colon cancer incidence in humans, and has been found to inhibit nitrosation. Results of a study measuring the overall tumor incidence of mice fed instant coffee at 5% of their total diet for two years indicated an inverse correlation between coffee intake and tumor incidence[82] Chlorogenic acid has also inhibited lipid peroxidation in rat liver induced by carbon tetrachloride, a potent liver carcinogen.[41]

Caffeic acid and ferulic acid have, like chlorogenic acid, inhibited nitrosamine formation. These three compounds have also detoxified ultimate carcinogenic metabolites of polycyclic aromatic hydrocarbons (PAH), which are the epoxides of PAH. Chlorogenic and ferulic acid have been shown to inhibit neoplasia in the forestomach of mice, colonic tumors in Syrian golden hamsters, liver tumors in rats, and tumor promotion in mouse epidermis when applied topically.[41]

Flavonoids. Flavonoids are polyphenols that are small, lipid soluble compounds. This group of phenols can be found in almost all plants and includes thousands of individual compounds.

Epidemiological studies in Finland and in the Netherlands have indicated an inverse relationship between flavonoid intake and coronary heart disease.[38,52,68,82] Flavonoids in the diet seemed to lower platelet aggregation, reduce thrombic tendencies, scavenge free radicals and modify various enzyme functions. A 26 year Finnish cohort study by Knekt *et al.*[52], concluded that foods containing flavonoids (especially quercetin) were the foods most strongly inversely related to mortality due to coronary heart disease.[68] A cohort study organized in Holland found an inverse relationship between the consumption of black tea, onions and apples (all sources of flavonoids) and coronary heart disease. The mean flavonoid intake of the subjects was 26.6 mg per day. Quercetin was the main flavonoid consumed.[38] Estimates for the average daily intake of flavonoids in the diets of Western countries ranges from 25 mg to one gram per person per day.[57,82]

Flavonoids are anticarcinogenic as well as protective against cardiovascular disease. The chemopreventive mechanisms exhibited by flavonoids, classified according to the classification system of Kelloff *et al.* is in Table 32.13.

TABLE 32.13
CLASSIFICATION OF CHEMOPREVENTIVE MECHANISMS BY
KELLOFF *ET AL.* APPLIED TO FLAVONOIDS[49]

Mechanism	Subset
Carcinogen Blocking Activities	Inhibit Formation or Activation of Carcinogen
Antioxidant Activities	Scavenge Oxygen Radicals[38,49,82]
Antioxidant Activities	Inhibit Arachidonic Acid Metabolism
Antiproliferation/Antiprogression	Modulate Signal Transduction
Antiproliferation/Antiprogression	Modulate Growth Factor Activity

Flavonoids apparently prevent the metabolic activation of a procarcinogen. These compounds exert their antioxidative activities via scavenging free radicals such as peroxy radicals, singlet oxygen and superoxide radicals, and also by inhibiting arachidonic acid metabolism, an action which prevents certain enzymes (lipoxygenase and prostaglandin H synthase) from catalyzing the activation of procarcinogens.[49]

Carcinogenesis can be defined as "a progressive series of disorders in the function of signal transduction."[49] Flavonoids may also modulate signal transduction by inhibiting protein kinase C (PKC) which is involved in one of the steps of signal transduction, the means of communication between cells. By halting communication, the cells will not be involved in activities that are out of control and normal cellular controls could be restored.

Flavonoids also modulate growth factor activity by inhibiting tyrosine kinase, an enzyme which plays a key role in cell proliferation and transformation. Another chemopreventive mechanism of flavonoids is the inhibition of polyamine metabolism, which is significant in cell proliferation, differentiation, and malignant transformation.[49]

Two types of flavonoids, the catechins and the flavonol quercetin have been widely studied. Catechins are found in many fruits and vegetables; however, they are most abundant in green and fermented teas. Quercetin is ubiquitous in fruits and vegetables and is the most biologically active flavonoid. They both exert protective effects against chronic diseases.

Catechins. Tea runs a close second to water as the most widely consumed beverage in the world. Nearly 2.5 million metric tons of tea are manufactured per year. The three main types of tea (black, green, and oolong) account for ~80%, ~20% and ~2% of all tea products, respectively.[85] The primary consumers of green tea reside in China, Japan, India, many North African nations and Middle Eastern countries. Western and some Asian populations prefer black tea. The manufacture of green tea involves steaming or drying fresh tea leaves at elevated temperatures. Precaution is taken in this process to avoid the oxidation of the polyphenolic compounds. Tea leaves are plucked, withered, macerated and dried in the production of black tea. During this process, the polyphenols undergo oxidative polymerization via fermentation which results in the conversion of catechins to aflavins and thearubigins. Ultimately, green tea contains a total of 35-52% (solid weight) catechins and flavonols. Black tea contains 3-10% catechins, 3-6% theaflavins, 12-18% thearubigins and other components.[85]

To date, green tea has been more widely studied than black tea. Recent studies show that black tea may have similar anticarcinogenic properties to green tea.[41,85] Of the 35-52% total green tea polyphenols (GTP) found in green tea leaves, epigallocatechin gallate (EGCG) is the main compound, comprising over 40% of the total phenolic mix.[85] This catechin is considered the principle antimutagenic and anticarcinogenic compound in green tea, although many other GTPs are antimutagens and anticarcinogens as well.[41,85]

Green tea polyphenols function via the anticarcinogenic mechanisms of flavonoids outlined in Table 32.13. Specifically, these compounds have been shown to be anticarcinogenic in the lung, forestomach, esophagus, colon, duodenum, liver, and mammary gland, in animal model studies.[62] The anticarcinogenic functions of GTP are listed in Fig. 32.2.

- Interact with cytochrome P-450 and inhibit associated monooxygenase activity
- Enhance antioxidant enzymes (glutathione peroxidase, catalase, quinone reductase)
- Enhance Phase II (glutathione-S-transferase) enzyme
- Inhibit chemically induced lipid peroxidation
- Exhibit anti-inflammatory activity
- Enhance gap junction intracellular communications
- Inhibit cellular proliferation
- Chemopreventive against initiation, promotion, and progression stages of carcinogenesis

---

FIG. 32.2. ANTICARCINOGENIC FUNCTIONS OF GTP[85]

EGCG has been demonstrated to have an effect on the promotion stage of carcinogenesis and DNA-adduct formation as well as antioxidative properties.[85] EGCG has also been shown to inhibit nitrosamine formation by reducing nitrite production caused by chronic inflammation. It has been postulated that simultaneous consumption of tea with foods that are undergoing nitrosation in the stomach of human subjects may be protective.[82] Vitamin C and alpha-tocopherol are also blocking agents for nitrosamine formation.[62] Green and black tea polyphenols have both shown to be protective at each stage of carcinogenesis in the skin.[85] It has been estimated that a heavy green tea drinker consumes one gram per day of EGCG.[16] However, to date, there is no information available concerning the pharmacokinetics of the uptake and distribution of phenols.

Quercetin. Quercetin is found in nearly all fruits and vegetables, tea, coffee, legumes, roots, and cereal grains. Estimated intake of this flavonol is 5% of the average 1 gram per day intake of flavonols.[41] Quercetin, which is mainly found in food in the less active glycosidic forms, is formed in the mouth and gut by bacterial hydrolysis of these glycosides.

Quercetin is anticarcinogenic as well as protective against cardiovascular disease. Bacterial assay studies had indicated that quercetin is mutagenic *in vitro*; however, many *in vitro* and *in vivo* animal model studies have demonstrated that quercetin is actually an anticarcinogen and exerts its protective effect via various chemopreventive mechanisms. The chemopreventive mechanisms of flavonoids in Table 32.13 also apply to quercetin. Many studies have concentrated on the anticarcinogenic effects of quercetin and have found that this compound is a potent inhibitor of carcinogenesis. Table 32.14 depicts the results of some of these investigations.

TABLE 32.14
THE RESULTS OF VARIOUS STUDIES PERTAINING TO THE ANTICARCINOGENIC
EFFECTS OF QUERCETIN[41]

Study Conditions	Result
in vitro	Significantly reduces the carcinogenic activity of several cooked food mutagens
in vitro	Inhibits binding of polycyclic hydrocarbons (PAH) to DNA
in vitro	Inhibits growth of colon cancer cells
in vivo (mice)	Depressed colon tumor incidence in mice fed 2% quercetin for 50 weeks
in vivo (mice)	Inhibits initiation and promotion of mouse skin tumor formation
in vivo (rats)	Inhibits binding of PAH to DNA in the epidermal and lung tissue of SENCAR rats
in vivo (rats)	Inhibits induction of mammary cancer by 7,12-dimethlyl-benz[a]anthracene and N-nitrosomethylurea in rats
in vivo and *in vitro*	Inhibit enzymes involved in tumor production (e.g. protein kinase C, lipoxygenase, ODC, cytochrome P-450/P-448 monooxygenases)

Quercetin has also inhibited the tumor promoters TPA, and aflatoxin B_1. These flavonols have inhibited allergic and inflammation responses of the immune system as well as the cyclooxygenase pathway to arachidonic acid metabolism to prostaglandins. In addition to its anticarcinogenic properties, quercetin is also protective against cardiovascular disease and is considered a vasoprotective and antithrombic agent.[41] It has been demonstrated that this compound blocks oxidative modification of LDL cholesterol; thus, protecting the body against heart disease.[82]

Indoles

Indoles are phytochemicals found in cruciferous vegetables, including Brussels sprouts, kale, cabbage, broccoli, cauliflower, and spinach. They exist as glucosinolates in the raw vegetable, which are hydrolyzed to their respective indoles via myrosinase after the plant cell is disrupted by cutting or chewing.[42] One of the main bioactive indoles studied is indole-3-carbinol (I3C) which exists as the glucosinolate 3-indolylmethyl glucosinolate (glucobrassicin).

Indoles have been reported to induce Phase I and Phase II hepatic enzymes[42] (Table 32.15). Evidence exists that indicates that I3C is a chemopreventive agent, especially for hormonal cancers involving estradiol, such as breast cancer. One of the several condensation products of I3C, indolo(3,2-b)carbazole (ICZ) is an acid derived condensation product formed in the stomach and is a potent antiestrogen.[59] The formation of 2-alpha-hydroxyestrone via 2-hydroxylation of estradiol has been reported to increase in the presence of I3C[42,49,66,82] by inducing the Phase I enzyme P450 1A2[42] in rodents and in humans. The reduction of mammary tumors in rats have been observed by a high cruciferous diet.[49,59] It has been observed that the acid condensation products of I3C are actually the chemopreventive compounds due to the conversion of I3C to these products upon digestion.[42] I3C has also been associated with decreased colon and lung cancers in humans.[66]

TABLE 32.15
CLASSIFICATION OF CHEMOPREVENTIVE MECHANISMS BY KELLOFF, *ET AL.*
APPLIED TO INDOLE-3- CARBINOL[49]

Mechanism	Subset
Carcinogen Blocking Activities	Deactivate/Detoxify Carcinogen

Michnovicz *et al.*[66] have shown that 350-500 mg (5-7 mg per kg body weight) of I3C from Brussels sprouts or raw cabbage consumed per day for seven days significantly increases estradiol metabolism in men and women. These researchers measured estradiol metabolism via the degree of total body

2-hydroxylation. Brussels sprouts and raw cabbage contain approximately 1000 ppm of I3C.[66] The effects were only observed when I3C was taken orally.

Indole-3-acetonitrile (IAN) is another indole present in cruciferous vegetables. This compound has also been reported to induce 2-hydroxylation of estradiol in rat liver cells; however, its potency is about threefold less than I3C.[65] IAN was also reported to be 50% less effective than I3C in the demethylation and consequent inactivation of the carcinogens 4-(methylnitrosamino)-1-(3-pyridyl)-1-butanone (NNK) and N-nitrosodimethylamine (NDMA) in rats.[42]

It has been reported that IAN levels increased during cooking cruciferous vegetables in water[80,81] and I3C levels decreased.[81] Slominksi and Campbell[81] observed that cabbage heated at 100C for zero to 30 minutes resulted in an increase in IAN formation and about a 50% loss of the formed IAN due to leaching into the water. Unprocessed cruciferous vegetables contained a higher concentration of I3C than IAN.[81] Reduced levels of indoles in cooked cruciferous vegetables are observed due to partial inactivation of the hydrolytic enzyme myrosinase.[42] A 50% reduction in indole glucosinolate was found to occur in cabbage cooked in boiling water. The level of indole glucosinolate was not affected by fermentation.[42] Since I3C is reported as the more potent indole, uncooked cruciferous vegetables, which contain a higher level of this compound, would seem to be more protective against hormonal cancers.

Another indole, the essential amino acid L-tryptophan (L-Trp) has been reported to inhibit the formation of aminoimidazo (IQ)-type and IQ-like mutagenic compounds, which are formed during the cooking, broiling or frying of protein-rich foods such as beef, fish, pork or lamb.[46] IQ-type mutagens have been reported to produce various cancers in rodent studies: liver, intestine, lung, stomach and skin cancers have been observed in mice, and cancers of the mammary gland, colon, small intestine, liver, pancreas and ear duct have been reported in rats. Creatinine has been identified as an essential compound for IQ formation. The proposed mechanism for antimutagenicity of L-Trp is competition with creatinine for aldehyde precursors required for IQ mutagen formation. By making these aldehydes unavailable to creatinine, L-Trp disrupts the pathway of mutagen formation and inhibits the production of IQ-type and IQ-like compounds.[46] However, L-Trp must be unheated in order to be protective. This amino acid, which is found in protein-rich foods, forms very mutagenic compounds, Trp-1 and Trp-2, when cooked, broiled, or heated in any other manner.[76]

Isothiocyanates

Isothiocyanates are, like indoles, found in cruciferous vegetables, such as cabbage, Brussels sprouts, turnips, radishes, watercress, broccoli, and cauliflower. These compounds are naturally present as glucosinolates, formed via glucosinolase enzymes or cooking, and are partly responsible for the pungent flavor of these vegetables.[36] When the vegetable cells are damaged, such as in mastication, the enzyme myrosinase is released. Myrosinase catalyzes the hydrolysis of glucosinolate and isothiocyanates are formed via Lossen rearrangement.[36] Isothiocyanates exist naturally in many structures which are structure-activity dependent; thus, they can inhibit a variety of cancers depending on the target tissues and isothiocyanate structure.[36]

Although a variety of isothiocyanates exists in nature, three compounds in particular have been studied extensively for their chemopreventive properties. These compounds include phenethyl isothiocyanate (PEITC), which is found in high quantities in Brussels sprouts, cauliflower, cabbage, kale and turnips, benzyl isothiocyanate (BITC) and sulforaphane, found mainly in broccoli and cabbage. PEITC is naturally present in cruciferous vegetables as the glucosinolate gluconasturtiin and BITC is found as the glucosinolate glucotropaeolin.[42]

The main effect on chronic diseases that isothiocyanates have exhibited is that they are anticarcinogenic compounds (Table 32.16). Isothiocyanates have been demonstrated to be potent inhibitors of nitrosamine-induced tumors[49] attributed to their inhibition of P450 enzymes. PEITC and BITC play a preventive role in lung cancer. The two major lung carcinogens found in cigarette smoke that are considered the main causes of lung cancer in smokers are nicotine derived nitrosaminoketone (NNK) and a polynuclear aromatic hydrocarbon benzo(a)pyrene (B-alpha-P). Rodent studies have indicated that PEITC inhibited lung tumor induction by NNK by blocking the metabolic activation of this carcinogen, and BITC inhibited lung tumor induction by B-alpha-P. A combination of PEITC and BITC inhibited tumorigenesis from the two carcinogens. These studies indicate that the isothiocyanates PEITC and BITC inhibit tumor promotion as well as enhance detoxification.[36]

TABLE 32.16
CLASSIFICATION OF CHEMOPREVENTIVE MECHANISMS BY KELLOFF *ET AL.*
APPLIED TO ISOTHIOCYANATES[49]

Mechanism	Subset
Carcinogen Blocking Activities	Inhibit Formation or Activation of Carcinogen
Carcinogen Blocking Activities	Deactivate/Detoxify Carcinogen[36,62]

A study was performed in humans by measuring the effect of PEITC from watercress on the amount of detoxified NNK metabolites excreted in the urine. The subjects (eleven smokers) chewed two ounces of watercress three times a day for three days (two ounces of watercress releases 5-10 mg of

PEITC) and a statistically significant increase in detoxified NNK metabolites were measured in the urine.[36] Thus, PEITC detoxified NNK either by preventing activation of the procarcinogen or by inducing Phase II enzymes to detoxify the compound.

All isothiocyanates have been shown to be very reactive inhibitors of cytochrome P-450 enzymes, which may activate carcinogens, and to enhance certain Phase II enzymes (such as UDP-glucuronoyl transferase) to detoxify carcinogens.[36,42] The isothiocyanate sulforaphane has been reported as a very potent monofunctional inducer, inducing the detoxifying Phase II enzymes but not affecting the Phase I P450 enzymes.[42] Mammary tumors in rats were reduced in number, size, incidence and rate of development when sulforaphane was administered.[62] The majority of studies suggest that isothiocyanates must be present at the same time as the carcinogen, or administered before carcinogen exposure, in order to inhibit tumor formation.[36,42]

Allylic Sulfur Compounds

Allylic sulfur compounds are responsible for the typical odor of *Allium* vegetables, a group which includes onions, garlic, leeks, shallots, chives, and scallions. These compounds have been detected in both cooked and raw *Allium* vegetables as well as in garlic supplements. Some allylic sulfur compounds which have exhibited chemopreventive qualities include diallyl sulfide, diallyl disulfide, diallyl polysulfide, and S-allyl cysteine. These phytochemicals have also been associated with a decrease in cardiovascular diseases.

A decreased risk for stomach cancer with increased *Allium* vegetable consumption has been reported in most case control studies. A study called the Netherlands Cohort Study (NLCS)[22] monitored the effects of onions, leeks and garlic supplements on the risk of stomach cancer. The study, which began in 1986 examined 120,852 subjects and ran for 3.3 years, found a significant decrease in the risk of carcinoma in all parts of the stomach except in the cardia with increased onion consumption.[22] In contrast, neither leeks nor garlic supplements were associated with decreased stomach cancer risk. Dorant *et al.*[22] report that an important risk factor for cancer of the non-cardia areas of the stomach is the bacteria *Helicobacter pylori*. Further investigation of the effect of *Allium* vegetables on this bacteria may reveal that the protective effects of allylic sulfur compounds against stomach cancer are due, in part, to their antibacterial properties. Allylic sulfur compounds have been observed to inhibit induced tumors in rodents of the stomach, skin, liver, colon, lung and cervix.[42]

Garlic, which contains a variety of allylic sulfur compounds, has been reported to inhibit stomach cancer.[31] The phytochemicals in garlic have been reported to prevent the formation of nitrosamines via three mechanisms. Garlic inhibits the synthesis of nitrite and nitrosamines by bacteria and fungi. The spontaneous syntheses of nitrosamines is directly inhibited by garlic, and nitrite has a greater affinity for sulphhydryl compounds than for amines; thus, allylic sulfur compounds competitively prevent the formation of nitrosamines.[31] Garlic is high in sulfur and contains 60% diallyl disulfide, 14% diallyl sulfide, 6% allyl propyl disulfide, and 4-10% allyl methyl trisulfide.[19] Other compounds found in garlic include dimethyldisulfide, dimethyltrisulfide, dipropyldisulfide, allylmercaptan, S-allylcystein and ajoene.

The reported chemopreventive mechanisms of allylic sulfides are that they modulate Phase I and Phase II enzymes.[49,19] Allylic sulfides prevent the metabolic activation of a procarcinogen (Table 32.17). The compounds also induce the carcinogen scavenger glutathione, and its catalyst, glutathione-S-transferase.[49] Some allylic sulfur compounds are unstable at temperatures above 60C.[42] The aroma is lost as well as the chemopreventive properties due to alteration of the compounds from heat.

TABLE 32.17
CLASSIFICATION OF CHEMOPREVENTIVE MECHANISMS BY KELLOFF *ET AL.*
APPLIED TO ALLYLIC SULFIDES[49]

Mechanism	Subset
Carcinogen Blocking Activities	Inhibit Formation or Activation of Carcinogen
Carcinogen Blocking Activities	Deactivate/Detoxify Carcinogen

Garlic has also been shown to exert a positive effect on cardiovascular disease. One study which examined the effects of garlic powder on cardiovascular disease in humans, reported that 900 mg of garlic powder tablets per day reduced LDL cholesterol by 11% while the HDL level was not changed. Garlic powder has been shown to reduce diastolic blood pressure and levels of serum cholesterol and triglycerides.[82] For a more extensive review, the reader is referred to the analyses by Lawson.[56]

Monoterpenes

Monoterpenes are compounds that are major constituents of plant essential oils. They occur in monocyclic, bicyclic and acyclic forms and are either simple or oxygenated hydrocarbons. These compounds are found in small quantities in citrus peel oil, dill weed oil, caraway oil, and food flavorings such as mint, and other natural plant oils. Orange peel and other citrus oils are the most abundant sources of these phytochemicals.[33,62,82]

One of the predominant monoterpenes studied is D-limonene, which is found in the oils of citrus peel, mint, myristica, caraway, thyme, cardamom, coriander, and celery, as well as many others, and is the simplest monocyclic monoterpene. Lemon and orange oils consist of over 90% D-limonene.[42] In rodent studies, this compound inhibited a variety of cancers including cancer of the mammary glands, stomach, lung, skin and liver.[33] D-limonene has been reported to induce Phase I and Phase II hepatic enzymes in rats.[42] In mice, D-limonene has been observed in induced liver cancer to increase glutathione transferase activity three times that of a control and 2.36 times in intestinal cancer.[90] Elegbede et al.[23] observed a significant reduction in mammary cancer in rats with a diet containing either 1000 ppm or 10,000 ppm of D-limonene one week prior to administration of the mammary tumor inducer 7,12-dimethylbenz-[a]anthracene (DMBA), and for 27 weeks after DMBA was introduced with the continuous addition of D-limonene in the diet at 1000 ppm or 10,000 ppm. No toxicity of D-limonene has been observed in rodents, dogs or humans.[23,42]

D-limonene has also been successfully used as a dissolving agent for gallstones in humans. The solution used by Igimi et al.[44] to dissolve the stones was 97.0 parts D-limonene, 2.1 parts polysorbate 80 and 0.9 part of sorbitan monoleate. The nonionic surfactants were necessary to create emulsions for dispersion of the bile surrounding the stones so D-limonene could come into contact with them.

Another monoterpene, D-carvone, found in caraway seed oil and dill weed oil, inhibited the activation of dimethylnitrosamine and in turn decreased the induction of forestomach tumors in a rodent model.[31]

Perryl alcohol is a compound closely related to monoterpenes. This phytochemical, found in cherries, has been reported to prevent cancer of mammary glands, lung, forestomach, liver, and skin in mice.[62] The potency of perryl alcohol has been demonstrated to be five times greater than limonene against mammary tumors in rats.[33]

Both monoterpenes and perryl alcohol have been shown to inhibit farnesyl-protein-transferase, an enzyme involved in the activation of ras oncogenes (Table 32.18). These oncogenes have been associated with mammary gland carcinogenesis.[49]

TABLE 32.18
CLASSIFICATION OF CHEMOPREVENTIVE MECHANISMS BY KELLOFF ET AL.
APPLIED TO MONOTERPENES AND PERRYL ALCOHOL[49]

Mechanism	Subset
Antiproliferation/Antiprogression	Inhibit Oncogene Activity

Carotenoids

Carotenoids are lipid soluble compounds which are responsible for the yellow to red pigments of plants. The phytochemicals in this group of 700 known compounds are classified as micronutrients due to the ability of 50 carotenoids to convert to vitamin A in the body. Beta-carotene is the most efficient pro-vitamin A carotenoid, and has been the most widely studied for its beneficial effects on health. Recently, researchers have been focusing more on some of the other carotenoids, mainly those which are predominant in human plasma. The carotenoids which account for 90% or more of the circulating carotenoids in humans are: alpha-carotene, alpha-cryptoxanthin, beta-cryptoxanthin, lutein, lycopene, zeaxanthin, and also beta carotene.[18,48] Alpha-carotene, beta-carotene and beta-cryptoxanthin are the pro-vitamin A carotenoids of these seven plasma related compounds while the remainder of them have no vitamin A activity.

The sources of these compounds are mainly fruits and vegetables. The significant sources for each of the seven carotenoids listed above are tabulated in Table 32.19.

TABLE 32.19
SOURCES OF THE PREDOMINANT CAROTENOIDS MEASURED IN HUMAN PLASMA

Carotenoid	Source
alpha-Carotene	Green beans, lima beans, carrots, green peas, prunes, pumpkin, winter squash
beta-Carotene	Apricot, green beans, lima beans, broccoli, Brussels sprouts, cabbage, cantaloupe, carrots, pink grapefruit, kale, kiwi, lettuce, mango, muskmelon (honeydew), oranges, papaya, peaches, green peas, sweet potato, prunes, pumpkin, acorn squash, winter squash, spinach, tomatoes and tomato-based products
alpha-Cryptoxanthin	Oranges
beta-Cryptoxanthin	Mango, oranges, papaya, peaches, prunes, winter squash
Lutein	Green beans, lima beans, broccoli, Brussels sprouts, cabbage, kale, kiwi, lettuce, mango, muskmelon (honeydew), oranges, papaya, peaches, green peas, sweet potato, prunes, pumpkin, acorn squash, winter squash, spinach
Lycopene	Apricot, pink grapefruit, tomatoes and tomato-based products
Zeaxanthin	Kale, kiwi, lettuce, mango, muskmelon (honeydew), oranges, papaya, peaches, green peas, prunes, pumpkin, acorn squash

Table adapted from Khachik, F. *et al.*[51]

Beta-Carotene. Beta-carotene, the most efficient pro-vitamin A carotenoid, is also the most widely studied carotenoid. Epidemiological studies have reported inverse correlations between beta-carotene and cancer, cardiovascular disease, and cataracts.[4] The average daily intake of beta-carotene of career age adults is 2.4-3.0 mg.[53] The ideal range of beta-carotene in the diet, calculated by Lachance, is 5.2 to 6.0 mg.[53] For "at risk" populations, recent clinical trial results reported that approximately 20 mg, or more, of dietary beta-carotene per day is required to exert protective effects and an estimated intake of 6 mg per day of beta-carotene is sufficient for those who are not in the "at risk" category.[4] The bioavailability of carotenoids has been measured at 10-30%.[75] To obtain 20 mg of beta-carotene from common vegetables in the U.S., assuming 30% bioavailability, 14 cups of spinach, 33 cups of baked winter squash, or 1041 cups of cabbage must be consumed per day (Table 32.20). For "healthy" individuals, about four cups of spinach, ten cups of baked winter squash or 312 cups of cabbage daily would add the minimum 6 mg of beta carotene to the diet. These portions are extremely high and the amount of beta-carotene recommended would be difficult to achieve for most Americans who consume a typical diet high in fat and protein and low in vegetables and fruits. These vegetables also contain other phytochemicals with which beta-carotene may act in synergism or possibly as a marker for more potent antioxidants; thus, possibly affecting the minimum serving sizes necessary for chemoprevention. Recommended doses of various phytochemicals for protection still need to be established for humans.

TABLE 32.20
APPROXIMATE VEGETABLE SERVING SIZES CONTAINING 20 MG AND 6MG OF
BETA CAROTENE AT 10%, 30%, AND 100% ABSORPTION

Vegetable	Beta-Carotene (mg/g vegetable)[1]	Raw Vegetable Containing 20mg Beta Carotene[1] (cups) with Absorption at:			Raw Vegetable Containing 6mg Beta Carotene[1] (cups) with Absorption at:		
		10%	30%	100%	10%	30%	100%
Broccoli	0.0233	57 (spears)*	19 (spears)	5.7 (spears)	17 (spears)	5.7 (spears)	1.7 (spears)
Cabbage	0.0008	3125	1041	313	940	312	94
Green beans	0.0047	410	136	41	123	41	12
Spinach	0.0890	40	14	4	12	4	1
Winter Squash (baked)	0.0099	98	33	10	30	10	3

* 1 broccoli spear is not equal to 1 cup.
[1] Beta-carotene values from Khachik, *et al.*[51]; weights used to calculate broccoli and green beans from *Krause's Food, Nutrition and Diet Therapy*[61]; the weights used for cabbage, spinach and winter squash calculations from *Understanding Nutrition*[71]

Some studies have reported that beta-carotene either has no effect on carcinogenesis or that the carotenoid actually increased the risk of cancer. In a mouse study conducted by Nishino[69], beta-carotene at a level of 0.05% administered in drinking water for 20 weeks did not significantly suppress lung tumors induced by the tumor initiator 4-nitro-quinoline-1-oxide and the tumor promoter glycerol. The highly publicized Finnish Alpha Tocopherol/Beta Carotene (ATBC) Cancer Prevention Study reported that the subjects who were given beta-carotene supplements had an 18% higher incidence of lung cancer[86]; however, of this group of subjects only those who consumed alcohol as well as smoked cigarettes were observed as having a higher risk of lung cancer.[54] Thus, the subjects in the beta-carotene supplement group who did not heavily consume alcohol were not at a higher risk for lung cancer.[54] This study involved 29,133 Finnish male smokers, ages 50-69, who were randomly treated with either 20 mg of beta-carotene and/or 50 mg of vitamin E daily.

It has been postulated that one of the reasons that the ATBC study is at variance with the reported beneficial effects of beta-carotene is that six years of treatment may not have been a long enough time to deter carcinogenesis, which progresses gradually over decades. Carcinogenesis may have already been on its malignant path in the smokers before the study began due to the potent carcinogens in tobacco smoke.[14]

Another possible cause for the discrepancy between the ATBC study and epidemiological studies involving beta-carotene is that beta-carotene supplements omit the other phytochemicals in fruits and vegetables that are chemopreventive. Epidemiological studies have focused mainly on the beta-carotene in green and yellow fruits and vegetables. The possibility exists that beta-carotene is a marker for other chemopreventive phytochemicals, such as other carotenoids, present in these foods as well.[69] Also, beta-carotene has been shown to work synergistically with other compounds to inhibit carcinogenesis, compounds which do not exist in beta-carotene supplements.[78,86] Beta-carotene supplements mainly consist of the commercially available beta-carotene stereoisomer, all-*trans*-beta-carotene, which is only one of 272 beta-carotene stereoisomers[20]; thus, aside from being devoid of other phytochemicals, beta-carotene supplements contain only a small fraction of the total stereoisomers of beta-carotene which also exert antioxidative activities.

A study conducted by Shklar et al.[78] reported that a mixture of beta-carotene, reduced glutathione, vitamin E, and vitamin C was more effective in preventing oral cancer in male hamsters than each of the compounds in the mixture was alone. The evidence in this study shows that the constituents work together synergistically, not merely in an additive manner. The chemopreventive effect of each individual constituent increased when in the mixture. Another study, conducted by Nishino[69], reported that a mixture of 60% beta-carotene, 30% alpha-carotene, and 10% of "other" carotenoids (e.g. gamma, carotene and lycopene) significantly reduced tumors of the skin, lung and duodenum in mice. The efficacy of this mixture for chemoprevention was higher than for beta-carotene alone. Esterbauer et al.[25] demonstrated that carotenoids work in a sequence and that beta-carotene is the last line of defense against oxidized LDL in vitro.

Lutein. Lutein is a carotenoid which, along with lycopene, is a much more effective antioxidant than the other carotenoids.[51] An extremely low rate of several cancers has been reported in the South Pacific Island nations (e.g., Cook, Fiji and Tahiti). A few of the many fruits and vegetables typically consumed in these nations (including Chinese cabbage, amaranthus leaves, fafa) were analyzed for their beta-carotene and lutein content. All but four of the 22 edible plants analyzed contained a higher concentration of lutein than beta-carotene.[51] These results suggest that beta-carotene is a marker for more potent chemopreventive compounds, such as lutein.

Lutein concentrations are also greater in vegetables commonly consumed in the U.S. (Table 32.21). To date, there are no recommended daily amounts of lutein for disease protection as there are for beta carotene even though most diets contain higher levels of lutein and lycopene.[51]

TABLE 32.21
COMPARISON OF THE LEVELS OF LUTEIN AND BETA-CAROTENE DETECTED
IN SELECT VEGETABLES

Vegetable	Lutein[1] (mg/g vegetable)	Beta-Carotene[1] (mg/g vegetable)	Amount of Carotenoid in Raw Vegetable[2]		
			Vegetable Amt.	Lutein (mg)	Beta-Carotene (mg)
Broccoli	0.0283	0.0233	1 spear*	4.3	3.5
Cabbage	0.0031	0.0008	1 cup	0.3	0.06
Green Beans	0.0059	0.0047	1 cup	0.6	0.5
Spinach	0.1250	0.0890	1 cup	6.9	4.9
Winter Squash (baked)	0.0700	0.0099	1 cup	14.4	2.0

* 1 broccoli spear is not equal to one cup
[1] Values from Khachik et al.[51]
[2] Weights used to calculate broccoli and green beans from *Krause's Food, Nutrition and Diet Therapy*[61]; weights used for cabbage, spinach and winter squash calculations from *Understanding Nutrition*[71]

Lycopene. Lycopene is a carotenoid found nearly exclusively in tomatoes and tomato-based products, such as tomato paste, tomato sauce, and tomato-based soups, and, in the tomato consumer, is an abundant carotenoid in human plasma and a variety of tissues.[30,51] Low lycopene levels are found in diets devoid of tomato-based products, even in diets high in fruits and vegetables.[30,75] It has been observed that the bioavailability of this lipophilic compound increases in the presence of heat and oil. Giovannucci et al.[30] reported that lycopene from tomato sauce correlated with increased plasma levels of lycopene while lycopene from tomato juice did not. Another study compared plasma levels of lycopene in subjects who consumed tomato juice versus an experimental recipe of tomato juice cooked in an oil medium and found concentrations of lycopene to be 2-3 times higher in subjects who consumed the tomato juice in oil.[30]

Carotenoids all exhibit antioxidative properties; however, lycopene is the most efficient quencher of singlet oxygen of all of these compounds.[30,58] In a comparison of the chemopreventive efficacy of lycopene, alpha-carotene, and beta-carotene, lycopene inhibited the cancer cell growth of the human endometrium, mammary, and lung, with greater potency than the other two carotenoids.[58] A prospective cohort study conducted in the U.S. by Giovannucci et al.[30] utilized subjects from an ongoing study of the causes of cancer and heart disease in men (The Health Professionals Follow-up Study) to evaluate the effect of diet on prostate cancer. This study, which followed the subjects over a period of 6 years, concluded that only 4 of 46 fruit and vegetable related products significantly correlated with decreased prostate cancer risk. Three of the four foods, tomato sauce, tomatoes, and pizza, were tomato-based foods and were primary sources of lycopene. (The fourth food, associated with lower prostate cancer risk, strawberries, is a source of ellagic acid (see section on Ellagic Acid) and does not contain lycopene.)

Summary of Carotenoids. All carotenoids are antioxidants and have been shown to be effective chemopreventors in many animal models and epidemiological studies. Carotenoids have also been reported to have other chemopreventive properties (Table 32.22). Lycopene suppressed insulin-like growth factor I (IGF-I) in human cells in vitro.[58] Cell replication in various tumors due to IGF-I stimulation has been reported, especially in human mammary cells because they have membrane receptors for, and excrete IGF-I.[49,58] Carotenoids have also been shown to increase intercellular communication via enhancing cell gap junctions[49,58]; thus, inhibiting carcinogenesis.

TABLE 32.22
CLASSIFICATION OF CHEMOPREVENTIVE MECHANISMS BY KELLOFF ET AL.
APPLIED TO CAROTENOIDS[49]

Mechanism	Subset
Antioxidant Activities	Scavenge Singlet Oxygen
Antiproliferation/Antiprogression	Modulate Hormonal/Growth Factor Activity[58]
Antiproliferation/Antiprogression	Increase Intercellular Communication[49,58]

Beta-carotene, particularly all-*trans*-beta-carotene, is the most extensively studied carotenoid mostly due to the commercial production of this beta-carotene stereoisomer, and its high availability.[20] However, some recent research has focused on other carotenoids present in vegetables and fruits which exhibit potent antioxidative and other chemopreventive properties. These compounds, such as lutein, lycopene, and alpha-carotene, coexist in certain natural foods, suggesting that beta-carotene is a secondary carotenoid in terms of anticarcinogenesis and/or a marker in epidemiological studies for these more effective compounds. Carotenoids and other phytochemicals have been shown to act synergistically, increasing the efficacy of the phytochemicals in chemoprevention.[69,78]

Esterbauer, *et al.*[25] reported that the carotenoids exert their antioxidative effects in a sequential manner. They observed that in copper oxidation of LDL *in vitro*, the antioxidant vitamin E was the first lipophilic antioxidant to react with the oxidant (i.e., in order of defense, alpha-tocopherol, then gamma tocopherol), lycopene was next in line followed by cryptoxanthin, lutein and zeaxanthin (which all acted at the same time), and, finally, beta-carotene was the last line of defense.[25] According to these findings, merely studying the effect of beta-carotene on carcinogenesis or cardiovascular disease omits the front-line of defense and concentrates only on the compound which acts last in the lipid antioxidant defense system.

The absorption of dietary carotenoids is low at 10-30% (see Table 32.20). The unabsorbed compounds are excreted in the feces. Dietary fiber, especially pectin, inhibits beta-carotene absorption. Dietary fat significantly enhances carotenoid absorption because carotenoids are only absorbed if bile salts are present.[75] Thus, in order to benefit from the protective effects of carotenoids, it is necessary to consume carotenoid-rich foods with a small amount of fat, preferably without increasing daily consumption of lipid-containing foods.

In general, carotenoids accumulate in serum and tissues, including fat, liver, skin, adrenal glands and testes.[48] The toxicity of carotenoids has been reported as very low in humans.[48] Carotenodermia (yellowing of the skin, a harmless condition) is the predominant effect of high carotenoid levels in humans.[48] Crystalline deposits on the retina have been observed with high doses of the carotenoid canthaxanthine. It is unknown whether these deposits affected retinal functions.[48]

Antioxidant Vitamins and Selenium

Aside from their nutritive contributions to the diet, the antioxidant vitamins E and C, and the trace mineral selenium, have been reported to inhibit carcinogenesis and protect against cardiovascular disease. The highest quantities of the lipophilic vitamin E in natural foods products are found in cereal oils and cereal products (especially wheat germ). Losses of vitamin E occur in vegetable oil processing such as the production of margarine and shortening and also high heat.[26] All animal and plant cells contain vitamin C. Rose hips, black and red currants, strawberries, parsley, oranges, lemons, grapefruit, cabbages and potatoes are especially high in vitamin C. The essential mineral selenium is most abundant in beef, egg yolk, seafood, chicken, whole grain cereals, and garlic.

Vitamin E. The lipophilic nature of alpha-tocopherol allows it to remain in the phospholipid layers of cell membranes, near the polyunsaturated fatty acids (PUFA). The potent antioxidant effects of vitamin E help protect PUFA from free radical attack[75,82], and also prevents oxidation of LDL[31], ultimately protecting against cardiovascular disease (see Cardiovascular Disease Section). An eight year prospective cohort study reported up to a 40% reduction in cardiovascular disease in men and women who took vitamin E supplements.[82]

Vitamin E has also been reported to reduce the risk of cancers of the breast, gastrointestinal tract, esophagus[82], mouth and pharynx.[82] Approximately 20% of vitamin E in the U.S. diet is supplied by fruits and vegetables.[75] The remaining 80% is provided by vegetable oils, margarine, whole grain and fortified cereals, nuts and eggs.[31] The daily recommended intake of this vitamin is 15 IU (or 10 mg) per day for adult men and women.[71] The half life of this vitamin is approximately 2 weeks and no known level of toxicity exists. The absorption of vitamin E has been estimated at 20-40% with the remaining 60-80% exiting as feces.[75] Vitamin E is spared by vitamin C, and both are destabilized by heat.

Vitamin C. Ascorbic acid is an aqueous antioxidant which scavenges water soluble superoxides and hydroxyl radicals. Vitamin C has also been reported to function as a chain-breaking antioxidant in lipid peroxidation.[75] These antioxidative mechanisms have associated the compound with reduced risk for cardiovascular disease and carcinogenesis.[31,75,82] Vitamin C has also been reported to reduce the risk of cataracts by reducing the level of oxidized proteins in the lens.[82] Vitamin C is very unstable in heat and light. The half life of this compound is sixteen days. The RDA for vitamin C is 60 mg to 70 mg per day for adult men and women.[71] An intake of a diet considered optimal to dietary guidelines delivers about 225 mg of vitamin C.

Interactions between vitamins C and E as well as between vitamin C and selenium exist. Vitamin C has been shown to spare vitamin E; thus, when vitamin E oxidizes to tocopheroxyl radical by giving up a hydrogen atom to a free radical, vitamin C oxidizes to ascorbyl radical and restores the hydrogen atom on vitamin E.[45] The spared vitamin E has restored antioxidative properties and is free to quench another free radical. In the presence of selenium, vitamin

C has been reported to hinder the effectiveness of the mineral against colon cancer in a rat study.[31]

Selenium. The trace mineral selenium is a cofactor for the Phase II enzyme glutathione peroxidase and has been reported to enhance vitamin E[11]; thus, this element has strong antioxidative properties and plays a role in preventing carcinogenesis as well as in lowering risk of cardiovascular disease. Selenium has been reported to reduce skin, colon, liver and mammary cancer in animals.[82] Carcinogenesis due to large doses of this compound have been observed; however, an abundance of evidence exists that implicates selenium as a potent anticarcinogenic agent.[31] The amount of selenium in vegetables is dependent upon the amount of the mineral in the soil; thus, soil that is rich in selenium will yield vegetables rich in selenium. The RDA for selenium is 0.055 mg per day for adult females and 0.07 mg per day for adult males.[71] Doses greater than 0.4 mg per day have been reported to be toxic.[26]

Other

The intention of this paper was to outline the main groups of phytochemicals; however, other compounds which have not been mentioned here exist and also have potential roles in disease prevention. A list of some of these compounds includes glycyrrhetinic acid, phytic acid, saponins and sesamol.

Summary of Active Compound Groups

The phytochemical groups discussed in this review are listed in Table 32.23, along with their sources and organoleptic/aesthetic properties. The general mechanisms against carcinogenesis of these compounds as well as an indication whether they are involved in cardiovascular disease prevention is listed in Table 32.24. A list of the reported levels for benefit from some of the active compounds is in Table 32.25.

REGULATORY ISSUES

Fruits, Vegetables and Grains

The NLEA (Nutrition Labeling and Education Act) restricts the labeling of the full benefits of fruits, vegetables, grains, and legumes in carcinogenesis and cardiovascular disease. Only three permissible health claims related to cancer and cardiovascular disease exist under this act. One health claim, CFR 101.76[17], requires that only fruits, vegetables and grains containing at least 10% of the daily requirements of dietary fiber can be labeled as foods which possibly

TABLE 32.23

A LIST OF SOME PHYTOCHEMICAL GROUPS, THEIR SOURCES AND
ORGANOLEPTIC/AESTHETIC, OR NUTRITIVE PROPERTIES

Phytochemical Group	Some Phytochemicals in the Phytochemical Group	Sources	Organoleptic/ Aesthetic or Nutritive Property
1. Phenols	Simple phenols, Phenolic acids, Hydroxycinnamic acid derivatives, Flavonoids	Almost all fresh fruits and vegetables, cereal grains, tea (black and green), nuts	Flavor, color, or aroma
2. Indoles	Indole-3-carbinol, Indole-3-acetonitrile, L-tryptophan	Cruciferous vegetables (including Brussels sprouts, kale, cabbage, broccoli, cauliflower, spinach, watercress, turnip, radish)	Pungent flavor
3. Isothiocyanates	Phenethyl isothiocyanate, Benzyl isothiocyanate, Sulforaphane	Cruciferous vegetables (including Brussels sprouts, kale, cabbage, broccoli, cauliflower, spinach, watercress, turnip, radish)	Pungent flavor
4. Allylic Sulfur Compounds	Diallyl sulfide, Diallyl disulfide, S-allyl cysteine, Allyl propyl disulfide, Ajoene	*Allium* vegetables (including garlic, onion, leek, shallot, chive, scallion)	Flavor
5. Monoterpenes	D-limonene, D-carvone	Citrus oils, vegetable oils, spice oils	Flavor, aroma
6. Monoterpene-like	Perryl alcohol	Cherries	Flavor
7. Carotenoids	Alpha-carotene, Beta-carotene, Alpha-cryptoxanthin, Beta-cryptoxanthin, Lutein, Lycopene, Zeaxanthin	Most red to yellow colored fruits and vegetables	Color
8. Antioxidant Vitamins	Vitamin C, Vitamin E	Fruits and vegetables, whole cereal grains	Nutrient
9. Antioxidant Mineral	Selenium	Garlic, whole cereal grains	Nutrient

TABLE 32.24
SOME PHYTOCHEMICALS AND THEIR CHEMOPREVENTIVE MECHANISMS
(ACCORDING TO THE CLASSIFICATION OF KELLOFF *ET AL*[49] AND EVIDENCE
OF CARDIOVASCULAR DISEASE (CVD) PREVENTION

Phytochemical Group	Specific Phytochemical(s) in the Phytochemical Group	Sources	Chemopreventive Mechanism	Evidence of CVD Prevention (Y=Yes, N/A = not available)
1. Simple Phenol	Curcumin	Turmeric, curry, mustard	Carcinogen Blocking Activity Antioxidant Activity Antiproliferation/Antiprogression	N/A
2. Diphenol: Isoflavonoid	Genistein, Daidzein	Soybeans	Antioxidant Activity Antiproliferation/Antiprogression	N/A
3. Phenolic Acid	Ellagic Acid	Grains, nuts, strawberry, cranberry, blackberry, blueberry, raspberry	Carcinogen Blocking Activity Antioxidant Activity	N/A
4. Phenol: Flavonoids	Thousands of compounds including Quercetin, Catechins	Almost all edible plants	Carcinogen Blocking Activity Antioxidant Activity Antiproliferation/Antiprogression	Y
5. Indoles	Indole-3-Carbinol	Cruciferous vegetables	Carcinogen Blocking Activity	N/A
6. Isothiocyanates	Phenethyl Isothiocyanate, Benzyl Isothiocyanate, Sulforaphane	Cruciferous vegetables	Carcinogen Blocking Activity	N/A
7. Allylic Sulfur Compounds	Allylic Sulfides	*Allium* vegetables	Carcinogen Blocking Activity	Y
8. Monoterpenes	D-Limonene	Citrus oils	Antiproliferation/Antiprogression	N/A
9. Monoterpene-like	Perryl Alcohol	Cherries	Antiproliferation/Antiprogression	N/A
10. Carotenoids	Alpha-carotene, beta-carotene, alpha-cryptoxanthin, beta-cryptoxanthin, lutein, lycopene, zeaxanthin	Most red to yellow colored fruits and vegetables	Antioxidant Activity Antiproliferation/Antiprogression	N/A
11. Antioxidant Vitamins	Vitamins C & E	Fruits and vegetables, whole cereal grains	Antioxidant Activity Carcinogen Blocking Activity Antiproliferation/Antiprogression (vitamin E)	Y
12. Antioxidant Mineral	Selenium	Garlic, whole cereal grains	Antioxidant Activity Antiproliferation/Antiprogression	Y

TABLE 32.25
SOME PHYTOCHEMICALS, THEIR SOURCES, REPORTED DOSES
FOR CHEMOPREVENTION AND STABILITY

Phytochemical Group	Specific Phytochemical(s) in the Phytochemical Group	Source(s)	Reported Level of Phytochemical for Chemoprevention*	Stability
1. Diphenol: Isoflavonoid	Genistein	Soybeans	50 mg/ day (10)	1. Heat resistant 2. Stable to water or acid washing. 3. Unstable in aqueous alcohol
2. Phenol: Flavonoid	Quercetin	All edible plants, including onions, apples, black and green teas	26.6 mg/ day (38)	N/A
3. Indoles	Indole-3-Carbinol	Cruciferous vegetables	5-7 mg/kg body weight (approx. 350-500 mg/day) (66)	1. Reduced levels when boiled in water. 2. Reduced levels when heated.
4. Allylic Sulfur Compounds	Allylic Sulfide	*Allium* vegetables	900 mg garlic powder per day reduced CVD in humans (82)	1. Unstable at temperatures > 60C
5. Carotenoids	Beta-carotene	Most red to yellow colored fruits and vegetables	6-20 mg (4)	1. Relatively stable at high temperatures.

N/A = Not available
*Allylic sulfides: Level reported is for reduction of cardiovascular disease (CVD).

reduce the risk of cancer. CFR 101.77[17] states that only those fruits, vegetables and grains that contain at least 0.6 grams of soluble fiber per serving can be claimed as foods which reduce the risk of coronary heart disease. The third legal health claim concerning fruits and vegetables, CFR 101.78[17], requires that the food in question must contain, without fortification, a minimum of 10% of the daily requirements of dietary fiber, or vitamin A, or vitamin C before a statement indicating that a possible reduction of cancer risk can be obtained by consumption of the food.

These three laws are limiting for phytochemical-rich natural foods which do not contain the minimum 10% of the daily requirements for vitamin A (essentially carotenoid content), vitamin C or fiber. Cabbage, string beans, and grapes, for instance, do not meet the minimum requirement of nutrient content for the health claims.[43] These foods; however, have a significant phytochemical content which has been reported to prevent certain types of cancers. Cabbage is rich in indoles, isothiocyanates, and flavonoids (see Phenols, Indoles and Isothiocyanates sections), string beans contain lutein and beta-carotene (see Carotenoids section), and grapes contain flavonoids and hydroxycinnamic acids (see Phenols section). Recently, the FDA has proposed to allow health claims for products which are strictly composed of fruits and vegetables, regardless of their vitamin A, vitamin C, and fiber content.[43] The alleviation of these strict rules on fruits and vegetables may increase consumer awareness of the benefits against certain chronic diseases that can be derived from all of these natural foods, not just a select few which meet the stringent FDA nutrient content requirement regulation.

Supplements

The Dietary Supplement Health and Education Act of 1994 (DSHEA)[87], permits manufacturers of supplements to make claims regarding the health benefits of these products. The FDA defines supplements as "a product (other than tobacco) intended to supplement the diet that bears or contains one or more of the following ingredients: (A) a vitamin, (B) a mineral, (C) an herb or other botanical, (D) an amino acid, (E) a dietary substance for use by man to supplement the diet by increasing the total dietary intake, or (F) a concentrate, metabolite, constituent, extract or combination of any ingredient in clause (A), (B), (C), (D), or (E)" [Section 3(1)(A-F)]. Under this definition, phytochemicals are considered to be supplements. The only limitation on the health claims permissible for supplements under this act is that a statement may not "claim to diagnose, mitigate, treat, cure, or prevent a specific disease or class of diseases" [Section 6(C)].

According to the DSHEA, the "burden of proof" for safety is on the FDA; therefore, it is up to the government to analyze each supplement that is introduced to the market and prove that it is either safe and the claims are correct, or that the product is unsafe. If proven deleterious to health, the FDA has the authority to discontinue production and sale of the supplement.

Although there is an abundance of data demonstrating that the various phytochemicals are beneficial for health, purifying these compounds and manufacturing them as supplements is premature at this point. Essential data about the toxicity (relative risk) of the compounds, levels of efficacy in humans, types or degrees of interactions between the compounds, absorption and bioavailability, and stability of these compounds must still be obtained and confirmed. The possibility exists that some compounds act as markers for, or are less potent than, other compounds which have not been as extensively studied present in the same food (for example, beta-carotene and lutein and lycopene). Supplements will most likely omit other more potent phytochemicals and some compounds which have not yet been identified. Universally approved analytical methods must be agreed upon so the actual quantity of each compound per supplement is measured accurately.

Until these questions are answered, consumption of phytochemical supplements will not offer the same benefits as obtaining these compounds from fruits, vegetables, legumes, and grains. When the majority of these unknown variables are eventually uncovered, phytochemical supplements may possibly offer more consistent benefits due to minimal variability in doses of the compound(s) and allow targeting for health. The levels of phytochemicals in plant foods are highly variable and are dependent upon many conditions including variety, season of the year, weather conditions during growth, geography, type of storage, and minerals in the soil.[7] Biotechnology applied to germ plasma and directed to controlling specific phytochemical entities could make a significant contribution to the standardization of supplement products.[55] Meanwhile, the consumer would be advised to adopt the daily practice of consuming no less than five servings (2 fruit and 3 vegetable) of fruits and vegetables.[55]

SUMMARY AND CONCLUSIONS

There is an abundance of evidence which shows that edible plant foods (i.e., fruits, vegetables, whole grains, nuts and teas) are protective against America's number one chronic diseases: cancer and cardiovascular disease; however, more organized research is needed. The studies completed to date have not been performed in a systematic manner. Instead, data concerning various phytochemicals exists as a result of epidemiological studies, *in vitro* assays, rodent studies and/or human metabolic studies which all utilize various sources, doses and methods of administration of the phytochemical being studied. Thus, there are few standardized source(s), dosage(s) or route of administration methods so that responses to variables could be compared accurately. More useful and correlating information would exist if the researchers performed phytochemical studies in a cooperative manner; thus, data from *in vitro/in vivo* bioassays and rodent studies could more realistically be extrapolated for use in human studies, including clinical trials. Standardized information concerning the results/risks of a compound for humans thus would be comparable and applicable.

A small percentage of the studies regarding phytochemicals report negative effects for some compounds. Although the amount of these studies are minor compared to the multitude of studies reporting the benefits of phytochemicals, they should not be overlooked. For example, a negative effect on health due to the flavonoid naringenin has been observed by Lee *et al.*[57] Naringenin is the major flavonoid in grapefruit juice and has been shown to inhibit the enzyme

11-B-hydroxysteroid dehydrogenase (11B-OHSD). Hypertension and hypokale-mia in children results from high cortisol levels in the kidney which occurs when the child is 11B-OHSD deficient.[57] Green tea catechins have been observed to enhance liver carcinogenesis in rats in a dose-dependent manner.[82] The flavonoid quercetin has been reported to cause nuclear DNA damage in rats.[77] In cattle, quercetin is a potent mutagen found in bracken fern and is a cofactor in bovine papillomavirus type 4 (BPV-4), which is associated with carcinogenesis of the upper alimentary canal.[72] The position taken by the American Dietetic Association (ADA) on the few contradictory findings is "these studies are in the minority in terms of total numbers; however, their clinical implication must be considered in the design of future clinical trials and in recommendations for intake, supplementation, and/or potential future fortification of the food supply."[4]

Eating fruits and vegetables, and other natural plant products, introduces many disease-fighting phytochemicals into the body; however, a person with a diet high in plant foods also obtains further benefits from such food choices because a diet high in plant foods is usually lower in fatty foods or other more caloric foods.[82] It has been reported that a reduction in calories is associated with a reduced risk of various cancers in humans.[82]

Consumption of fruits, vegetables, whole grains, nuts, and legumes has been shown by various studies to reduce the risk of chronic diseases. It is difficult to pinpoint exactly which compound is most beneficial. Research has shown that there may be a synergistic effect of two or more phytochemical compounds in the food. Optimal combinations and levels of certain compounds need to be determined. Until the exact mechanisms of the compounds are clear, the "consumption of supplements will only provide selected components in a concentrated form, not the diversity of phytochemicals that occur naturally in food."[4] The best recommendation at the current time is to consume at least 5-9 servings of fruits and vegetables per day, as recommended by the Food Guide Pyramid, preferably in a variety of colors reflecting an array of phytochemicals, as well as nutrients and sources of food energy.

FUTURE WORK

An overwhelming amount of data shows that there is a link between diet and chronic diseases; however, the following issues still need to be more thoroughly addressed.

- Pre-use characterization of the chemistry and concentrations of the com-pounds.
- Relative toxicity of the compounds.

- Bioavailability (i.e., whether the compounds are absorbed and utilized, or function beneficially if they are not absorbed).
- Half life of each compound in the body.
- Development of a standard for each compound, because the amount of phytochemicals in foods is highly variable and dependent upon many conditions including edible plant variety, season of the year, weather conditions during growth, geography, type of storage, and minerals in the soil.
- Whether the compounds act additively, synergistically, or independently.
- The interaction of nutrient and non-nutrient chemicals.
- The mechanisms of action of the compounds in the body (i.e., the metabolism of the compounds).
- Stability (temperature, time, processing)
- Doses required for effective outcomes.
- Acceptable universal methods of analyses.
- Negative effects (if any) of the compounds.

REFERENCES

[1] Adlercreutz, H. 1995. Phytoestrogens: epidemiology and a possible role in cancer protection. Environmental Health Perspectives 103(Suppl. 17): 103–112.

[2] Alberts, D. and Garcia, D. 1995. An overview of clinical cancer chemoprevention studies with emphasis on positive Phase III studies. J. Nutr. 125: 692S–697S.

[3] The American Dietetic Association. Nutrition Trends Survey, conducted by the Wirthlin Group, Underwritten by Kraft Foods. 1995: Chicago, pp. 1–7.

[4] The American Dietetic Association. 1995. Position of American Dietetic Association: phytochemicals and functional foods. J. Am. Dietet. Assoc. 95(4):493–496.

[5] Ames, B.N, Shigenaga, M.K. and Hagen, T.M. 1993. Oxidants, antioxidants, and the degenerative diseases of aging. Proc. Nat. Acad. Sci., USA 90:7915–7922.

[6] Anderson, R. and Wolf, W. 1995. Compositional changes in trypsin inhibitors, phytic acid, saponins and isoflavones related to soybean processing. J. Nutr. 125:581S–588S.

[7] Anon. 1996. What's the threat of chemicals in food? Consumers' Res. 79(4):28–32.

[8] Ashendel, C.L. 1995. Diet, signal transduction and carcinogenesis. J. Nutr. 125:686S–691S.

[9] Barnes, S. 1995. Effect of genistein on in vitro and in vivo models of cancer.

J. Nutr. 125:777S–783S.

[10] Barnes, S., Peterson, G. and Coward, L. 1995. Rationale for the use of genistein-containing soy matrices in chemoprevention trials for breast and prostate cancer. J. Cellular Biochem., Suppl. 22:181–187.

[11] Belitz, H.D. and Grosch, W. (ed.). 1987. Food Chemistry. Heidelberg: Springer-Verlag Berlin.

[12] Block, G., Paterson, B. and Subar, A. 1992. Fruit, vegetables, and cancer prevention: a review of the epidemiological evidence. Nutr. & Cancer 18(1): 1–29.

[13] Brown, W.V. 1990. Dietary recommendations to prevent coronary heart disease. Annals N.Y. Acad. Sci. 598:376–388.

[14] Buring, J.E. and Hennekens, G.H. 1995. Beta-carotene and cancer prevention. J. Cell. Biochem. Suppl. 22:226–230.

[15] Byers, T., Lachance, P.A. and Pierson, H.F. 1990. New directions: the diet-cancer link. Patient Care 34–38.

[16] Chan, M.M.-Y., Ho, C.-T. and Huang, H.-I. 1995. Effects of three dietary phytochemicals from tea, rosemary, and turmeric on inflammation-induced nitrite production. Cancer Letters 96:23–29.

[17] Code of Federal Regulations (CFR). 1996. Volume 21: Food and Drugs. Part 101, Food Labeling, Subpart E, Health Claims. Published by Office of Federal Register National Archives and Records Administration.

[18] Cozzi, R., Ricordy, R., Bartolini, F., Ramadori, L. Perticone, P. and DeSalvia, R. 1995. Taurine and ellagic acid: two differently-acting natural antioxidants. Environmental Molecular Mutagenesis 26:248–254.

[19] Dausch, J.G. and Nixon, D.W. 1990. Garlic: a review of its relationship to malignant disease. Preventive Med. 19:346–361.

[20] Doering, W.v.E. 1996. Antioxidant vitamins, cancer, and cardiovascular disease. The New England J. Med. 335(14):1065.

[21] Doll, R. and Peto, R. 1981. The causes of cancer: quantitative estimates of avoidable risks of cancer in the United States today. J. Nat. Cancer Inst. 66:1192–1308.

[22] Dorant, E., Van den Brandt, P.A., Goldbohm, R.A. and Sturmans, F. 1996. Consumption of onions and a reduced risk of stomach carcinoma. Gastroenterology 110:12–20.

[23] Elegbede, J.A., Elson, C.E., Qureshi, A., Tanner, M.A. and Gould, M.N. 1984. Inhibition of DMBA-induced mammary cancer by the monterpene d-limonene. Carcinogenesis 5(5):661–664.

[24] Enstrom, J.E., Kamin, L.E. and Kelin, M.A. 1992. Vitamin C intake and mortality among a sample of the United States population. Epidemiology 3:194–202.

[25] Esterbauer, H., Gebecki, J. Puhl, H. and Jurgens, G. 1992. The role of lipid peroxidation and antioxidants in oxidative modification of LDL. Free Radical

Biol. & Med. 13:341–390.

[26] Fennema, O. (ed.) 1985. Food Chemistry. Marcel Dekker, New York.

[27] Fernandez, M.L., Sun, D.M., Montano, C. and McNamara, D.J. 1995. Carbohydrate-fat exchange and regulation of hepatic cholesterol and plasma lipoprotein metabolism in the guinea pig. Metabolism 44(7):855–864.

[28] Fotsis, T., Pepper, M., Adlercreutz, H., Hase, T., Montesano, R. and Schweigerer, L. 1995. Genistein, a dietary ingested isoflavonoid, inhibits cell proliferation and in vitro angiogenesis. J. Nutr. 125:790S–797S.

[29] Galvez, J., de la Cruz, J.P., Zarzuelo, A. and de la Cuesta, F.S. 1995. Flavonoid inhibition of enzymic and nonenzymic lipid peroxidation in rat liver differs from its influence on the glutathione-related enzymes. Pharmacology 51:127–133.

[30] Giovannucci, E., Ascherio, A., Rimm, E.B., Stampfer, M.J., Colditz, G.A. and Willett, W.C. 1995. Intake of carotenoids and retinol in relation to risk of prostate cancer. J. Nat. Cancer Inst. 87:1767–1776.

[31] Golberg, Israel (ed.). 1994. Functional Foods, Designer Foods, Pharmafoods, Nutraceuticals. Chapman and Hall, New York.

[32] Golbitz, P. 1995. Traditional soyfoods: processing and products. J. Nutr. 125:570S–572S.

[33] Gould, M.N. 1995. Prevention and therapy of mammary cancer by monoterpenes. J. Cell. Biochem., Suppl. 22:139–144.

[34] Grisham, M.B. 1992. Reactive Metabolites of Oxygen and Nitrogen, pp. 4–10. R.G. Landes Company, Austin, TX.

[35] Hawrylewicz, E.J., Zapata, J.J. and Blair, W.H. 1995. Soy and experimental cancer, animal studies. J. Nutr. 125:698S–708S.

[36] Hecht, S.S. 1995. Chemoprevention by isothiocyanates. J. Cell. Biochem., Suppl. 22:195–209.

[37] Herman, C., Adlercreutz, T., Goldin, B.R., Gorback, S.L., Hockerstedt, K.A., Watanabe, S., Hamalainen, E.K., Markkanene, M.H., Makela, T.H., Wahala, K.T., Hase, T.A. and Fotsis, T. 1995. Soybean phytoestrogen intake and cancer risk. J. Nutr. 125:757S–770S.

[38] Hertog, M.G.L., Feskens, E., Hollman, P., Katan, M. and Kromhout, D. 1993. Dietary antioxidant flavonoids and risk of coronary heart disease: the Zutphen Elderly Study. Lancet 342:1007–1011.

[39] Ho, C.-T., Lee, C.Y. and Huang, M.-T. (ed.). 1992. Phenolic Compounds in Food and Their Effects on Health. ACS Symp. Ser. 506. American Chemical Society, Washington, D.C.

[40] Holzman, D. 1995. Whatever happened to beta-carotene? J. Nat. Cancer Inst. 87(23):739–741.

[41] Huang, M.-T., Ho, C.-T. and Lee, C.Y. (ed.). 1992. Phenolic Compounds in Food and Their Effects on Health. ACS Symposium Series 507. American Chemical Society, Washington, D.C.

[42] Huang, M.T., Osawa, T., Ho, C.-T. and Rosen, R.T. (ed.) 1994. Food Phytochemicals for Cancer Prevention. ACS Symp. Ser. 546. American Chemical Society, Washington, D.C.

[43] Hunter, B.T. 1996. Relaxing the rules on food labels. Consumers' Res. 79(4):18–21.

[44] Igimi, H., Hisatsugu, T. and Nishimura, M. 1976. The use of d-limonene preparation as a dissolving agent of gallstones. Digestive Diseases 21(11):926–939.

[45] Jacob, R.A. 1995. Nutr. Res. 15:755–766.

[46] Jones, R.C. 1987. Studies on the inhibition of the formation of aminoimidazo-quinoline and -quinoxaline type mutagens/carcinogens. Ph.D. Thesis, Rutgers, the State University, New Brunswick, NJ.

[47] Katzenellenbogen, J.A. 1995. The structural pervasiveness of estrogenic activity. Environmental Health Perspectives 103 (Suppl. 7):99–101.

[48] Kelloff, G.J., Boone, C.W., Steele, V.E., Fay, J.R., Lubet, R.A., Crowell, J.A. and Sigman, C.C. 1994. Clinical development plan: beta-carotene and other carotenoids. J. Cell. Biochem. Suppl. 20:110–115.

[49] Kelloff, G.J., Boone, C.W., Steele, V.E, Fay, J.R., Lubet, R.A., Crowell, J.A. and Sigman, C.C. 1994. Mechanistic considerations in chemopreventive drug development. J. Cell. Biochem. Suppl. 20:1–24.

[50] Kennedy, A.R. 1995. The evidence for soybean products as cancer preventive agents. J. Nutr. 125:733S–743S.

[51] Khachik, F., Beecher, G.R. and Smith, J.C., Jr. 1995. Lutein, lycopene, and their oxidative metabolites in chemoprevention of cancer. J. Cell. Biochem. Suppl. 22:236–246.

[52] Knekt, P., Jarvinen, R., Reunanen, A. and Maatela, J. 1996. Flavonoid intake and coronary mortality in Finland: a cohort study. British Med. J. 312:478–481.

[53] Lachance, P.A. Micronutrients in cancer prevention. In Food Phytochemicals for Cancer Prevention (Huang, M.T., Osawa, T., Ho, C.-T. and Rosen, R.T., eds.) ACS Symp. Ser. No. 546. American Chemical Society, 1994:49–64, Washington, D.C.

[54] Lachance, P.A. 1996. "Natural" cancer prevention. Science 272:1860–1861.

[55] Lachance, P.A. Unpublished data. Oct. 1996.

[56] Lawson, L.D. Bioactive organosulfur compounds of garlic and garlic products. In Human Medicinal Agents from Plants, (Kinghorn, A.D. and Balandrin, M.F., eds.) ACS Symp. Ser. No. 534. American Chemical Society, 1993:306–330, Washington, D.C.

[57] Lee, Y.S., Lorenzo, B.J., Koufis, T. and Reidenberg, M. 1996. Grapefruit juice and its flavonoids inhibit 1-B-hydroxysteroid dehydrogenase. Clin. Pharmacol. Therapeutics 59(1):62–71.

[58] Levy, J., Bosin, E., Feldman, B., Giat, Y., Munster, A., Danilenko, M. and

Sharoni, Y. 1995. Lycopene is a more potent inhibitor of human cancer cell proliferation than either alpha-carotene or beta-carotene. Nutr. & Cancer 24:257–266.

[59] Liu, H., Wormke, M., Safe, S.H. and Bjeldanes, L.F. 1994. Indolo [3,2-b]carbazole: a dietary-derived factor that exhibits both antiestrogenic and estrogenic activity. J. Nat. Cancer Inst. 86(23):1758–1765.

[60] McNamara, D.J. and Howell, W.H. 1992. Epidemiological data linking diet to hyperlipidemia and arteriosclerosis. Semin-Liver-Dis. 12(4):347–355.

[61] Mahan, K.L. and Arlin, M. (ed.). 1992. Krause's Food, Nutrition and Diet Therapy, 8th Ed., p. 764. W.B. Saunders Company, Philadelphia, PA.

[62] Marwick, C. 1995. Learning how phytochemicals help fight disease. J. Am. Med. Assoc. 274(17):1328–1330.

[63] Messina, M. 1995. Modern applications for an ancient bean: soybeans and the prevention and treatment of chronic disease. J. Nutr. 125:567S–569S.

[64] Messina, M.J., Persky, V., Setchell, K.D.R. and Barnes, S. 1994. Soy intake and cancer risk: a review of the in vitro and in vivo data. Nutr.& Cancer 21:113–131.

[65] Michnovicz, J.J. and Bradlow, H.L. 1990. Induction of estradiol metabolism by dietary indole-3-carbinol in humans. J. Nat. Cancer Inst. 11:947–949.

[66] Michnovicz, J.J. and Bradlow, H.L. 1991. Altered estrogen metabolism and excretion in humans following consumption of indole-3-carbinol. Nutr. & Cancer 16:59–66.

[67] Molteni, A., Brizio-Molteni, L. and Persky, V. 1995. In vitro hormonal effects of soybean isoflavones. J. Nutr. 125:751S–756S.

[68] Muldoon, M.F. and Kritchevsky, S.B. 1996. Flavonoids and heart disease. British Med. J. 312:458–459.

[69] Nishino, H. 1995. Cancer chemoprevention by natural carotenoids and their related compounds. J. Cell. Biochem. Suppl. 22:231–235.

[70] Nishino, H., Iwashima, A., Hakura, Y., Matsuura, H. and Fuwa, T. 1989. Antitumor-promoting activity of garlic extracts. Oncology 46:277–280.

[71] Whitney, E.N. and Hamilton, E.M.N. (eds.). 1984. Understanding Nutrition, 3rd Ed., pp. H98-H102. West Publishing Company, Philadelphia.

[72] Pennie, W.D. and Campo, M.S. 1992. Synergism between bovine papilloma-virus type 4 and the flavonoid quercetin in cell transformation in vitro. Virology, 190(2):861–865.

[73] Persky, V. and Van Horn, L. 1995. Epidemiology of soy and cancer: perspectives and directions. J. Nutr. 125:709S–712S.

[74] Peterson, G. 1995. Evaluation of the biochemical targets of genistein in tumor cells. J. Nutr. 125:784S–789S.

[75] Rock, C.L., Jacob, R.A. and Bowen, P.E. 1996. Update on the biological characteristics of the antioxidant micronutrients: vitamin C, vitamin E, and the carotenoids. J. Am. Dietet. Assoc. 96:693–702.

[76] Rosen, J.D. Unpublished data. Sept. 1994.

[77] Sahu, S.C. and Washington, M.C. 1992. Effect of ascorbic acid and curcumin on quercetin-induced nuclear DNA damage, lipid peroxidation and protein degradation. Cancer Letters 63(3):237–241.

[78] Shklar, G., Schwartz, J., Trickler, D. and Cheverie, S.R. 1993. The effectiveness of a mixture of beta-carotene, alpha-tocopherol, glutathione, and ascorbic acid for cancer prevention. Nutr. & Cancer 20:145–151.

[79] Sirtori, C., Lovati, M., Manzoni, C., Monetti, M., Pazzucconi, F. and Gatti, E. 1995. Soy and cholesterol reduction: clinical experience. J. Nutr. 125:598S–605S.

[80] Slominski, B.A. and Campbell, L.D. 1988. Gas chromatographic determination of indole acetonitriles in rapeseed and Brassica vegetables. J. Chromatography 454:285–291.

[81] Slominski, B.A. and Campbell, L.D. 1989. Formation of indole glucosinolate breakdown products in autolyzed, steamed and cooked Brassica vegetables. J. Agri. Food Chem. 37:1297–1302.

[82] Stavric, B. 1994. Role of chemopreventers in the human diet. Clin. Biochemistry 27:319–327.

[83] Steele, V.E., Periera, M.A., Sigman, C.C. and Kelloff, G.J. 1995. Cancer chemoprevention agent development strategies for genistein. J. Nutr. 125:713S–716S.

[84] Stone, N.J. 1990. Diet, lipids, and coronary heart disease. Endocrin-Metab-Clin-North Am. 19(2):321–344.

[85] Stoner, G.D. and Mukhtar, H. 1995. Polyphenols as cancer chemopreventive agents. J. Biochem. Suppl. 22:169–180.

[86] The Alpha-Tocopherol, Beta-Carotene Cancer Prevention Study Group. The effect of vitamin E and beta-carotene in the incidence of lung cancer and other cancers in male smokers. 1994. New England J. Med. 330:1029–1035.

[87] United States Code Service Advance Legislative Service. 1994. Public Law 103–417. Dietary Supplement Health and Education Act. Lawyers Cooperative Publishing.

[88] Wattenberg, L.W. 1985. Chemoprevention of cancer. Cancer Res. 45:1–8.

[89] Weisburger, J.H. 1991. Nutritional approach to cancer prevention with emphasis on vitamins, antioxidants, and carotenoids. Am. J. Clin. Nutr. 53: 226S–237S.

[90] Zheng, G., Kenney, P.M. and Lam, L.T. 1993. Potential anticarcinogenic natural products isolated from lemongrass oil and galanga root oil. J. Agri. Food Chem. 41(2):153–156.

INDEX